Fast and Accurate Finite-Element Multigrid Solvers for PDE Simulations on GPU Clusters

---

Dissertation
zur Erlangung des Grades eines
Doktors der Naturwissenschaften

Der Fakultät für Mathematik der
Technischen Universität Dortmund
vorgelegt von

Dominik Göddeke

Bibliografische Information der Deutschen Nationalbibliothek

Die Deutsche Nationalbibliothek verzeichnet diese Publikation in der Deutschen Nationalbibliografie; detaillierte bibliografische Daten sind im Internet über http://dnb.d-nb.de abrufbar.

ISBN 978-3-8325-2768-6

Logos Verlag Berlin GmbH
Comeniushof, Gubener Str. 47,
10243 Berlin
Tel.: +49 (0)30 42 85 10 90
Fax: +49 (0)30 42 85 10 92
INTERNET: http://www.logos-verlag.de

[Homer] D'oh, I need to dedicate this to someone.[/Homer]

Fast and Accurate Finite Element Multigrid Solvers for PDE Simulations on GPU Clusters
Dominik Göddeke

Dissertation eingereicht am: 25. 2. 2010
Tag der mündlichen Prüfung: 10. 5. 2010

Mitglieder der Prüfungskommission

Prof. Dr. Stefan Turek (1. Gutachter, Betreuer)
Prof. Dr. Heinrich Müller (2. Gutachter)
Prof. Dr. Joachim Stöckler
Prof. Dr. Rudolf Scharlau
Dr. Matthias Möller

Acknowledgements

It is impossible to acknowledge everyone who directly or indirectly contributed to the realisation of this thesis. I apologise to all those not mentioned by name: Your help and support is really appreciated.

In particular, I would like to express my sincere gratitude to the following people: Professors Stefan Turek and Heinrich Müller guided my scientific career from the first contact to Numerical Mathematics and Computer Graphics as an undergraduate student to the strongly interdisciplinary field of scientific computing in which I finally settled with this thesis. I am deeply indebted to them for giving me the opportunity to work in the friendly atmosphere at the Institute of Applied Mathematics and the Chair of Computer Graphics at TU Dortmund. I want to thank them for their ideas, fruitful discussions and encouraging criticisms, their belief in my work when I got lost, and last but not least for taking the risk and having the vision to support this endeavour into an emerging research area that at the beginning no one really expected to have such a big impact.

I am very grateful to my fellow PhD students and friends in the FEAST group, Christian Becker, Sven Buijssen, Matthias Grajewski and Hilmar Wobker. Developing such a large-scale package like FEAST is always a group effort, and I believe we did extremely well sharing the tedious task of implementing features for the common good even if it meant devoting the 'occasional' week for non-thesis work, while still pursuing our separate research projects. Knowing that I could always rely on our group for discussing design decisions, toying with ideas, going through painful debugging sessions, solving hardware, cluster and scheduling issues and related nightmares in parallel computing, and last but not least, sharing a beer if all else failed cannot be appreciated enough. The conference proceedings and journal papers we wrote together clearly do not tell the entire story. While not being technically members of the FEAST group, Matthias Möller, Thomas Rohkämper and Michael Köster are acknowledged for the same reasons.

I am very grateful to Robert Strzodka for our collaboration during the past years. What started out from his talk at my 'inauguration workshop' in early 2005 soon became a common vision that we pursued relentlessly. Without Robert, I would never have met Patrick McCormick and Jamaludin Mohd-Yusof from Los Alamos, and the research the four of us pursued together (and the papers we wrote) would not have been possible. In fact, Robert deserves the same credits for shaping me into a scientist as Stefan Turek and Heinrich Müller do.

Matthias Möller, Hilmar Wobker, Robert Strzodka and Jamaludin Mohd-Yusof deserve special credits for proofreading sections of this thesis and discussing questions that arose. Their constructive criticism helped a lot to improve clarity of this thesis and to make the presentation more concise.

In my time at TU Dortmund, I helped supervising quite a few Diploma theses and student project groups. Discussing ideas with 'my' students cannot be overestimated, and I believe this has always been fruitful for both sides. Among the works I supervised, I want to acknowledge Hendrik Becker, Markus Geveler, Dirk Ribbrock and Mirko Sykorra. One project group that I launched was so successful that we published the results in Computer Physics Communications, and I am indebted to my co-authors Danny van Dyk, Markus Geveler, Sven Mallach, Dirk Ribbrock and Carsten Gutwenger for the opportunity to work together productively for more than two years.

NVIDIA and AMD/ATI supported my research through generous hardware donations, and I owe gratitude to Mark Harris, Mike Houston, David Luebke, Simon Green, Christian Sigg, Dominik Behr and Sumit Gupta for fruitful discussions and untiring assistance. The same is true for collaborators and colleagues I discussed many scientific and non-scientific ideas with at conferences or otherwise, in particular John Owens, Aaron Lefohn, Dimitri Komatitsch, Gordon Erlebacher, David Michéa, Cliff Woolley, Daniel Nogradi, Jonathan Cohen, Yao Zhang, Jens Krüger and Rüdiger Westermann; and virtually everyone I collaborated and exchanged ideas with in one form or another.

Finally, I am very grateful to my parents Michaela and Nikolaus, my sisters Maria and Anne and my brother Niklas for encouragement and support. My godchild Linda Lübbers, and Hannah, Jannis and Pia Imöhl made sure to keep me sane and focused through all these years.

Support of this work by the DFG (Deutsche Forschungsgemeinschaft) through the grants TU102/22-1 and TU102/22-2 is gratefully acknowledged.

Dortmund, February 25, 2010

Dominik Göddeke

Contents

1

Introduction

This chapter introduces and motivates the (wider) context of this thesis, establishes and summarises its contribution, and provides an outline of the following chapters.

1.1. Introduction and Motivation

Scientific Computing is the field of research within applied and numerical mathematics that is concerned with the development, analysis *and* efficient realisation and implementation of state-of-the-art numerical algorithms on computer systems. In practice, it is very interdisciplinary in nature, because it constitutes an integral part of *computational science and engineering (CSE)*, which bridges the gap between (numerical) mathematics and computer science on the one hand, and on the other hand, all kinds of engineering disciplines. In the context of this thesis, *continuum mechanics* is of particular interest, i. e., fluid dynamics and solid mechanics. Nonetheless, in academia and in the industry alike, a wide range of application domains are encountered in CSE, ranging from computational chemistry and computational biology (molecular dynamics, genome sequence alignment, drug design) over geophysics (seismic wave propagation, earthquake predictions) to computational finance (risk assessment and trade predictions), to name just a few. Computational science adds *simulation* as the third pillar to science and engineering, between theoretical analysis and experimental verification: Computer systems are used to analyse, to gain understanding and to solve problems from a wide range of application domains. Numerical simulations are often used for economical reasons (when exhaustive experimental analysis is too expensive), or to study phenomena which cannot be accessed in other ways, for instance for the analysis of hazardous substances when experiments are impossible.

In continuum mechanics, natural phenomena are modeled with fundamental physical conservation laws (e. g., conservation of mass, energy and momentum), expressed mathematically as (systems of) *partial differential equations (PDEs)* with initial and boundary conditions that stem from experimental measurements or theoretical analysis. The first step in the numerical simulation is the selection of a suitable mathematical model for the problem at hand, for instance the Navier-Stokes equations in the case of fluid flow. Starting from this initial model and an associated set of parameters, a *discretisation* process is applied to transform the problem from a continuous representation to a finite approximation, in both space and time. After a solution of the discrete problem has been computed, features of interest are extracted either from a visualisation of the results, or by computing characteristic parameters from the solution. Examples in the scope of this thesis include von Mises stresses in solid mechanics (which are an important measure for predicting material failure), or drag and lift coefficients in fluid dynamics. These parameters are then used to refine the initial model, and the process is repeated iteratively.

The wide variety of numerical techniques to solve PDEs is clearly beyond the scope of even these generic introductory paragraphs. In the context of this thesis, it suffices to state that many techniques boil down to *linear equation systems* that need to be solved repeatedly, and this task is often the most computationally demanding one in the course of a simulation. This is true for instance in the case of implicit time discretisation techniques, or in the case of the Newton-Raphson scheme as an exemplary algorithm for the linearisation of nonlinear problems. We only consider grid-based discretisations like finite elements, differences and volumes, and owing to their strictly local support, the system matrices of these linear systems are very *sparse*, i. e., only a very small number of entries per row is nonzero.

It is obvious that the *computational demands* of CSE workflows of this kind are insatiable. It is not uncommon to face linear systems with billions of unknowns, that need to be solved for thousands of timesteps. Due to the sheer size of the problems, a parallel solution on large supercomputer systems is often unavoidable. *Efficiency* is of utmost importance, in all stages of the solution process:

- *Numerical efficiency* affects both the discretisation and the solution. Here, we only elaborate upon efficiency aspects of spatial discretisations and the solution of sparse linear systems. *Finite element methods* are advantageous over finite difference or finite volume approaches in terms of flexibility and accuracy. Furthermore, finite element techniques provide a complete and well-understood theoretical framework which for instance enables rigorous a posteriori error control and thus adaptivity: Less overall unknowns have to be invested for a given target accuracy. For the linear systems of equations encountered throughout this thesis, *(geometric) multigrid methods* are the most efficient solvers (see Section 2.2): They combine the smoothing properties of inexpensive iterative methods with corrections on a hierarchy of nested coarser grids (coarser representations of the discretised domain), and they converge asymptotically optimal: Computing the solution requires a linear amount of calculations in the number of unknowns, and the convergence rates are provably independent of the mesh width h and thus of the problem size. When equipped with powerful smoothing operators, convergence rates of 0.1 are common, which means that only a small amount of iterations are needed to compute the solution. These favourable theoretical properties should be preserved in practice, no matter if the computation is carried out serially or in parallel. In the latter case, the domain is partitioned into a number of *subdomains*, and *multilevel domain decomposition methods* are often applied: For an ideal parallelisation, the convergence rates should be independent of the partitioning of a given domain into subdomains, see Section 2.3.

- The term *parallel efficiency* refers to how well a given algorithm can be parallelised: In the scope of this thesis, three levels of parallelism are encountered, the coarse-grained parallelism between the nodes in a compute cluster, the medium-grained parallelism between the CPU cores in each node, and the fine-grained parallelism within many-core processors. Each level of parallelism has its own associated communication model, from message passing between distributed memories over (explicit or implicit) locks and mutexes on shared data in global off-chip memory, down to blocks of concurrent threads synchronising via small, on-chip memories. Two important aspects in this context are *weak and strong scalability* of an algorithm. The former means that when doubling the problem size and the number of compute resources, the time to solution should not change, and the latter means that doubling the amount of resources but leaving the problem size unchanged should lead to a halving of the solution time, i. e., the parallel speedup is linear in the amount of added compute resources. Different levels of parallelism exhibit different communication characteristics, and for perfect both strong and weak scalability, communication must be fully overlapped with independent computations. One may argue that weak scalability can be identified, in terms of convergence rates rather than time to solution, with *parallel (numeri-*

cal) efficiency: Doubling the number of subdomains and the number of compute nodes but refining each subdomain by the same amount should not change convergence rates.

- Finally, good *hardware efficiency* requires highly optimised implementations, with the goal to extract a significant fraction of the peak performance of modern computer architectures: Hardware efficiency affects both the single-processor performance and the communication characteristics of a given architecture on all levels of parallelism. In practice this means that *high performance computing* techniques, efficient data structures, data layouts and blocking techniques adapted to the memory hierarchy of each architecture must be developed and applied: Different strategies are required for different computer systems and different levels of parallelism.

In practice, these three aspects of efficiency are often mutually exclusive. Algorithms that maximise the numerical efficiency do not necessarily scale well in parallel, and solution techniques exhibiting good strong scaling may converge too slowly for reasonable weak scaling. Different tradeoffs must be taken into consideration for different levels of parallelism, and if hardware efficiency is also taken into account, the search space for an optimally suited algorithm becomes very high-dimensional. Throughout this thesis, we often use the term *computational efficiency* to refer to parallel and hardware efficiency simultaneously. In our group, this field of study is termed *hardware-oriented numerics*, and we discuss its importance in the outset of this thesis in Section 1.1.1.

Obviously, these considerations imply that the different aspects of computational science are strongly interwoven, and scientific computing constitutes more than the sum of its parts from (natural) sciences, computer science, computer architecture and (applied and numerical) mathematics. Consequently, this thesis is also interdisciplinary in nature, and it contributes techniques to improve not only one aspect of efficiency, but techniques to (simultaneously) maximise numerical, parallel and hardware efficiency. In the remainder of this section, we present several instructive examples that help us to motivate the concrete contributions of this thesis which are presented in Section 1.2.

1.1.1. Hardware-Oriented Numerics

Hardware-oriented numerics is, broadly speaking, the combination of high performance numerics and high performance computing, and therefore a subset of scientific computing. We consider a very important example—the solution of a Poisson problem in two space dimensions—to illustrate the core idea of hardware-oriented numerics to *simultaneously maximise numerical and computational efficiency*. This scalar elliptic PDE is a fundamental prototype for all problems encountered and examined in the (wider) scope of this thesis: Poisson problems are encountered in fluid dynamics (pressure-Poisson problem in operator splitting approaches) and solid mechanics (Lamé equations discretised with separate displacement ordering), but for instance also in electrostatics (potential calculations). They are known to be hard to solve, because the condition number of the corresponding discrete linear system depends on the mesh width of the discretisation, which means that they are harder to solve the finer the discretisation becomes (the more accurate the continuous PDE is approximated).

To emphasise the importance of asymptotics, a simple back-of-the-envelope calculation is very instructive; it should of course be taken with a grain of salt: Let us consider a Poisson problem, discretised with bilinear finite elements for a problem size of 1.28 billion ($1.28 \cdot 10^9$) unknowns (this is the largest problem solved in this thesis). If one used standard Gaussian elimination to solve the linear system resulting from the discretisation of this problem, and if the system matrix were stored densely, then approximately $1.4 \cdot 10^{27}$ floating point operations would be needed for the concrete example at hand: Gaussian elimination to solve a dense linear system with N unknowns executes asymptotically $2N^3/3 + \mathcal{O}(N^2)$ floating point operations. The RoadRunner system installed at Los Alamos National Laboratory has been the world's first supercomputer to break the petaflops barrier, it is capable of achieving 1.105 PFLOP/s (10^{15}) floating point operations per second in the Linpack test.[1] Even with this tremendous level of performance, it would still take more than 40 175 years to solve the problem using Gaussian elimination. The fastest multigrid solver we present for this problem in Chapter 6, which furthermore exploits the sparsity pattern of the system matrix, computes the solution in less than one minute.

In the following, we consider a prototypical situation that can easily be computed with a conventional, serial code. In Section 2.3 we discuss the aspects of parallel hardware-oriented numerics that are related to distributed memory clusters in detail. Figure 1.1 depicts the (unstructured) coarse grid of this test problem comprising eight elements, and the numerical solution computed on a refined grid. Bilinear quadrilateral conforming finite elements are employed for the discretisation, and uniform refinement is applied so that a *refinement level* L corresponds to $N = (2^L + 1)^2$ unknowns: For $L = 10$, the problem size considered in the following, the linear system to be solved has eight million unknowns.

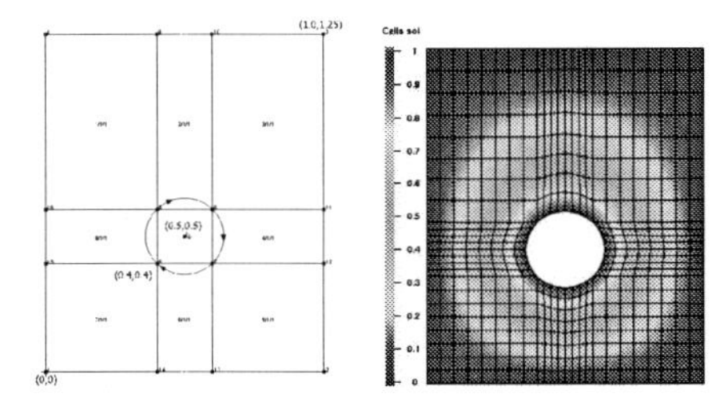

Figure 1.1: Test domain and coarse grid ($L = 0$, left) and computed solution (right) for a simple experiment to demonstrate the ideas of hardware-oriented numerics.

This experiment has been performed in joint work with fellow PhD students [137], and two software packages are evaluated, FEAT and FEAST.[2] FEAT is an established and actively used finite element analysis software package primarily aiming at numerical flexibility and efficiency [27]. FEAST is a next-generation toolkit that prototypically implements many concepts of hardware-oriented numerics. Both packages are actively being developed in our group, and the author of this thesis has contributed significantly to FEAST during the pursuit of this thesis.

To solve the problem depicted in Figure 1.1, FEAT assembles a global matrix and right hand side, the matrix is stored in the standard *compressed sparse row (CSR)* format,[3] which involves indirect memory accesses to iterate over the nonzero entries. We consider two numbering schemes for the unknowns, a lexicographical enumeration first along the x-axis and then along the y-axis (denoted as 'XYZ'), and a fully stochastic enumeration. The latter is often used to emulate computations on fully adaptive, highly unstructured grids. We refer to the underlying technical report [137] and to Turek [224] for more details on different numbering schemes, and to the textbook by Saad [191] for a discussion of various sparse matrix storage formats. In contrast, FEAST uses a block-structured approach with (sub-) matrices that correspond to each refined coarse element, and a linewise numbering in each of these patches (see Section 2.3). The benefit of this approach

[1] See `http://top500.org/lists/2008/11` for details.

[2] see `http://www.featflow.de` and `http://www.feast.tu-dortmund.de`

[3] In the literature, this format is often also referred to as *compact row storage*, *compressed sparse row* and other acronym permutations thereof.

is that every submatrix has a banded structure, and this special structure is exploited in the design and implementation of numerically and computationally efficient data structures, preconditioners and other operations from numerical linear algebra. In particular, each band is stored as an individual vector, and matrix-vector multiplication does not require 'pointer chasing' through memory. Furthermore, blocking to efficiently exploit cache hierarchies and other aspects of modern processors is enabled by FEAST's approach. In summary, FEAST's discretisation technique is designed around *locally structured, globally unstructured grids* (which does not prevent adaptivity per se), and we refer to Section 2.3 for more background on FEAST.

Two different solvers are evaluated, conjugate gradients as a representative of simple Krylov subspace schemes, and multigrid, which is asymptotically optimal for this elliptic problem. In both cases, we employ either a Jacobi preconditioner (labeled `simple`), or the most numerically and computationally efficient preconditioner available. For the CSR matrices that can incorporate changes of the nonzero pattern easily, preconditioners based on incomplete LU factorisations (ILU) are commonly considered to be most efficient, at least among the techniques that have some kind of black-box character. Preconditioners that maintain the computationally favourable band format are harder to design, and we choose an alternating direction implicit combination of Gauß-Seidel and line-relaxation (ADITRIGS, see Section 2.3 for details). All measurements are obtained on a single core of a compute server equipped with Opteron 2214 processors and fast DDR2-667 memory, a machine that is used frequently throughout this thesis.

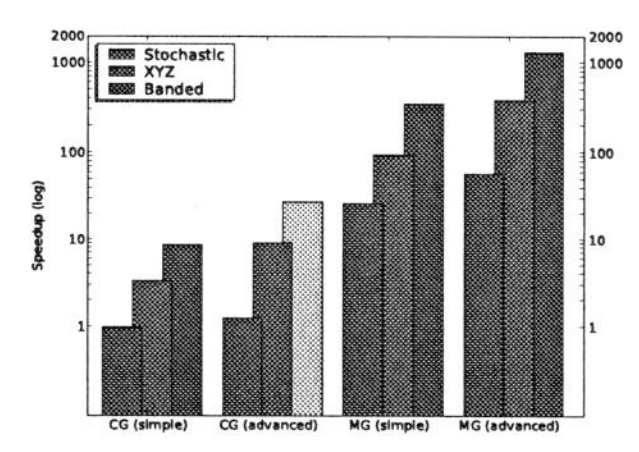

Figure 1.2: Relative performance improvement obtained by hardware-oriented numerics for a Poisson problem, normalised timings. The y-axis uses a logarithmic scale.

Figure 1.2 illustrates the performance results; as only relative improvements are relevant for the argument in this introductory section, all measurements are normalised to the slowest configuration. The data point for the conjugate gradient solver using an advanced preconditioner for the banded data layout is estimated, because at the time these tests have been performed, a symmetric advanced preconditioner has not been available in FEAST. We first observe that the data layout and the data access patterns are of great importance for the efficiency of algorithms running on modern computer hardware. A simple reordering of the unknowns improves performance by a factor of 3 without changes to the data layout, and better data layouts can gain a factor of 10–30. Very 'bad' data layouts (e. g., stochastic numbering) let the performance drop to a mere 1 % of the theoretical peak floating point throughput of the Opteron processor. The technical report lists all numbers in detail [137]. On the other hand, it is essential that the structure of the mathematical problem does not get lost in the design of the solvers. Utilising a hierarchical solver that exploits multigrid structures quickly pays off and may lead—in combination with the 'correct' data layout—to a speedup by a factor of 1 300 and even more in computation time compared to a straightforward implementation of a simple numerical scheme. An extremely tuned implementation of a numerically poor algorithm is easily outperformed by a more advanced numerical scheme, so the importance of advanced numerical methodology clearly dominates all implementational aspects.

But on the other hand, advanced algorithms implemented without awareness of the underlying machine architecture exhibit degrading performance, as we demonstrate below.

Consequently, our hypothesis is that in the field of high performance finite element simulations, significant performance improvements can only be achieved by maximising computational and numerical efficiency simultaneously: Numerical and algorithmic foundation research must accompany (long-term) technology trends; and prospective hardware trends must enforce research into novel numerical techniques that are in turn better suited for the hardware. This view is shared by many other researchers: Keyes [126] and Colella et al. [50] survey trends towards terascale computing for a wide range of applications including finite element software and conclude that only a combination of techniques from computer architecture, software engineering, numerical modelling and numerical analysis will enable a satisfactory scale-out on the application level, see also the bibliography of work related to FEAST in Section 2.3. The following observation, which motivates hardware-oriented numerics, thus summarises the context of this thesis:

> *Substantial performance improvements can only be achieved by hardware-oriented numerics, i. e., by simultaneously maximising numerical and computational efficiency.*

While these examples and conclusions are all based on serial code, the same line of argument holds for large-scale parallel computations executing on distributed memory clusters. This is an important aspect in view of the contribution of this thesis, but to keep this introduction concise, we do not argue along these lines. The interested reader is referred to the introductory chapters of previous theses on FEAST for motivations which focus on this aspect [15, 130, 242], and to Section 2.3 and the references therein for an overview of FEAST.

1.1.2. The Memory Wall Problem

The sustained bandwidth and the latency to access data in off-chip memory is much more important for grid-based codes than the raw compute performance, as the discretisation process leads to large, but very sparse matrices. The *arithmetic intensity*, defined as the ratio of floating point operations per memory access, of computations involving such matrices is very low, commonly less than one. This is problematic, because an access to off-chip memory can cost hundreds of clock cycles, while modern processors are able to execute several arithmetic operations per cycle in their vector units: *Computing is cheap, and data movement is expensive.* This problem is known as the *memory wall problem*, and is further aggravated by the fact that peak processor performance continues to increase at a much faster pace than peak memory bandwidth: Over the past decade, processor speed has been improving at 60 % per year (this is not to be confused with Moore's Law), whereas DRAM access is only getting faster by 10 % anually [97, 134].

The common approach in computer architecture and processor design to alleviate the memory wall problem is to employ ever larger hierarchies of fast, on-chip cache memories. In order to exploit these caches, *data reuse* must be maximised on the software side. In view of hardware-oriented numerics, techniques like temporal and spatial blocking have been developed to keep data in-cache for as long as possible. In combination with vectorisation techniques, efficient block-memory transfers that better exploit the available bandwidth are enabled. Similarly, current processors support asynchronous prefetching of data between different memory levels. Section 2.3 provides references to early and recent work on the topic. These observations lead to the second fundamental motivation for the work presented in this thesis:

> *Memory performance and memory performance optimisations are much more important than (peak) compute performance for sparse codes.*

1.1.3. Multicore Architectures

Software development for finite element problems, driven by the mathematical community, is traditionally focused on the improvement of the numerical methodology. For instance, adaptivity techniques, higher order discretisations, and a wide range of stabilisation approaches are being used, developed and refined. Hardware aspects used to play only a minor role, since codes automatically ran faster with each new generation of processors during the past decade. This trend has come to an end, because physical limitations have led to a *paradigm change in processor architectures*: Aggressively optimising circuits, deep processor pipelines and in particular large on-chip cache memories have increased the power requirements and consequently the heat dissipation beyond acceptable scales, both for supercomputing architectures and for commodity processors. Furthermore, leaking voltage in high-end chip designs has become a problem that is increasingly hard to alleviate. These aspects are commonly summarised with the term *power wall*, hardware designers are able to put more transistors on a chip than they can afford to turn on. As a consequence, frequency scaling is limited to 'natural frequency scaling' by process shrinking; the reduction in feature size is, roughly speaking, proportional to the number of admissible transistors under a fixed power budget. Performance can no longer be improved by simply increasing the CPU clock frequency. On the other hand, ever wider superscalar architectures with more aggressive out-of-order execution have reached their limits as well, the hardware is no longer able to extract enough instruction-level parallelism to hide latencies (*ILP wall*). An important consequence is:

> *Together, the power wall, the memory wall and the ILP wall form an 'impenetrable brick wall' for serial legacy codes [8].*

Processor manufacturers have begun to migrate towards on-chip parallelism in the form of *chip multiprocessors (CMP)*, these designs are more commonly known as *multicore architectures*: The (approximately) doubled amount of transistors per chip generation (Moore's Law) is invested into a doubling of processing cores. This trend has reached the mass market in 2005, when both AMD and Intel introduced commodity multicore CPUs with their Athlon 64 X2 and Pentium D chips. At the time of writing, quadcore processors are standard, and the first six- and eight-core designs are appearing. As a consequence, the performance of each individual core is remaining constant or even degrades, which leads us to the next important motivation of this thesis:

> *Single-threaded legacy codes no longer run faster automatically.*

We illustrate this important observation based on FEATFLOW [225], a CFD solver built on top of FEAT which has already been used for the experiment presented in Section 1.1.1. FEATFLOW draws its strength from state-of-the-art numerical techniques and a flexible design to facilitate the straightforward addition of novel finite-element related functionality, but its implementation is 'typical' for the mathematical community in the sense that it does not include hardware-specific optimisations at all: Standard data structures like CSR to store sparse matrices are employed, and all optimisations are left in the responsibility of the compiler. In addition, FEATFLOW is strictly single-threaded. Since the early days of its development 15 years ago, FEATFLOW contained a suite of well-defined CFD benchmarks, and the project website[4] lists timing results on workstations constituting the state of the art during these years. Figure 1.3 depicts the relative performance gain based on this benchmark data. In the covered time frame from 1993 to 2008, hardware progress (and associated improvements in compiler technology) delivered a speedup by a factor of 80, essentially for free. While this number is indeed impressive, CPU peak performance in these 15 years increased more than 1 000-fold. The reason for this discrepancy is the memory

[4] http://www.featflow.de

wall problem, see Section 1.1.2. However, the important observation in the outset of this thesis is that the *free* performance gain has begun to stagnate with the advent of multicore processors. We observe this postulated consequence in an established code already, and not in (artificially constructed) microbenchmarks.

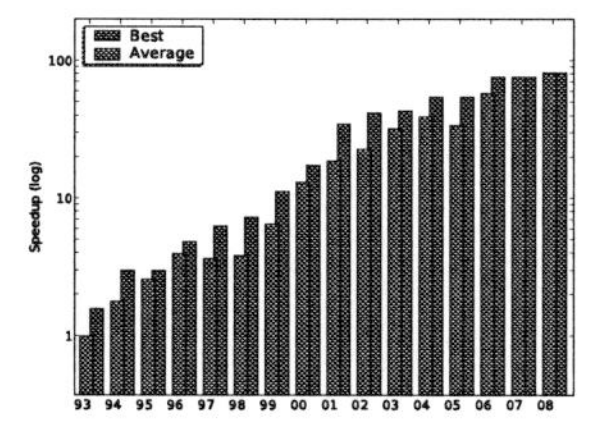

Figure 1.3: Normalised performance of the single-threaded FEATFLOW benchmark, 1993–2008. The bars labeled 'best' represent the fastest machine up to a given year, and the bars labeled 'average' correspond to the averaged performance of all machines evaluated in one year. The y-axis uses a logarithmic scale.

Consequences for Numerical Codes

This development shifts most of the responsibilities to maximise efficiency and, more importantly, *scalability and usability of a given software on future hardware generations* from the hardware architects and the compiler community to the software developer: Task and data parallelism must be exploited not on the level of coarse grained parallelism (compute nodes in a distributed memory cluster), but explicitly at a finer scale:

> *The evolution towards on-chip parallelism is evident, and exploiting it is in the responsibility of the software developer.*

Multicore designs significantly increase the peak compute performance from one hardware generation to the next, because the number of cores increases exponentially. At the same time, memory performance continues to improve at a significantly slower rate. Due to *pin limits* (the limited amount of physical data paths to move data from off-chip memory to the processing cores), the memory wall problem is even worsened by multicore architectures: With memory controllers moving on-chip and core counts increasing rapidly, more and more parallel computations have to compete for the limited bandwidth to off-chip memory, which often scales with the number of sockets per compute node rather than the number of cores per processor [241]. In addition, memory access speed becomes increasingly non-uniform, as on-chip cache memories used for fast data exchange between cores are not shared by all cores, and on-chip memory controllers maintain only their local portion of main memory. This leads us to the final observation to be taken into account in the work presented in this thesis:

> *In conventional general-purpose architectures, the trend towards on-chip parallelism aggravates the memory wall problem and prevents future scalability of legacy finite element codes.*

In high-performance computing, innovations stemming from research on *symmetric multiprocessing (SMP)* shared memory systems during the last decade have lead to the development of libraries and programming environments to exploit the parallelism within a compute node, the most prominent example is OpenMP.[5] OpenMP enables the programmer to declare parallelism by explicitly inserting certain preprocessor statements into the code. Many current research efforts aim at bridging the gap between SMP and CMP systems, in particular in view of the non-uniform memory characteristics mentioned above. It is important to note that at the time of writing, compiler support to automatically parallelise a given finite element code using threads is still in its infancies because this task requires global knowledge of the data flow, which is hard to extract by compilers acting locally on the instruction stream. Efficient autoparallelisation is not expected in the medium term [8].

1.1.4. Emerging Many-Core Architectures

In this thesis, we go one step further and do not consider multicore CMP parallelism, but rather the massively fine-grained parallelism exposed by unconventional yet promising *emerging manycore architectures*. Currently available multimedia processors such as the Cell BE processor and—in particular—*graphics processing units (GPUs)* have gained increasing interest in view of the above mentioned trends and issues. In fact, they are considered as forerunners of these trends, and are commonly labeled emerging many-core architectures, a nomenclature we follow throughout this thesis. This observation makes these chip designs very attractive outside their traditional application domains, for general-purpose and (in the scope of this thesis) high performance numerical computations. The author of this thesis has actively researched the potential of these architecture for finite element computations since 2005, see also Section 1.2. GPUs in particular alleviate many of the issues by their fundamentally different processor design, because they exploit parallelism to a much greater extent than conventional CPUs, and because they are optimised for maximising throughput of a given computation rather than latency of an individual operation. This is achieved by *massive hardware multithreading*, in combination with a parallel memory subsystem.

As a concrete motivation, we list several important features of the NVIDIA GeForce GTX 280 graphics accelerator, the most powerful GPU employed throughout this thesis: Its 30 so-called multiprocessors ('cores') and its 512 bit interface to main memory result in rather impressive peak numbers of over 1 TFLOP/s raw compute performance and 140 GB/s of bandwidth to fast off-chip memory, at least an order of magnitude higher than conventional designs. A single such graphics board consumes 236 W under full load, and costs less than 400 Euro, placing it at the 'bleeding edge' not only in raw peak numbers, but also in increasingly important metrics such as performance/Watt and performance/Euro. In view of the memory wall problem, it still costs approximately 600 clock cycles to access data in off-chip memory, but these GPUs can 'hide' this latency via a very efficient lightweight thread scheduling mechanism implemented directly in hardware, keeping more than 30 000 threads 'in-flight' simultaneously. For comparison, HyperThreading technology in current quadcore CPUs maintains two threads per core. Stalled threads waiting for memory transactions are suspended, and automatically resumed once the data is available. In essence, GPUs are specialised for high throughput rather than minimisation of latencies for random access, they excel at *data-parallel computations*. It is worth noting that the multi-billion dollar market of video games entails rapid evolution and superlinear performance improvements across chip generations at comparatively low acquisition costs, in contrast to more specialised solutions aiming at the HPC market alone.

In the next section, we describe how the development towards fine-grained parallelism lead to the contribution of this thesis.

[5] `http://openmp.org/`

1.2. Thesis Contributions

The main contribution of this thesis, in view of the paradigm shift outlined in Section 1.1, is to demonstrate that *GPUs as current representatives of emerging many-core architectures are very well-suited for the fast and accurate solution of PDE problems*, in particular for the solution of large sparse linear systems of equations on large compute clusters: We report on at least one order of magnitude speedup over highly-tuned conventional CPU solutions, *without sacrificing neither accuracy nor functionality*. This improvement in performance is achieved over an already highly optimised, numerically and computationally efficient implementation that follows the concept of hardware-oriented numerics, see Section 1.1.1. Speedups of this order are highly relevant in practice, because the asymptotical 'free' speedup for bandwidth-limited codes is a mere 10 % per year [97], see also Section 1.1. This contribution is achieved by combining techniques from numerical mathematics *and* computer science to design algorithms and data layouts that efficiently exploit fine-grained parallelism, and to design numerical schemes that scale perfectly with thousands of concurrently active threads. Consequently, this thesis as a whole is interdisciplinary in nature.

The individual contributions of this thesis are summarised thematically below. In addition, we provide pointers to later sections which explain the results in detail. Section 1.3 presents our contributions in a chronological fashion by listing the (peer-reviewed) articles and conference proceedings that have been published in the pursuit of this thesis to disseminate results and novel contributions in a timely manner. Finally, Section 1.4 provides a linear sorting of the topics covered in this thesis. We summarise this thesis at three different angles to make the contents as accessible as possible.

1.2.1. Mixed Precision Iterative Refinement

Floating point arithmetic on computer systems provides finite, limited precision, and computational precision and result accuracy are related in a highly unintuitive, 'nonlinear' way. An important theoretical result states that the condition number of a linear system essentially determines the (minimally) required computational precision to deliver an accurate result. Due to truncation and cancellation errors, this theoretical result however does not tell the whole story. For all problems considered in this thesis, i. e., for the prototypical Poisson problem, for linearised elasticity and for stationary laminar flow, the condition number of the system matrix depends on the discretisation mesh width, and thus deteriorates with increasing problem size. We demonstrate that already for simple test cases that do not involve complex geometries, anisotropic and/or locally adapted grids, standard single precision floating point arithmetic may be insufficient and lead to wrong solutions, see Section 4.5.

Mixed precision methods have been known for more than 100 years already. They have gained rapid interest with the arrival of computer systems in the 1940s to 1960s. At that time, floating point computations had been much more expensive than memory accesses. The idea of these methods is to iteratively improve a solution computed in low precision, by using a high precision defect correction loop: The majority of the computations are carried out in cheaper low precision. To maximise efficiency, one obviously wants to execute as many operations as possible in a floating point format that is *natively* supported in hardware. If the problem at hand is 'not too ill-conditioned' in low precision, iterative refinement provably converges to high precision accuracy. Not natively available higher precision formats can also be emulated in software, for instance with the so-called double-single technique (native pair arithmetic), which uses only natively available arithmetic operations.

Since that time, mixed precision iterative refinement has received little attention in the community. The commonly applied procedure has become, in the past 20 years, to execute all computations entirely in double precision, the highest precision natively available in commodity hardware. This is an immediate consequence of the fact that, for a wide range of practical problems, double precision yields sufficient accuracy.

However, double precision may not be available on some hardware, or may be significantly slower. In view of the memory wall problem, using single precision is also advantageous, because the bandwidth requirements are halved, and twice the amount of data can be stored at each level of the memory hierarchy. In this thesis, we therefore revisit these techniques from a memory performance perspective, which has not commonly been done in the past years. Chapter 4 is dedicated to this contribution: The theoretical foundation of mixed precision methods is based on the exact solution of the auxiliary low precision problems, which is not possible in our context where approximate, iterative methods for the large, sparse systems have to be employed. After reviewing related work and the theoretical background, we present a mixed precision iterative refinement framework that is specifically designed for this situation: An outer iterative refinement loop is preconditioned with an inner multigrid or Krylov subspace solver.

We thoroughly analyse our method with a wide range of numerical experiments in a variety of different contexts in Sections 5.6–5.8 and Sections 6.3–6.5. In particular, we demonstrate the following important findings: Even for severely ill-conditioned problems (condition numbers on the order of 10^{15}), our mixed precision solvers always compute the same result as if computing entirely in double precision. This is not the case when using an emulated precision format based on the unevaluated sum of two native single precision values. When employing multigrid to approximately solve the auxiliary problems to no more than two digits accuracy, we obtain a speedup of 1.7–2.1 over executing entirely in double precision, on CPUs and on GPUs. The combination of these two results means that mixed precision iterative refinement, executing hybridly on the CPU and on accelerators that do not support double precision at all, is the only viable approach to exploit the performance of such hardware without sacrificing accuracy for the problems examined in this thesis. Reducing the residuals further is, in contrast to theory that is based on exact solutions, not beneficial because the inner solvers converge to wrong solutions in single precision or may not converge at all. The speedup is in line with the bandwidth considerations outlined above, and underlines the increased efficiency of our solvers.

Furthermore, even though we do not examine this aspect in detail, the mixed precision approach is very beneficial for problems that are too ill-conditioned even for (native) double precision. A higher-precision format has to be emulated expensively in software, and analogously to the case of double and single precision examined in this thesis, minimising the amount of arithmetic work performed in the expensive format is beneficial for the overall execution time.

1.2.2. Co-Processor Acceleration of Existing Large-Scale Codes

The parallelisation and specialisation of resources as observed in many-core processor designs results in a major challenge for the programming model, in particular for clusters and supercomputers where the coarse-grained parallelism (handled by message passing among distributed memory nodes) must interact with the fine-grained parallelism in co-processors and the increasing heterogeneity (and medium-grained parallelism) within each compute node. In Section 6.1 we argue that re-implementations of codes for each emerging new architecture are prohibitively expensive.

In view of these challenges, an important contribution of this thesis is a *minimally invasive* method to incorporate unconventional and emerging hardware into the existing large-scale finite element toolkit FEAST, which we consider typical in this context (even though it is, at this point, 'only' a research code). This contribution bridges the gap between numerical mathematics and software engineering. FEAST [15] comprises more than 100 000 lines of code in its core modules already; Section 2.3 describes its design and implementation in detail. It is designed around a multilevel, multilayer domain decomposition approach called ScaRC [130] and executes on large commodity based clusters and supercomputers. In two other theses, FEAST has been extended by applications for the challenging large-scale solution of solid mechanics [242] and fluid dynamics [44] problems.

The core idea of our approach is to take advantage of the multilayer ScaRC scheme that clearly separates the 'outer' MPI layer from the 'inner' local layer: We offload the entire local work of the Schwarz preconditioner onto co-processors, and leave the 'outer' layer unchanged. The proposed methodology is described in Chapter 6, and has two important advantages: First, it encapsulates the heterogeneities within each compute node so that MPI sees a globally homogeneous system; and second, applications built on top of FEAST benefit from the acceleration through (unconventional) fine-grained parallel co-processor hardware without any code changes.

We have implemented this approach, prototypically using GPUs as co-processors (see also the next Section 1.2.3). The resulting 'FEASTGPU' extension of FEAST enables the execution of large-scale computations on heterogeneous GPU-enhanced clusters. We present substantial performance improvements (in terms of time to solution and energy efficiency) for the prototypical Poisson problem, and the two FEAST applications FEASTSOLID and FEASTFLOW (cf. Sections 6.3–6.5), for computations on clusters ranging from 4 to 160 GPUs. Furthermore, we analyse hybrid solvers that simultaneously perform work on the CPUs and the GPUs. Excellent weak scalability is demonstrated on a cluster of 160 GPU-enhanced nodes. We show that the mixed-precision two-layer ScaRC scheme, analogous to the small-scale experiments summarised in Section 1.2.1, is faster and delivers exactly the same accuracy as if computing in double precision alone, for large-scale computations with up to one billion unknowns for challenging problems from linearised elasticity. As we only accelerate portions of the entire solver, the speedup is limited by the fraction of time spent in the unaccelerated parts. Depending on the concrete configuration, we accelerate between 50 and 90 % of the entire solution scheme, so the total speedup is bounded between factors of 2 and 10. We develop an analytical model similar to Amdahl's Law to predict the total speedup based on this accelerable fraction (which is a property of the unmodified CPU-based code) and the local speedup (which can be measured in local solver benchmarks, see Section 1.2.3), and show that the predicted values are recovered in the experimental results.

1.2.3. GPU-Based Multigrid Solvers on Generalised Tensor Product Domains

The third contribution of this thesis is concerned with scalable fine-grained numerical and computational parallelisation techniques for emerging many-core architectures. We develop algorithms to efficiently solve sparse linear systems using multigrid and Krylov subspace methods on GPUs, specifically tailored to finite element discretisations on generalised tensor product domains. These solvers constitute the main building block in the scope of our minimally invasive co-processor acceleration outlined in the previous section. The presented techniques are not limited to finite element based discretisations, but rather applicable for similar grid-based techniques like finite differences or finite volumes.

Many of the required operations parallelise in a straightforward manner, at least after suitable, nontrivial reformulations, or by applying reduction techniques. We describe these algorithms for both the 'legacy' GPU programming model of using graphics APIs, and for the 'modern', more powerful NVIDIA CUDA architecture. However, numerically strong preconditioners and multigrid smoothing operators, which are required for challenging problems and high degrees of mesh or operator anisotropy, exhibit recursive, inherently sequential data dependencies, and their parallelisation is thus very challenging. In Section 5.4 we contrast (exact) wavefront with (inexact) multicolouring techniques to parallelise such preconditioners. Only the latter are well-suited for the GPU architecture in our case because they exhibit a sufficient amount of independent parallelism. We develop a favourable renumbering technique for Gauß-Seidel type preconditioners that alleviates many of the disadvantages induced by the lesser degree of coupling inherent to multi-colouring. Furthermore, we present an efficient implementation of cyclic reduction for parallel line relaxation preconditioning (the solution of tridiagonal linear systems).

We thoroughly analyse the numerical and computational efficiency of these solvers in Sections 5.5–5.8, in combination with mixed precision iterative refinement methods. Our most important results are: Multicolouring reduces the numerical efficiency of a multigrid solver, but our favourable renumbering scheme almost recovers the convergence rates of the sequential case. For 'simple' preconditioners, we achieve speedup factors of at least one order of magnitude over optimised CPU code, and in many cases, the speedup is even 20- to 40-fold. F and in particular W cycle multigrid reduces the achievable speedup significantly, due to the amount of time spent on coarser levels of refinement where simply not enough independent parallelism is available to saturate the GPU. We concisely summarise our findings by comparing (well-known) CPU performance characteristics and guidelines with the GPU case in Section 5.9.

1.2.4. A Short History of GPU Computing

During the pursuit of this thesis since early 2005, the view of GPUs as many-core processors has changed frequently, and the underlying programming model to exploit fine-grained parallelism has been continuously adapted to technological advancements. When the author started his venture into GPU computing (also known as *general purpose computations on GPUs (GPGPU)*), this field of research had been obscure, and only few researchers had driven it forward. In these days, GPU computing was commonly identified with a rather ad-hoc, 'hacky' approach and a general 'proof of concept' attitude, mostly because fundamental algorithmic research had to be 'hidden' behind using graphics APIs to indirectly program the device. Nonetheless, GPU computing has gained significant momentum since these early days, and has evolved into an established research area. Hardware vendors have recognised the benefits of GPU computing and have provided high-level programming environments to express parallelism more efficiently. GPU computing is nowadays no longer restricted to academia; commercial products and applications are beginning to appear. A noteworthy contribution of this thesis is thus our historical survey of the evolution of GPU hardware and GPU computing in Chapter 3, with a special focus on the numerical simulation of PDE problems and sparse problems in general. In this chapter, we also explain architectural differences between CPU and GPU processor designs, and discuss why GPUs are a well-suited architecture for data-parallel numerical computations.

1.3. List of Publications

During the pursuit of this thesis, the author has published and contributed to a number of peer-reviewed book chapters and journal articles, and intermediate results have been presented at international conferences and published in peer-reviewed proceedings volumes. As GPU computing evolved rapidly during the pursuit of this thesis, early dissemination of novel contributions has been and remains very important. In the following, we list these publications chronologically, ordered by publication type, and link them to the various sections (and covered topics) of this thesis.

The author of this thesis is convinced that research (in scientific computing and elsewhere) is fostered *only* by group efforts, and consequently, all publications listed below have been written in close collaboration with a number of co-authors. This is particularly true for large-scale software projects like FEAST, which involve a large amount of cooperative programming effort: Different publications concerning different aspects and features take different angles, and the order in which authors are listed varies. Lead-authorship in each of the publications below implies that the main novel contribution is attributed to the author of this thesis. But, and this is very important, without the preceeding and joint work within the FEAST group (Susanne Kilian, Christian Becker, Sven Buijssen and Hilmar Wobker) and the collaborations with colleagues (Robert Strzodka, Jamaludin Mohd-Yusof and Patrick McCormick, and Dimitri Komatitsch, Gordon Erlebacher and David Michéa), most of the publications would not have been possible at all. We owe gratitude to all collaborators, and their contribution is gratefully acknowledged.

Book Chapters

In an upcoming textbook tentatively titled *Scientific Computing with Multicore and Accelerators* and scheduled for publication in late 2010, one chapter will describe FEAST and the GPU-acceleration of FEAST developed by the author of this thesis (see Chapter 6). This chapter has been co-written by the entire currently active FEAST group at TU Dortmund: Stefan Turek, Sven Buijssen, Hilmar Wobker and the author of this thesis [234]. A second chapter, co-authored with Robert Strzodka [80], describes fine-grained parallelisation techniques for numerically strong preconditioners in mixed precision multigrid solvers. The contents of this chapter can best be summarised as a 'didactically enhanced' version of Section 5.4 of this thesis.

Journal Articles

At the time of writing, five journal articles have been published, and three additional ones have been accepted but have not been published yet.

Our first journal article appeared in the *International Journal of Parallel, Emergent and Distributed Systems* in early 2007, co-authored with Robert Strzodka and Stefan Turek. It surveys hardware accelerators, and evaluates native, emulated and mixed precision iterative solvers for sparse linear systems [85]. It constitutes a significantly extended and expanded update of our very first conference proceedings paper [82] which has been, to the best of our knowledge, the first publication that reported on double precision accuracy on GPUs that natively only support single precision (at that time), with a significant speedup over double precision on the CPU. Chapter 4 and several sections in Chapter 5 are concerned with an update of this publication.

Our concept of the minimally invasive integration of hardware accelerators into FEAST (see Chapter 6) has been published in a series of three articles. The first one, in the *International Journal of Computational Science and Engineering* [87], presents the general concept, and its evaluation for the prototypical Poisson problem. The second one appeared in *Parallel Computing* in 2007 and extends the analysis to weak scalability and energy efficiency on a cluster of 160

GPUs [84]. Due to delays in the publication process, this paper finally appeared before the first one. The third paper in the series, again in the *International Journal of Computational Science and Engineering* [90], applies our concept to the acceleration of FEASTSOLID, an application built on top of FEAST for the simulation of large-scale solid mechanics problems (developed by Wobker [242]). These articles have been co-authored with Robert Strzodka, Hilmar Wobker, Jamaludin Mohd-Yusof, Patrick McCormick and other members of the FEAST group.

The author of this thesis also contributed significantly to another journal paper by the FEAST group, which emphasises the concepts of hardware-oriented numerics and compares the performance of FEAST on supercomputers and GPU-accelerated clusters. This article is accepted for publication in *Concurrency and Computation* [233]. Software design issues in heterogeneous high performance computing with accelerator hardware have been pursued in the HONEI project at TU Dortmund, the design approach and preliminary results have been published in *Computer Physics Communications* [236].

Our efficient implementation of cyclic reduction for the parallel solution of tridiagonal systems (cf. Section 5.4), co-developed with Robert Strzodka, has been written up for publication in the *IEEE Transactions on Parallel and Distributed Systems*, and recently got accepted [81].

The results from recent joint work with Dimitri Komatitsch, Gordon Erlebacher and David Michéa on accelerating a finite element seismic wave propagation simulation on a large cluster of 192 GPUs have been accepted for publication in the *Journal of Computational Physics* in April 2010 [136].

Conference Proceedings

The early dissemination of intermediate results at international conferences, and the resulting discussions with expert practitioners in the field, is extremely important. Highlights during the pursuit of this thesis include:

The very first paper by the author of this thesis, presented at a small conference in Germany, has been to the best of our knowledge the first to achieve double precision accuracy on single precision GPUs by revisiting mixed precision iterative refinement methods from a performance perspective [82]. This work has gained quite some interest in the community, and as a consequence, we have been invited to present our work in a minisymposium held in conjunction with the renowned *ACM/IEEE Supercomputing conference* in November 2006.

The minimally invasive co-processor acceleration, applied to the solution of large-scale linearised elasticity problems, has been accepted as a poster presentation a year later [86].

A submission to the *International Conference on Supercomputing (ISC)* in 2008 reporting on hardware-oriented numerics as implemented in FEAST and applied to our approach to integrate GPU acceleration (co-authored by the entire FEAST group), has received the PRACE award [6] which recognises contributions of young researchers towards petascale computing.

Our development of mixed precision schemes has been recognised by NVIDIA, and we have been able to experiment on early engineering samples of their first GPU architecture that supports double precision natively in hardware [79]. The results have been presented in an invited talk at their first NVISION conference on visual computing.

In a contribution to the *IEEE High Performance Computing & Simulation* conference in 2009, we have explored the limitations of our minimally invasive integration approach by applying it to the acceleration of stationary laminar flow simulation, i. e., by accelerating the application FEAST-FLOW [88].

Finally, the *SIAM Computational Science and Engineering Conference 2009* is worth noting because of its impact factor [89].

[6] http://www.prace-project.eu/news/prace-award-presented-to-young-scientist-at-isc201908

1.4. Thesis Outline

This thesis is written so that all chapters are self-contained, i. e., each chapter provides its own motivation and summary. Nonetheless, the chapters all build on each other. In the following, we provide a high-level overview of each chapter, and refer to their respective introductory sections for a more detailed outline.

In Chapter 2 we summarise the necessary background. Section 2.1 presents, exemplarily for the Poisson problem, the core ideas of finite element discretisation. Section 2.2 introduces iterative solvers for sparse linear systems of equations, in particular Krylov subspace and multigrid methods. FEAST as the current prototypical implementation of a wide range of concepts from hardware-oriented numerics is described in detail in Section 2.3.

Chapter 3 provides an overview of GPU hardware and GPU computing. We take up a historical perspective and present the (concurrent) evolution of the hardware and the programming model in detail. Section 3.1 is concerned with the 'legacy GPGPU' approach to program graphics hardware through graphics APIs, while Section 3.2 covers 'modern GPGPU' through dedicated compute APIs such as NVIDIA CUDA or AMD Stream. In these sections, we present both hardware and software aspects; and furthermore provide a bibliography of related work which underlines the evolution of GPU computing, with a special focus on the solution of PDE problems, since its early, obscure beginnings. In Section 3.3 we discuss GPUs in terms of parallel computing taxonomies and parallel programming models. The chapter concludes with an outlook on future programming APIs, future graphics architectures and the convergence of CPU and GPU designs in Section 3.4.

Chapter 4 is concerned with native, emulated and mixed precision methods. Section 4.1 motivates their use and briefly revisits floating point arithmetic, using a number of instructive examples that highlight common pitfalls in floating point computations. Section 4.2 explains techniques to emulate high precision formats using only (native) low precision data and operations, and thus techniques that are required when no native high precision format is available. Section 4.3 presents mixed precision methods from a historical and numerical perspective, and provides the necessary theoretical background and a bibliography of related work. Here, we also introduce our mixed precision iterative refinement solver tailored to the solution of large sparse linear systems of equations. Mixed and emulated precision techniques are discussed from a hardware efficiency and hardware design point of view in Section 4.4. Finally, we exemplarily analyse the precision requirements for the accurate solution of Poisson problems (prototypical for all applications considered in this thesis) in much detail in Section 4.5.

In Chapter 5, we describe, analyse and evaluate our GPU implementation of efficient linear solvers specifically tailored to systems stemming from the discretisation of scalar PDEs on generalised tensor product grids. These meshes correspond to individual subdomains in FEAST, and furthermore, the solvers we present and evaluate in this chapter constitute the local solvers (parts of the Schwarz preconditioner) of the two-layer SCARC approach in FEAST. We are interested in gaining insight into GPU performance characteristics for iterative linear solvers and sparse equation systems in contrast to conventional CPU implementations, the various tradeoffs in parallelism and the performance improvements of GPUs over CPUs. Section 5.1 provides a detailed overview of the various contributions of this chapter. In Section 5.2 and 5.3 we describe the efficient implementation of numerical linear algebra operations and multigrid solver components, both in terms of 'legacy GPGPU' programming techniques based on graphics APIs and for the CUDA architecture. Section 5.4 is then devoted to the efficient parallelisation of preconditioners and multigrid smoothers specifically tailored to systems stemming from generalised tensor product patches. Sections 5.5–5.8 present thorough numerical experiments, we evaluate performance and accuracy of our solvers in detail. Special focus is placed on the numerical performance of mixed precision methods in case of very ill-conditioned problems. In Section 5.9 we conclude with a list of guidelines for GPU performance characteristics, in contrast to CPUs.

Chapter 6 is devoted to our minimally invasive integration of hardware accelerators into FEAST. Section 6.1 summarises this chapter, and provides a detailed motivation of the two different approaches, acceleration and re-implementation, to adapt software packages to novel hardware architectures. Our integration procedure and the possibility to execute two-layer SCARC solvers hybridly on CPUs and GPUs are described in Section 6.2. In Section 6.3 we analyse and evaluate our approach with detailed numerical experiments. We consider the Poisson problem as an important building block, and discuss the accuracy and efficiency of the accelerated mixed precision two-layer SCARC solvers. We are interested in performance, speedup, weak scalability and energy efficiency of computations executing on large-scale GPU-enhanced clusters. Here, we also develop an analytical performance model to assess and predict how the local speedup achieved in the portions that are offloaded to accelerator hardware translate into global speedup of the entire solver. Section 6.4 and Section 6.5 assess the applicability and performance of the minimally invasive integration technique, using the two applications FEASTSOLID and FEASTFLOW. Again, our focus is on performance and accuracy, and we present benchmark computations for various ill-conditioned problems from the field of solid mechanics, for instance a cantilever beam configuration, and for laminar flow simulations. This chapter concludes with a summary and a critical discussion including preliminary experiments that examine the feasibility of several avenues for future work, in Section 6.6.

2

Mathematical Background

This chapter covers the mathematical background of this thesis. Section 2.1 briefly outlines the most important concepts of finite element discretisations, using the Poisson problem as an example. Section 2.2 explains iterative solution techniques for sparse linear systems of equations, in particular Krylov subspace methods and the multigrid algorithm. Finally, Section 2.3 explains FEAST, the finite element and solver toolkit that the author of this thesis has significantly contributed to, and based on which all implementations for this thesis have been performed.

2.1. Finite Element Discretisation

From an engineering and mathematical point of view, *finite element* (FE) methods are well-suited and very favourable for the numerical treatment of PDE problems in solid mechanics and fluid dynamics, due to their flexibility and accuracy. As a thorough introduction to the field is clearly beyond the scope of this thesis, we restrict ourselves to the discretisation of the two-dimensional Poisson equation with conforming bilinear quadrilateral finite elements, an important model problem which we also examine in many numerical tests throughout this thesis. We refer to the textbooks by Donea and Huerta [61] and Löhner [151] for more detailed, practically oriented introductions to the topic.

Given a domain $\Omega \subset \mathbb{R}^2$ with boundary Γ, the Poisson equation with homogeneous Dirichlet boundary conditions reads:

$$-\Delta u = f \text{ in } \Omega, \qquad u = 0 \text{ on } \Gamma$$

2.1.1. Variational Formulation of the Poisson Problem

In contrast to finite difference methods which require pointwise validity, the idea of finite element methods is to formulate an equivalent *variational equation*, also termed *weak formulation*, which is required only to be valid in some mean. The *method of weighted residuals* is used to construct this integral formulation by multiplying the residual of the strong form with a *test* or *weight function* w and integrating over the domain ($\mathbf{x} \in \Omega$):

$$(w, -\Delta u - f) = 0 \quad \forall w \in V \qquad \Leftrightarrow \qquad -\int_\Omega w\,\Delta u\, d\mathbf{x} = \int_\Omega wf\, d\mathbf{x} \quad \forall w \in V$$

As a minimal requirement to the space of test functions V, the inner product has to exist. The weak formulation imposes less regularity restrictions on the involved data, and we refer to the literature for more details on the exact requirements. Integration by parts yields *Green's formula*:

$$-\int_\Omega w\,\Delta u\, d\mathbf{x} = -\int_\Gamma w\,\frac{\partial u}{\partial \boldsymbol{n}}\, d\Gamma + \int_\Omega \operatorname{grad} w \operatorname{grad} u\, d\mathbf{x}$$

That is, one derivative is shifted from u to w and an additional boundary integral emerges, $\boldsymbol{n}$ denoting the outer normal unit vector of Γ. We assume zero boundary conditions for simplicity, so it immediately vanishes:

$$-\int_{\Omega} w\, \Delta u\, d\mathbf{x} = \int_{\Omega} \operatorname{grad} w \operatorname{grad} u\, d\mathbf{x}$$

Substitution into the weak formulation leads to:

$$\int_{\Omega} \operatorname{grad} w \operatorname{grad} u\, d\mathbf{x} = \int_{\Omega} w f\, d\mathbf{x}$$

So far, the space of test functions V is infinite dimensional. The next step is to seek an approximate solution only, by restricting V to a finite dimensional subspace $V_h \subset V$ with $N := \dim V_h < \infty$. For instance, V_h may be the space of all (piecewise) polynomials up to a certain degree. The approximate weak formulation then reads:

$$\int_{\Omega} \operatorname{grad} \mathrm{w}_h \operatorname{grad} \mathrm{u}_h\, d\mathbf{x} = \int_{\Omega} \mathrm{w}_h \mathrm{f}_h\, d\mathbf{x} \quad \forall \mathrm{w}_h \in \mathrm{V}_h \tag{2.1.1}$$

As V_h is finite dimensional, it has a finite basis $\Phi := \{\phi_i \in V_h, i = 1,\dots,N\}$:

$$V_h = \{\mathrm{w}_h, \quad w_h(x) = \sum_{i=1}^{N} w_i \phi_i(x), \quad w_i \in \mathbb{R}\}$$

Taking into account the linearity of the operators and the right hand side data f, Equation (2.1.1) is equivalent to:

$$\int_{\Omega} \operatorname{grad} \phi_i \operatorname{grad} \mathrm{u}_h\, d\mathbf{x} = \int_{\Omega} \phi_i f\, d\mathbf{x}, \quad i = 1,\dots,N$$

As the approximate solution u_h is sought in V_h, it can be written in terms of the basis Φ as well:

$$\mathrm{u}_h(x) = \sum_{j=1}^{N} \phi_j u_j(x)$$

Substituting and extracting the sum yields:

$$\sum_{j=1}^{N} \left(\int_{\Omega} \operatorname{grad} \phi_i \operatorname{grad} \phi_j\, d\mathbf{x} \right) u_j = \int_{\Omega} \phi_i f\, d\mathbf{x}, \quad i = 1,\dots,N \tag{2.1.2}$$

The functions $\{\phi_i\}$ are referred to as *trial functions* or ansatz functions. As the test and trial spaces are identical, we actually describe the *Galerkin weighted residual method.* Equation (2.1.2) is in fact a linear system of equations of the form $\mathbf{Au} = \mathbf{b}$ with the unknowns $\mathbf{u} = (u_1, \dots, u_N)$, the matrix entries $a_{ij} = \int_{\Omega} \operatorname{grad} \phi_i \operatorname{grad} \phi_j\, d\mathbf{x}$ and the right hand side $\mathbf{b} = (\int_{\Omega} \phi_1 f\, d\mathbf{x}, \dots, \int_{\Omega} \phi_N f\, d\mathbf{x})$.

In the following, we discuss the remaining open questions, namely, the choice of the trial functions ϕ_i, the subdivision of the domain Ω into actual elements, and finally in Section 2.2 the efficient solution of the arising linear system. We do not present details on numerical quadrature to evaluate the integrals needed in the assembly of the matrix and right hand side, as they are typically covered in any undergraduate level textbook on numerical mathematics.

2.1.2. Choice of Basis Functions and Subdivision of the Domain into Elements

In approximation theory, global trial functions are widely used, for instance Fourier series or Legendre polynomials. However, global trial functions exhibit a number of severe drawbacks:

First, determining an appropriate set of trial functions is very challenging except for the simplest geometries in two or three space dimensions. The resulting system matrix $\mathbf{A}$ is dense, and can easily become ill-conditioned except when using strongly orthogonal polynomials. Finally, the matrix coefficients have no physical significance.

The finite element method circumvents these difficulties by switching from global to *local* trial functions that commonly exhibit the following useful properties:

- *Interpolation property*: $\phi_i(x_j) = \delta_{ij}$, that is, at the discrete location i all but the i-th trial function vanish.
- *Constant sum property*: The trial functions form a partition of unity.
- *Conservation property*: The sum of all derivatives of all trial functions at any given location in the element vanishes as a direct consequence of the constant sum property.

The domain Ω is subdivided into a set of nonoverlapping subdomains Ω_i, $\Omega = \bigcup \Omega_i$. The Ω_i are called *elements* or cells, and their entire collection *grid* or *mesh*. A multitude of different geometric shapes of the elements are conceivable, like triangles or quadrilaterals in two space dimensions, or tetrahedrals, hexahedrals, prisms or pyramids in three space dimensions. The grids can be cartesian (i. e., all elements are orthogonal, *uniformly structured grids*), *structured* (i. e., comprise only one element type and the same connectivity for all inner elements, say for a quadrilateral grid, four elements meet at each inner grid point), or *unstructured* (i. e., element types may be mixed, and connectivity varies).

The approximate solution u_h and hence the test and trial functions are then defined on each element separately. Additionally, certain transition conditions between neighbouring elements are imposed to guarantee global properties like a continuous solution.

Concrete elements are characterised by their geometry, and the number, location and type parameters of the ansatz functions ϕ_i. This information is called *degrees of freedom*, based on the roots of the finite element method in mechanics. As an example of the 'zoo' of elements, we consider the quadrilateral bilinear conforming element in two space dimensions. The degrees of freedom for this element coincide with the vertices, meaning that on the *reference element*, the unit square $[-1,1]^2$, bilinear trial functions are used ($\xi,\eta \in [-1,1]$):

$$
\begin{aligned}
\phi_1 &= \tfrac{1}{4}(1-\xi)(1-\eta),\\
\phi_2 &= \tfrac{1}{4}(1+\xi)(1-\eta),\\
\phi_3 &= \tfrac{1}{4}(1+\xi)(1+\eta),\\
\phi_4 &= \tfrac{1}{4}(1-\xi)(1+\eta).
\end{aligned}
$$

The i-th function takes the value 1 at the i-th vertex and zero at the other vertices. Requiring a continuous transition between neighbouring elements in the vertices (and due to the linearity on the entire edge) results in the finite element space Q_1. In practice, elements are often transformed to the reference element via a suitable bijective mapping.

2.1.3. Mesh Refinement and Error Analysis

The goal of *refinement* of a given finite element grid is to achieve successively better approximations, with the goal that $\mathrm{u_h} \to u$ for $h \to 0$. Starting from a single Q_1 element (refinement level $L = 0$, mesh width $h = 1$), finer grids are iteratively created by subdividing each element into four new ones with mesh width $h/2$. In regular refinement, this is achieved by connecting opposing edge midpoints with new edges. Figure 2.1 shows a grid hierarchy for refinement levels $L = 1,2,3$. An alternative possibility that we occasionally make use of in this thesis is anisotropic refinement in x and y direction, e. g., to accurately resolve boundary layers or singularities. We refer to Section 2.3 for examples and the thesis by Kilian [130] for a formal definition. The refinement level L and the total number of unknowns N are related via $N=(2^L+1)^2$, for a scalar problem. Obviously, refinement implies a natural hierarchy, which is exploited in geometric multigrid solvers (see Section 2.2.3).

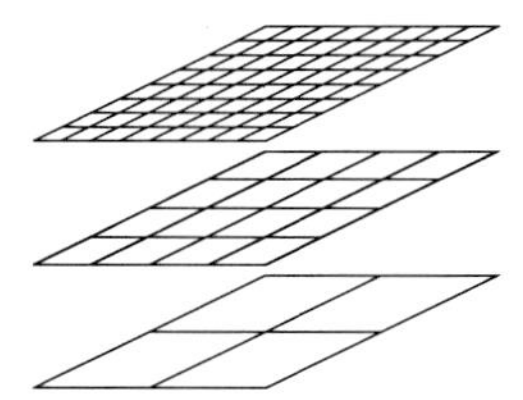

Figure 2.1: Mesh hierarchy for refinement levels 1,2 and 3.

Due to the local support of the trial functions, each inner grid point is coupled with itself and the eight vertices of its surrounding four elements, see Equation (2.1.2). The resulting system matrix is thus very sparse, containing nine entries for each row corresponding to an inner degree of freedom, and less for boundary vertices. Section 2.3 explains how a linewise numbering of the degrees of freedom within each refined element leads to a band-structured matrix, which can be exploited in the implementation of numerical linear algebra operations to improve performance.

A common approach to assess the correctness of computed solutions u_h, also extensively used thoughout this thesis, is to use an analytical function v, and its derivatives to construct a test problem with analytically known solution: The analytical Laplacian of v is used to define a Poisson problem $-\Delta u = f$ with $f := -\Delta v$, by evaluating the analytical formula in each grid point during the assembly of the right hand side. The absolute and relative approximation errors e_{abs} and e_{rel} of a computed solution u_h of this Poisson problem against the analytically known solution v are then measured in the L_2 norm:

$$e_{\text{abs}} = \sqrt{\int_\Omega (\mathrm{u}_h - v)^2 \, d\mathbf{x}} \qquad e_{\text{rel}} = \frac{\sqrt{\int_\Omega (\mathrm{u}_h - v)^2 \, d\mathbf{x}}}{\sqrt{\int_\Omega v^2 \, d\mathbf{x}}}$$

In this norm, the error reduces by a factor of four in each refinement step (i. e., halving of the mesh width h), as the approximation order of the Q_1 element space is $\mathcal{O}(h^2)$.

2.2. Iterative Solvers for Sparse Linear Systems

For the large sparse linear systems of equations we encounter in this thesis, direct methods are prohibitively expensive due to their asymptotically cubic runtime. This section briefly reviews two classes of iterative solution schemes, *conjugate gradients* as a representative of Krylov subspace methods, and *multigrid* as our favourite and highly efficient defect correction scheme. We restrict ourselves to the necessary algorithmic aspects of the schemes, and refer to the literature in particular for all convergence proofs. Recommended reading material includes the 'classic' textbook by Barrett et al. [13] on which most of our notation is based, and the textbooks by Saad [191], Hackbusch [103] (an English version is also available [104]) and Wesseling [238].

2.2.1. Basic Stationary Defect Correction Methods

The starting point for all *defect correction methods (relaxation methods)* to solve a linear system of equations $\mathbf{Ax} = \mathbf{b}$ is the following algebraic manipulation:

$$\begin{aligned}\mathbf{x} &= \mathbf{A}^{-1}\mathbf{b}\\ &= \mathbf{x} - \mathbf{x} + \mathbf{A}^{-1}\mathbf{b}\\ &= \mathbf{x} - \mathbf{A}^{-1}\mathbf{Ax} + \mathbf{A}^{-1}\mathbf{b}\\ &= \mathbf{x} + \mathbf{A}^{-1}(\mathbf{b} - \mathbf{Ax})\end{aligned}$$

Replacing the inverse of $\mathbf{A}$ with a suitable approximation, formulating the scheme in terms of an iteration, and adding a *damping parameter (relaxation parameter)* ω leads to the basic *preconditioned Richardson method*:

$$\mathbf{x}^{(k+1)} = \mathbf{x}^{(k)} + \omega\mathbf{C}^{-1}(\mathbf{b} - \mathbf{Ax}^{(k)})$$

The choice of $\mathbf{C}$, the so-called *preconditioner*, is crucial. On the one hand, its inverse should be a (spectrally) good approximation to the inverse of $\mathbf{A}$, optionally with amenable properties that A lacks. On the other hand, it should be much easier to invert than $\mathbf{A}$. Here, we list only the most basic choices, writing the coefficient matrix as a matrix sum of its diagonal and lower and upper triangular parts: $\mathbf{A} = \mathbf{L} + \mathbf{D} + \mathbf{U}$.

The *Jacobi* method couples each unknown only with itself, in more formal notation, $\mathbf{C}$ is the diagonal of $\mathbf{A}$:

$$\mathbf{C}^{\mathrm{JAC}} := \mathbf{D}$$

The application of the Jacobi preconditioner reduces to a simple scaling of the current defect with the inverse of the main diagonal. The order in which the equations in the linear system are examined is irrelevant, and the iteration is independent of the underlying numbering of the unknowns. In contrast, the *Gauß-Seidel* method examines them one at a time, in sequence, meaning that computed results are used as soon as they are available. The application of the Gauß-Seidel preconditioner is thus the inversion of the lower triangular part of the system matrix:

$$\mathbf{C}^{\mathrm{GS}} := (\mathbf{L} + \mathbf{D})$$

The Gauß-Seidel method is not independent of the numbering scheme; depending on the ordering, families of Gauß-Seidel preconditioners with different numerical properties can be constructed. The *SOR (successive overrelaxation)* method uses an additional relaxation parameter $\tilde{\omega}$ only for the lower triangular part, its optimal value can be determined in some special cases.

$$\mathbf{C}^{\mathrm{SOR}} := (\tilde{\omega}\mathbf{L} + \mathbf{D})$$

In practice, the Gauß-Seidel method typically needs half the number of iterations than the Jacobi method. All three methods however converge much too slow to be used as stand-alone solvers, their convergence rates depend on the mesh width h, and an optimal choice of the damping parameter ω is impossible in practice. Their main purpose (and practical relevance) stems from being suitable preconditioners in advanced methods such as Krylov subspace schemes, and in particular as *smoothers* in multigrid solvers.

2.2.2. Krylov Subspace Methods

Given a linear system $\mathbf{Ax} = \mathbf{b}$, an initial guess $\mathbf{x}^{(0)}$ and the initial residual $\mathbf{r}^{(0)} = \mathbf{b} - \mathbf{Ax}^{(0)}$, the *Krylov subspace* is the m-dimensional vector space $\mathrm{span}\{\mathbf{r}^{(0)}, \mathbf{Ar}^{(0)}, \ldots, \mathbf{A}^{m-1}\mathbf{r}^{(0)}\}$. All Krylov subspace methods iteratively construct such a vector space based on $\mathbf{A}$-orthogonalisation.

Conjugate Gradients

The basic idea of the *conjugate gradient (CG)* scheme is to construct an $\mathbf{A}$-*orthogonal* or *conjugate* basis $\{\mathbf{p}_j\}$ of $\mathbb{R}^N$ iteratively:

$$\mathbf{Ap}_i \cdot \mathbf{p}_j = \delta_{ij} \quad \forall i, j = 1, \ldots, N$$

When expressing the (unknown) solution $\mathbf{x}$ in terms of this basis,

$$\mathbf{x} = \sum_{j=1}^{N} \eta_j \mathbf{p}_j,$$

the coefficients η_j can be explicitly calculated due to the $\mathbf{A}$-orthogonality:

$$\eta_j = \frac{\mathbf{b} \cdot \mathbf{p}_j}{\mathbf{Ap}_j \cdot \mathbf{p}_j} \quad j = 1, \ldots, N$$

This basis is constructed via orthogonalisation, in fact, two coupled two-term recurrences suffice. The search directions computed in these recurrences are chosen to minimise the energy norm $(\mathbf{x}^{(k)} - \hat{\mathbf{x}})^{\mathsf{T}}\mathbf{A}(\mathbf{x}^{(k)} - \hat{\mathbf{x}})$. For the unpreconditioned CG variant, this minimum is guaranteed to exist if the coefficient matrix is symmetric and positive definite. For the preconditioned variant, in addition the preconditioner is required to be symmetric and positive definite. The steplength parameter is computed so that it minimises ${\mathbf{r}^{(k)}}^{\mathsf{T}}\mathbf{A}^{-1}\mathbf{r}^{(k)}$ for the current residual $\mathbf{r}^{(k)} = \mathbf{b} - \mathbf{Ax}^{(k)}$. Figuratively speaking, the algorithm computes (locally) optimal search directions towards the exact solution, and follows them for an optimal distance. Hence, assuming exact arithmetic, the CG algorithm computes the exact solution in at most N iterations, i. e., once the entire $\mathbf{A}$-orthogonal basis has been constructed. The pseudocode for the resulting algorithm is given in Figure 2.2.

The CG solver involves one matrix-vector product, three vector updates and two dot products per iteration, not taking into account the application of the preconditioner. In general, the number of iterations required until convergence can be bounded by the square root of spectral condition number of $\mathbf{C}^{-1}\mathbf{A}$.[1] For the Poisson problem discretised with a mesh width h (cf. Section 2.1), this means that the number of iterations is proportional to h^{-1}, see also Section 4.5.1. Consequently, each refinement step of a local tensor product patch (see Section 2.3.1) incurs approximately a doubling of iterations. The scheme requires, in addition to a few scalars, storage for the four auxiliary vectors $\mathbf{r}, \mathbf{z}, \mathbf{p}$ and $\mathbf{q}$.

Biconjugate Gradients Stabilised

In practice, the requirements of the conjugate gradient algorithm are often not satisfied, in particular the symmetry condition. As a remedy, several advanced and less constrained methods have

[1] Here we assume that $\mathbf{C}$ is one of the elementary, symmetric preconditioners presented in Section 2.2.1.

Compute $\mathbf{r}^{(0)} = \mathbf{b} - \mathbf{A}\mathbf{x}^{(0)}$ for some initial guess $\mathbf{x}^{(0)}$
Loop over $k = 1, 2, \dots$
  Apply preconditioner: $\mathbf{z}^{(k-1)} = \mathbf{C}^{-1}\mathbf{r}^{(k-1)}$
  $\rho_{k-1} = {\mathbf{r}^{(k-1)}}^{\mathsf{T}}\mathbf{z}^{(k-1)}$
  if $k = 1$
    $\mathbf{p}^{(1)} = \mathbf{z}^{(0)}$
  else
    $\beta_{k-1} = \rho_{k-1}/\rho_{k-2}$
    $\mathbf{p}^{(k)} = \mathbf{z}^{(k-1)} + \beta_{k-1}\mathbf{p}^{(k-1)}$
  endif
  $\mathbf{q}^{(k)} = \mathbf{A}\mathbf{p}^{(k)}$
  $\alpha_k = \rho_{k-1}/({\mathbf{p}^{(k)}}^{\mathsf{T}}\mathbf{q}^{(k)})$
  $\mathbf{x}^{(k)} = \mathbf{x}^{(k-1)} + \alpha_k\mathbf{p}^{(k)}$
  $\mathbf{r}^{(k)} = \mathbf{r}^{(k-1)} - \alpha_k\mathbf{q}^{(k)}$
  Check convergence: $\|\mathbf{r}^{(k)}/\mathbf{r}^{(0)}\| < \varepsilon$

Figure 2.2: The preconditioned conjugate gradient algorithm.

been devised. In this thesis, we make use of the *biconjugate gradients stabilised (BiCGStab)* method, which uses two short-term recurrences and works well for unsymmetric systems. The algorithm needs six auxiliary vectors, and performs two preconditioning steps per iteration. Roughly speaking, one BiCGStab step corresponds to two CG steps. We refer to the literature for a formal description, in particular to the textbook by Barrett et al. [13].

2.2.3. Geometric Multigrid Solvers

Multigrid (MG) methods belong to the most efficient solvers for sparse linear systems arising in the discretisation of (elliptic) partial differential equations (see Section 2.1). In this thesis, we only consider *geometric multigrid methods*, i. e., solvers acting on an existing hierarchy stemming from mesh refinement as discussed in Section 2.1.3.

Illustrative Motivation

Detailed Fourier analysis of the basic relaxation schemes reveals, that high frequency error components are annihilated quickly and efficiently, in only a few iterations. In contrast, low frequency components are more persistent and consequently determine the overall slow convergence of the basic relaxation schemes. If the error is already sufficiently smooth (i. e., it comprises low frequency components only), then the differences between nodal values associated with neighbouring vertices are small. All relaxation schemes act locally as they basically average neighbouring values in some order, which, figuratively speaking, explains the slow convergence: Many iterations are needed to propagate corrections through the entire grid.

Assuming that all high frequency error components have been 'smoothed out' already, then the remaining low frequency components on the current, fine grid appear much more oscillating on a coarser scale; where they can again be smoothed efficiently: Low frequency errors on a fine scale are high frequency errors on a coarser scale. Applying this concept on a sequence of successively coarser grids leads to a relatively small problem, which can be solved exactly in very little time using a tuned direct method like sparse LU decomposition. Obviously, the smoothing is cheaper the coarser the grid becomes. Each of the original low frequency components of the error can be reduced efficiently in this process, always at the appropriate scale.

The second phase then comprises transferring the computed correction term back to the original fine grid where it is used to update the previous iterate. On each intermediate scale, the corrected approximations are again smoothed, to eliminate high frequency error components introduced by the transfer operations. The whole procedure is called a *multigrid cycle*, more precisely, a V cycle; and is repeated until some convergence criterion holds. It has been shown that the convergence behaviour of the algorithm is independent of the mesh width h, called h-independency, and that the convergence rates of the multigrid algorithm are typically very good, given a sufficiently strong smoothing operator. Due to the hierarchical nature, the entire scheme has optimal linear complexity in the number of unknowns, with a reasonably small constant: Depending on the exact problem definition, dimensionality and discretisation, estimations in the literature vary between 100 and 2 000 floating point operations per unknown for a standard Poisson problem.

In summary, the core idea of *multigrid solvers* is: Defects are smoothed on consecutively coarser grids, at some point, an exact defect correction is computed, which is pushed back through the hierarchy to advance the solution on the original, finest grid. In other words, the multigrid method is a sequence of defect correction methods on meshes with different resolutions, and can thus be interpreted as a defect correction method on its own. We refer to the textbooks by Hackbusch [103, 104]) and Wesseling [238] for more details, and in particular, for all convergence proofs we intentionally omit.

Multigrid Algorithm

In contrast to many textbooks, we present the multigrid algorithm in its non-recursive form. This has several advantages: First, we are convinced that the challenges in an efficient implementation are more pronounced, an important aspect in the scope of this thesis.[2] Second, the presentation becomes much clearer; in particular, the notation is formally correct without reverting to long parameter lists in the otherwise commonly used recursive subroutine formulation.

The multigrid algorithm has the following input parameters:

- $\mathbf{A}^{(L)},\dots,\mathbf{A}^{(1)}$: system matrices individually assembled on the L levels, $\mathbf{A}^{(L)} := \mathbf{A}$
- $\mathbf{b}^{(L)} := \mathbf{b}$: right hand side vector on the finest level
- $\mathbf{x}^{(L)} := \mathbf{x}$: initial guess on the finest level, and solution after completion of the multigrid algorithm

Furthermore, four sets of auxiliary vectors are required, allocated on all levels of refinement: iteration and correction vectors $\mathbf{x}^{(L-1)},\dots,\mathbf{x}^{(1)}$ and $\mathbf{c}^{(L)},\dots,\mathbf{c}^{(1)}$; and right hand side and defect vectors $\mathbf{b}^{(L-1)},\dots,\mathbf{b}^{(1)}$ and $\mathbf{d}^{(L)},\dots,\mathbf{d}^{(1)}$. In our case (see Section 2.1.3), a full 'hierarchical' vector only requires $4/3$ of the storage than on the finest refinement level alone, the same holds true for the matrices. Similarly, the factor is $8/7$ for a Q_1 discretisation in three space dimensions. So the storage requirements are small compared to other iterative schemes like Krylov subspace solvers, as long as the smoother does not require too much additional memory. Figure 2.3 depicts the multigrid algorithm in pseudocode.

Multigrid Solver Components

Grid transfer operators are commonly referred to as *restriction* (from fine to coarse) and *prolongation* (from coarse to fine). Previous experience has shown that even for severely anisotropic (local) grids, the error introduced by using bilinear interpolation with constant coefficient weights for the entire grid can be neglected during grid transfers [225]. Consequently, it is not necessary to store a matrix representation of the transfer operators, the corresponding stencil values suffice.

[2] Also, long chains of recursive subroutine calls are less efficient per se, due to contexts being pushed onto stacks.

- Compute initial defect: $\mathbf{d}^{(L)} = \mathbf{b}^{(L)} - \mathbf{A}^{(L)}\mathbf{x}^{(L)}$
- Compute initial norm for convergence control: $d = \|\mathbf{d}^{(L)}\|$
- Loop until convergence or up to a maximal admissible number of iterations:
  - Set current level: $L_c = L$
  - Loop until $L_c = 1$ (restriction loop)
    - Presmoothing: smooth $\mathbf{A}^{(L_c)}\mathbf{c}^{(L_c)} = \mathbf{d}^{(L_c)}$ with initial $\mathbf{c}^{(L_c)} = \mathbf{0}$
    - $\mathbf{x}^{(L_c)} = \mathbf{x}^{(L_c)} + \mathbf{c}^{(L_c)}$
    - Compute new defect: $\mathbf{d}^{(L_c)} = \mathbf{b}^{(L_c)} - \mathbf{A}^{(L_c)}\mathbf{x}^{(L_c)}$
    - Restrict $\mathbf{d}^{(L_c)}$ to $\mathbf{d}^{(L_c-1)}$
    - Copy $\mathbf{d}^{(L_c-1)}$ to $\mathbf{b}^{(L_c-1)}$
    - Set current level: $L_c = L_c - 1$
  - Solve coarse grid problem $\mathbf{A}^{(1)}\mathbf{x}^{(1)} = \mathbf{d}^{(1)}$ with initial $\mathbf{x}^{(1)} = \mathbf{0}$
  - Loop until $L_c = L$ (prolongation loop)
    - Set current level: $L_c = L_c + 1$
    - Prolongate $\mathbf{x}^{(L_c-1)}$ to $\mathbf{c}^{(L_c)}$
    - Apply homogeneous Dirichlet boundary conditions to $\mathbf{c}^{(L_c)}$
    - Add correction to current solution: $\mathbf{x}^{(L_c)} = \mathbf{x}^{(L_c)} + \mathbf{c}^{(L_c)}$
    - Postsmoothing: smooth $\mathbf{A}^{(L_c)}\mathbf{x}^{(L_c)} = \mathbf{b}^{(L_c)}$
  - Compute final defect: $\mathbf{d}^{(L)} = \mathbf{b}^{(L)} - \mathbf{A}^{(L)}\mathbf{x}^{(L)}$
  - Perform convergence control with d and $\|\mathbf{d}^{(L)}\|$

Figure 2.3: Geometric multigrid algorithm.

This significantly increases performance, as grid transfer operations are reduced to efficient vector scaling, reduction and expansion operations as shown in Figure 2.4. We continue to use this approach throughout this thesis.

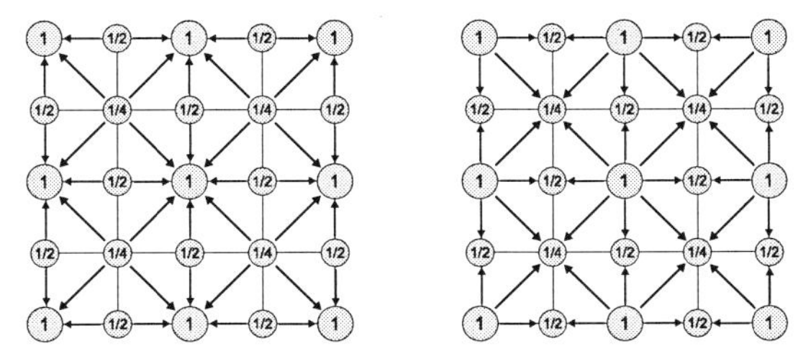

Figure 2.4: Grid transfer operations and corresponding weights: Restriction (left) and prolongation (right).

The restriction step is essentially a *gather* operation; for each coarse mesh point, the contributions from the neighbouring fine mesh points are summed with appropriate weights for direct and diagonal neighbours. In contrast, prolongation is a natural *scatter* operation, values from coarse mesh points are copied to the coinciding fine points, and accumulated for the mesh points not present in the coarse grid. On CPUs, both transfer operators can be implemented efficiently, without conditionals, using sequences of strided `axpy`-operations. In Section 5.2.3 we reformulate the prolongation as a gather operation which is executed independently for each fine point. This is an example of specifically re-formulating an operation for data-parallel architectures.

Formally, smoothers—denoted as S_h—are operators on the ansatz space V_h (see Section 2.1.1) that fulfil the fixed point property for the exact solution $\hat{\mathrm{u}}_h$ (i. e., $\mathrm{S}_h\hat{\mathrm{u}}_h = \hat{\mathrm{u}}_h$) and are stable (i. e.,

$\|u_h - S_h v_h\| \leq c\|u_h - v_h\|$ for a constant $c > 0$ and an arbitrary test function v_h). The constant c may depend on the mesh width h. In practice, smoothers are not necessarily defined in closed form, but as relaxation schemes configured to perform a fixed number of iterations. The Jacobi and Gauß-Seidel schemes, as defined above in discrete form, are frequently used. We present parallelised variants of smoothing operators tailored to the generalised tensor product structure of the mesh in Section 5.4. *Presmoothing* denotes the smoothing of the defect in the restriction phase, *postsmoothing* treats the correction term after prolongation. The practical efficiency depends to a great extent on the choice of the smoothing operator, which should be tailored to the problem at hand, in particular due to the ill-conditioning induced by anisotropies in the differential operators and the underlying mesh.

To solve the problems on the coarsest grid level, either a direct or an iterative solver can be used. In this background section, it suffices to view the coarse grid solver as a black box tool that computes an exact solution (or a reasonably well-converged approximation to it). The coarse grid problem is also sparse, which is typically exploited in 'sparse LU decomposition' based methods such as multifrontal techniques [57], that exhibit superlinear, but significantly less than quadratic runtime in the number of unknowns. We discuss the exact implementation in the presentation of FEAST (see Section 2.3) and in the discussion of our numerical experiments in Chapter 5 and Chapter 6.

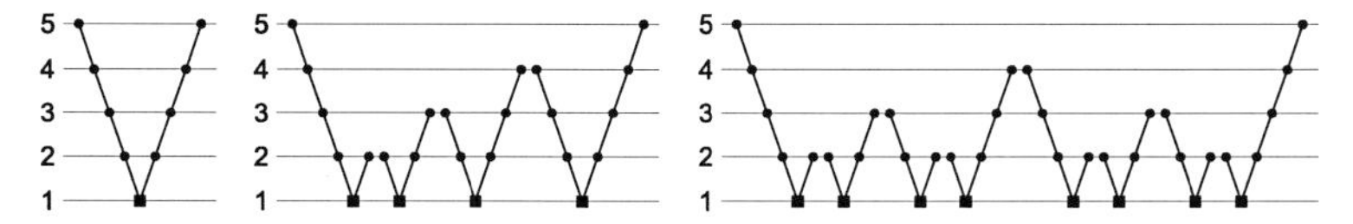

Figure 2.5: Multigrid cycle types V, F and W on five refinement levels. • denotes smoothing and ■ coarse grid corrections.

The order in which the mesh hierarchy is traversed is determined by so-called *multigrid cycles*. In the algorithmic description above, we described a V cycle, which traverses the hierarchy exactly once from fine to coarse and back. More powerful cycles are distinguished by the order in which they visit coarser levels repeatedly to increase robustness and to reduce the overall number of iterations until convergence, see Figure 2.5 for an illustration of the three cycle types V, F and W. For the advanced cycles, the majority of additional work is spent on coarser refinement levels, e. g., a W cycle is asymptotically twice as expensive as a V cycle.

Cycle control can be implemented efficiently into the non-recursive algorithm by keeping track of the 'direction' in which to proceed after each prolongation on intermediate levels.

2.3. FEAST – High Performance Finite Element Multigrid Toolkit

The numerical solution of elliptic PDEs is one of the most common, yet most challenging tasks in many practical scenarios. Examples of elliptic problems considered in this thesis are the Poisson equation and the equations of linearised elasticity for compressible material. Similarly, in the treatment of saddlepoint problems such as the (linear) Stokes and the (nonlinear) Navier-Stokes equations in fluid dynamics, or the elasticity equations for (nearly) incompressible materials in solid mechanics, elliptic subproblems have to be solved repeatedly, i. e., in each outer or nonlinear iteration, and this step often constitutes the most time-consuming subtask in the course of an entire simulation. It is not uncommon to be faced with linear systems comprising billions of unknowns from the spatial discretisation, and for instance for transient flows, such subproblems need to be solved in each timestep. The solution on large-scale distributed memory computer systems is inevitable.

Efficiency is thus of utmost importance in this situation, as motivated in the introduction. For elliptic PDEs, multilevel methods are obligatory from a numerical point of view. FEAST (*Finite Element Analysis and Solution Toolkit*) is our parallel finite element solver toolkit for the solution of these kinds of problems on HPC systems, actively developed at TU Dortmund under the supervision of Prof. Turek. FEAST and its underlying solution strategy SCARC (*Scalable Recursive Clustering*) realise many concepts of *hardware-oriented numerics* to simultaneously maximise numerical and computational efficiency, as motivated in Section 1.1.1. FEAST and SCARC have been presented in a number of PhD theses already [15, 98, 130, 242] (a fifth one is in preparation [44]), and we only briefly summarise the most relevant aspects for this work. Additional information can be found in book chapters [80, 234], journal articles [81, 84, 85, 87, 90, 228–230, 233] and conference proceedings [2, 16–20, 45, 82, 83, 86, 88, 131, 227, 231]. The author of this thesis has contributed to a significant amount of the more recent publications, and to the actual implementation. All numerical and computational experiments in Chapters 5 and 6 have been realised in the FEAST framework.

2.3.1. Separation of Structured and Unstructured Data

FEAST covers the computational domain with a collection of m quadrilateral subdomains. These patches form an unstructured coarse mesh (cf. Figures 2.6 and 2.7), and each subdomain is refined in a regular, generalised tensor product fashion so that N_{loc} grid points per patch are arranged in an $M \times M$ grid, $M = \sqrt{N_{\text{loc}}}$. Adaptive refinement strategies, that maintain the local tensor product property of the mesh, are realised via grid deformation techniques (r-adaptivity, see also Figure 2.7, right), anisotropic refinement within each subdomain to, e. g., capture boundary layers in CFD (Figure 2.6, top left), hanging nodes on subdomain edges (patchwise adaptivity), and any combination thereof. We refer to the thesis by Grajewski [98] for more details on adaptivity in the FEAST context. The resulting mesh is used to discretise the domain with finite elements, currently only the bilinear conforming Q_1 element is supported.

This approach caters to the contradictory needs of flexibility in the discretisation and efficient implementation: Instead of keeping all data in one general, homogeneous data structure, FEAST stores only local finite element matrices and vectors (corresponding to subdomains, as usual for parallel approaches) and thus maintains a clear separation of structured and unstructured parts of the domain. In short, FEAST grids are locally structured and globally unstructured. The latter is extremely important, as in realistic configurations, a globally structured grid is simply not feasible due to its enormous storage requirements, especially in three space dimensions. The tensor product property of the local meshes is crucial to achieving good performance in two ways, affecting both linear algebra and optimised multigrid components.

From a domain decomposition point of view, global computations are performed by a series

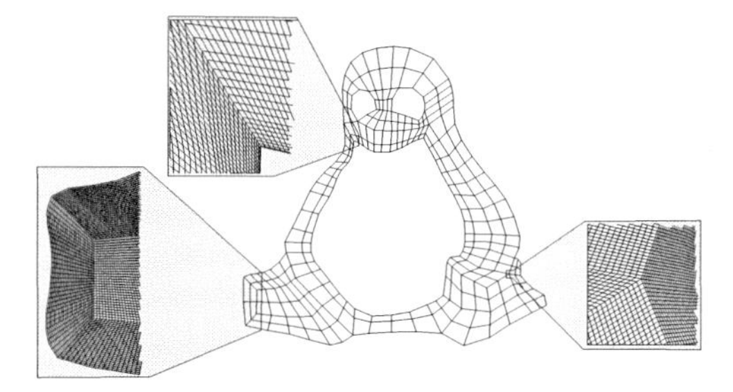

Figure 2.6: TUX example geometry. Important features of FEAST's discretisation approach are highlighted: globally unstructured, locally structured grids (bottom left, bottom right); anisotropic refinement (microscopic anisotropies, top left); boundary adaptation during refinement (bottom left); and coverage of large regions of minor interest with large, regularly refined subdomains (macroscopic feature size variations, bottom right).

of local operations on matrices representing the restriction of the 'virtual' global matrix on each subdomain. Local information is exchanged only over boundaries of neighbouring subdomains, with a minimal overlap. Several subdomains can be grouped together and treated within one MPI process. Details of this general sketch are explained in the following.

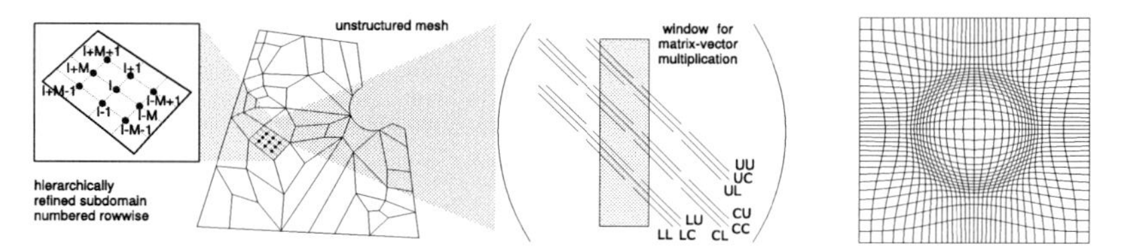

Figure 2.7: Left: Globally unstructured, locally structured mesh and corresponding banded matrix structure for one subdomain (macro). Right: Deformed patch that maintains the tensor product property.

Exploiting the Tensor Product Property in Linear Algebra Components

Fully adaptive, highly unstructured grids entail maximum flexibility in resolving singularities or quantities of interest in the domain. However, the resulting basic linear algebra components such as sparse matrix vector multiply (SPMV) perform very poorly on modern computer architectures [137]. The reason for this is that the corresponding matrices exhibit an unstructured, in the limit case even stochastic distribution of the nonzero entries. A multitude of storage techniques and data structures for such matrices have been proposed, we refer to the books by Barrett et al. [13] and Saad [191] for a comprehensive overview and only mention the *compressed sparse row* (CSR[3]) format as a well-known example.

Common to all these formats is that they generate coherency in the matrix entries due to the contiguous storage, but accesses to the coefficient vector are performed indirectly, requiring

[3]In the literature, this format is often also referred to as *compact row storage*, *compressed sparse row* and other acronym permutations thereof.

pointer chasing through caches and main memory. Even with established resorting techniques such as Cuthill-McKee renumbering, it is still challenging to achieve more than 1–5 % of the processor's peak performance, let alone peak memory bandwidth of the machine [93, 137]. This is particularly true in multi-threaded implementations as demonstrated by Frigo and Strumpen [75] and Williams et al. [241].

On the other hand, FEAST's regular refinement and the patchwise approach allow a linewise numbering of the unknowns, cf. Figure 2.7 (left). Consequently, the nonzero pattern of the local matrices enables a highly efficient DIA-like [191] storage format in which only the matrix bands containing nonzero entries are stored as individual vectors. Offdiagonal bands are zero-padded so that all bands have equal size, and the bands are stored contiguously in memory, ordered from bottom left to top right, with the historical exception of the main diagonal being stored first, as in the DIA format. It is instructive to note that no single variant of the DIA format is part of the established BLAS or LAPACK libraries except the special case of tridiagonal matrices. Banded matrices in the BLAS context are always less storage-efficient, storing all elements up to the matrix bandwidth; and in general, little standardisation exists for FE codes and data structures.

In Sections 5.2–5.4 we return to these questions, taking up the more general position of designing parameterised algorithms for many-core architectures. As part of his thesis, Becker [15] implemented optimised linear algebra components in the FEAST context, and collected the resulting routines in the SBBLAS (*'sparse banded BLAS'*) library.

The work by Douglas, Rüde, Hu and others, in particular in the earlier years of the DiME[4] project, provides an excellent overview of cache-oriented multigrid solver components, and we refer to the overview article by Douglas et al. [63] and the references therein. Similarly, Sellappa and Chatterjee [195] present many performance experiments for low-level tuning of common smoothing operators, down to the level of blocking for the TLB. Kowarschik and Weiß [138] summarise general cache optimisation techniques such as loop transformations, blocking and data structure rearrangements with a more hands-on approach.

Exploiting the Tensor Product Property in Multigrid Components

The second aspect in *exploiting locally structured parts* affects the design and implementation of grid transfer and smoothing operators within the multigrid solvers. As explained in Section 2.2.3 on page 26, the linewise numbering reduces grid transfer operators to the application of a stencil-based interpolation. This approach does not lead to interpolation errors that affect the convergence of the entire multigrid cycle.

The starting point to construct numerically efficient smoothing operators is a formal decomposition of the system matrix into its nine bands, see Figure 2.7 for notation:

$$\mathbf{A} = (\mathbf{LL}+\mathbf{LC}+\mathbf{LU})+(\mathbf{CL}+\mathbf{CC}+\mathbf{CU})+(\mathbf{UL}+\mathbf{UC}+\mathbf{UU})$$

Adopting this notation, the JACOBI- and GSROW smoothers, i. e., the preconditioners of the corresponding basic iteration scheme (cf. Section 2.2.1), read:

$$\mathbf{C}^{\text{JAC}} := \mathbf{CC} \qquad \mathbf{C}^{\text{GSROW}} := (\mathbf{LL}+\mathbf{LC}+\mathbf{LU}+\mathbf{CL}+\mathbf{CC})$$

The suffix 'row' is used to indicate that the smoothers correspond to a linewise numbering of the unknowns. The TRIDIROW smoother applies the inverse of the centre tridiagonal matrix:

$$\mathbf{C}^{\text{TRIDIROW}} := (\mathbf{CL}+\mathbf{CC}+\mathbf{CU})$$

[4]Data Local Iterative Methods for the Efficient Solution of Partial Differential Equations, `http://www10.informatik.uni-erlangen.de/Research/Projects/DiME`

The TRIGSROW smoother combines the TRIDIROW and GSROW smoothers:

$$\mathbf{C}^{\text{TRIGSROW}} := (\mathbf{LL} + \mathbf{LC} + \mathbf{LU} + \mathbf{CL} + \mathbf{CC} + \mathbf{CU})$$

Due to the rowwise numbering, the dependencies within each row (TRIDIROW), and to the left, right and bottom (TRIGSROW) of the underlying mesh are implicit (i. e., the application of the smoother involves formally inverting a system) but other dependencies are explicit. The same idea can be formulated for a columnwise numbering, so that the dependencies along the columns are implicit, resulting in the elementary smoothers TRIDICOL and TRIGSCOL. Alternating between which of the directions is treated implicitly leads to the class of *alternating direction implicit (ADI)* methods [179], and we denote the resulting elementary smoothers by ADITRIDI and ADITRIGS. We refer to the Diploma thesis by Altieri [1] for a more detailed derivation.

We finally note that these operators have been implemented very efficiently on commodity processors in the SBBLAS library. On data-parallel architectures however, the inherent recursion of these powerful, specifically tailored smoothers pose nontrivial challenges, and we return to this topic in Section 5.4.

2.3.2. Parallel Multilevel Solvers

The parallelisation of multigrid/multilevel solvers is in general a very challenging task:

- All prerequisites regarding computational efficiency and single-processor performance as developed in the previous section must be maintained.

- For parallel methods in general, *scalability* is essential. *Strong scalability* means that a doubling of resources (i. e., processors, compute nodes etc.) should result in a halving of execution time. *Weak scaling* on the other hand means that simultaneously doubling resources and problem size should not change the execution time. All parallel multigrid solvers must scale well, and both aspects of scalability are equally important in practice.

- Communication in parallel (HPC) compute clusters is much slower than computation. Network technology is rapidly improving, in particular, Infiniband technology delivers more bandwidth and less latency than standard ethernet interconnects. Improvements however proceed much slower than those of peak processor and even memory capabilities [97], which adds an additional layer of complexity to the memory wall problem: For instance, the USC cluster on which several experiments presented in this thesis have been performed employs 4x DDR Infiniband interconnects with a measured bandwidth of only 1/10 of available bandwidth between main memory and the CPUs, and latency even worse. This is the best interconnect of all clusters we have used. Furthermore, in practice, interconnects are often cascaded (for instance in a fat-tree topology), as extremely wide full bandwidth crossbar switches add significantly to the acquisition costs of HPC installations. To achieve good scalability, (global) communication obviously has to be minimised, and all locality requirements as outlined in the previous section apply.

- The numerical capacities of a serial multigrid scheme should be transferred to the corresponding realisation in parallel. This includes the independence of the convergence rates of the refinement level, the invariance of (local and global) anisotropies given a sufficiently robust and powerful smoothing operator, and of course the independence of the convergence behaviour from the partitioning of the domain into subdomains.

To alleviate these conflicting goals, SCARC, the solver concept at the core of FEAST, generalises techniques from *multilevel domain decomposition* (MLDD) and parallel multigrid (MG);

combining their respective advantages into a very robust, and (numerically *and* computationally) efficient parallel solution scheme for (scalar) elliptic PDEs.

A thorough presentation of the differences between these parallelisation techniques is clearly beyond the scope of this work, and we refer to the textbooks by Smith et al. [201] and Toselli and Widlund [223] for detailed coverage. The survey article by Xu [245] presents a general framework for MLDD and parallel MG methods which enables a number of fundamental convergence proofs.

Minimal Overlap

In domain decomposition, two separate classes are distinguished, *overlapping Schwarz* methods and *non-overlapping Schur complement* or *substructuring* techniques. The latter schemes typically involve solving dedicated interface problems to transfer information between subdomains, while the former have to perform communication between neighbouring subdomains to synchronise shared data due to overlapping subdomains. When designing the SCARC scheme, Kilian thoroughly evaluated different overlaps, and we refer to her thesis for an in-depth comparison in the context of finite element simulations [130]. She concludes, in line with Smith et al. [201], that a *minimal, nonzero overlap* is the necessary minimum from a numerical point of view, while larger overlaps typically incur worse scalability.

The decomposition as realised in the SCARC solvers works as follows: Geometrically speaking, the local grids on each subdomain are extended by one element layer from the neighbouring subdomains. In the limit case of identically shaped neighbouring subdomains, this approach reduces to commonly applied *ghost cell* or *halo* techniques. In the more practically relevant case of shape changes between subdomains, the approach is actually much better, because it incorporates anisotropies from neighbours that affect even orthogonally-shaped subdomains, as illustrated in Figure 2.8. The local refined grids on the subdomains are always true subsets of the global refined grid and consequently invariant to the decomposition into subdomains.

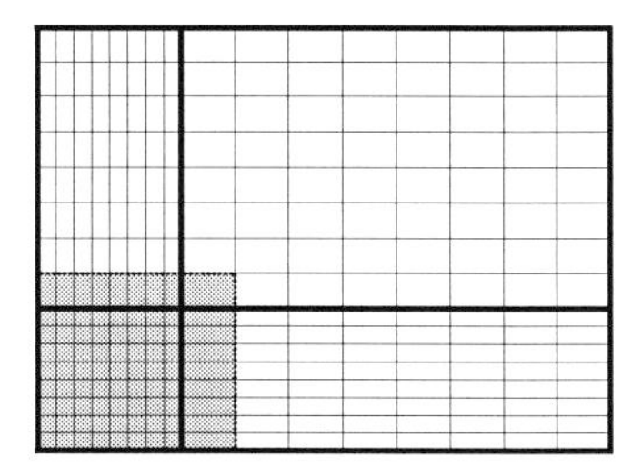

Figure 2.8: Minimal overlap, highlighted for the bottom left subdomain, refinement level 3.

In FEAST (as in all distributed memory schemes), global data exist only 'virtually', and operations involving global matrices or vectors are realised by combining the corresponding local contributions via communication of the overlapping data. Once the topological information of the overlapping element layer has been communicated in a preprocessing stage, the actual overlap reduces to few shared degrees of freedom, typically much less data than that associated with the entire element layer, e. g. cubature points etc. In case of the Q_1 element, only data associated with grid points on subdomain edges actually need to be shared. Additionally, the concept of *minimal overlap* does not create artificial discontinuities on subdomain boundaries, which constitutes a significant advantage in terms of global solver convergence: The solution on a subdomain is always a subset of the global solution.

During the assembly of the local matrices (see Section 2.1.1 on page 19), it makes no difference if a given degree of freedom lies in the interior of the domain or on a subdomain boundary, integration is always performed over all adjacent elements. For true geometric borders, the corresponding boundary conditions are included in the matrix as usual. On inner boundaries however, the respective local matrices contain the full entries of the virtual global matrix. Thus, no boundary conditions are prescribed, nor does the matrix correspond to a pure Neumann matrix as it would be the case if integration was performed only over the local elements, a technique used for instance in substructuring Neumann-Neumann methods [201]. Rather, the local matrix corresponds to a fictitious Dirichlet matrix on the extended element layer termed *extended Dirichlet boundaries*, realised by prescribing homogeneous Dirichlet boundary conditions on the extended boundary and immediately removing the resulting unit rows and columns. As a consequence of this construction process, which becomes important in Section 2.3.3, the local matrices are always invertible.

In the SCARC framework, the terminology of *full entries* (or *local state*) is used to denote this fact. In order to perform global operations, the matrices are switched to their *global state*, which means that matrix entries corresponding to subdomain boundaries only include local element contributions, cf. Figure 2.9. In the global state, the (formal) sum of all subdomain matrices thus yields the global matrix.

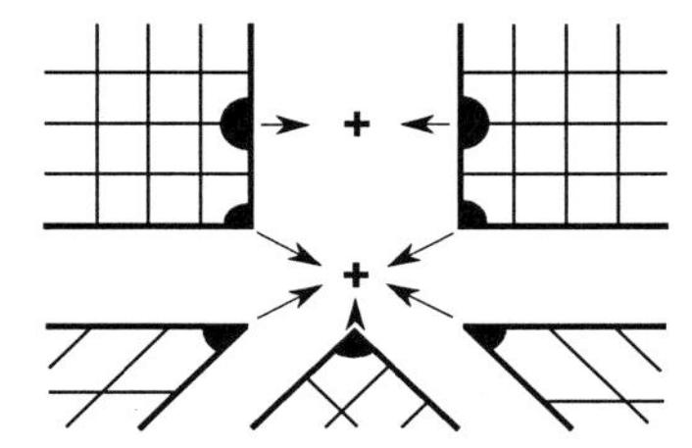

Figure 2.9: Illustration of global matrix state, contributions from the local matrices sum up to the virtual global matrix.

A similar distinction is made for vectors. (Pointwise) global vector-vector operations do not change the state, whereas global matrix-vector multiplications with a matrix in global state and a vector in local state yield a vector in global state, which has to be transformed (via communication with neighbouring subdomains) back to local state for subsequent global operations, at the latest prior to the next matrix-vector multiplication.

This distinction between the *local state* (full entries, regular matrix) and the *global state* (additive construction of the global matrix) constitutes an exact approximation and not just a mere median calculation on subdomain boundaries, because each patch knows exactly as much information from its neighbours as necessary. In view of the memory wall problem, the concept of *minimal overlap* thus exhibits perfect locality, as only a minimal amount of data has to be communicated.

In summary, the invariance to the decomposition transfers from the grid to the matrices and vectors, by keeping track of the *state* of the data. Becker [15] has implemented this coupling of subdomains very efficiently into the FEAST toolkit. His code stores both states of the subdomain matrices and allows for switching states in zero time and without communication at the cost of $\mathcal{O}(M)$ increased storage requirements. In the case of vectors, only communication between neighbouring subdomains is required to exchange and perform averaging of quantities to keep the values on subdomain boundaries synchronised. In any case, only $\mathcal{O}(M)$ data is communicated, i. e., the data associated with subdomain *edges*, which can be transferred in one block, alleviating the latency aspect of the memory wall problem for interconnects. The approach is slightly more difficult to implement efficiently for subdomains that do not share an entire edge but only meet at a single grid point due to the increased bookkeeping requirements, but conceptually identical.

In view of the conflicting goals listed above, the described concept of *minimal overlap* balances numerical (as the approach constitutes an exact approximation over subdomain boundaries) and scalability (as communication is never global except for global reductions such as norms and dot products) requirements. The (theoretical) property of perfect scaling for operations requiring only communication between neighbouring subdomains (e. g., matrix-vector and vector-vector operations) is also achieved in practice due to a carefully tuned implementation. Section 6.3.4 and Section 6.4.3 present scalability experiments on up to 160 nodes of a commodity based GPU-enhanced cluster, for the Poisson problem and a linearised elasticity solver.

Multilevel Domain Decomposition

In *multiplicative* domain decomposition methods, information is passed between subdomains as soon as it is available, so that local updates on one subdomain can immediately be used to advance the solution on other subdomains. This is typically achieved by incorporating available updates into the right hand sides prior to advancing to the next subproblem. In contrast, *additive* techniques treat all subdomains order-independently, and synchronisation is only performed once after all subdomains have been processed.

The dominant drawback of standard Schwarz methods is that their convergence depends on the number of subdomains, due to the nature of elliptic problems where the solution in an interior point depends on all boundary values: Information can only travel 'one subdomain at a time' but has to pass at least once through the entire domain. This drawback can be overcome by introducing a second, coarser grid and by solving the problem also on this grid, combining the solutions again either in a multiplicative or additive way via suitably defined prolongation and restriction operators. It can be shown that the resulting *two-level* Schwarz method converges independently of the number of subdomains [201]. Applying this idea recursively in case the coarser grid is still too large leads to *multilevel Schwarz methods*.

FEAST is rooted in serial geometric multigrid and each subdomain comprises $(2^L+1)^2$ degrees of freedom for scalar equations and the Q_1 element, L denoting the level of refinement. Hence, a natural hierarchy of different mesh resolutions and matching grid transfer operations are readily available.

In parallel, the strong recursive character of 'numerically optimal' serial multigrid smoothers such as incomplete LU decomposition (ILU(k)) leads to impractically high communication requirements. Consequently, the global smoothing is usually relaxed to a blockwise application of the smoothing operator. A blockwise treatment means that the smoother is applied on each subdomain independently either in a Jacobi or Gauß-Seidel fashion, and synchronisation is achieved by computing global defects intermittently, and in case of the Gauß-Seidel coupling, by incorporating updates into the local right hand sides as soon as they become available. As Smith et al. [201] point out, such a blocked approach is—in terms of domain decomposition—in fact an additive or multiplicative Schwarz preconditioner with zero overlap. In the literature, the terminology *block-Jacobi* and *block-Gauß-Seidel*, or *parallel subspace correction* and *successive subspace correction* is also commonly encountered [245].

In SCARC, smoothing is applied in an additive patchwise manner, exhibiting global block-Jacobi character in the sense that macroscopic anisotropies have potentially significant impact on the numerical efficiency of the parallel solver [130, 201]. As a remedy, the global multigrid is not used as a standalone solver, but as a preconditioner inside a global Krylov subspace iteration. Combined with the coarse grid problem, this provides additional global coupling of the unknowns and together with 'optimal step lengths' that are calculated within the Krylov-space method, helps to alleviate the negative effects of macro anisotropies inherent to the block-Jacobi approach. The resulting scheme is thus significantly more robust [15, 130]. As a consequence the number of global iterations decreases, which typically more than outweighs the additional computation and communication requirements, in particular the global communication needed to compute global dot products within the Krylov scheme.

In summary, we denote that SCARC is strictly speaking a *hybrid multilevel domain decomposition* method, or (slightly less exact, but more intuitive) a *generalised domain decomposition multigrid method*, trying to combine the respective advantages in view of the challenges listed at the beginning of this section. The coupling between the different mesh resolutions is realised in a multiplicative way via prolongation and restriction operators specifically tailored to the underlying tensor product meshes, while the smoothing and preconditioning is performed additively in a block-Jacobi fashion (with minimal instead of zero overlap, see previous paragraph). In short, SCARC is multiplicative 'vertically' and additive 'horizontally'.

Solving Coarse Grid Problems

In FEAST, the global coarse grid problems are solved on a master process, while all compute processes are idle. This approach is generally prone to suffer from Amdahl's Law, which basically states that the maximum attainable speedup of any parallel method and consequently its (strong) scalability is limited by its *sequential* portion.

FEAST either uses a Krylov subspace scheme or the highly optimised sparse LU decomposition implementation from the UMFPACK library [57] to solve the coarse grid problems, algebraic multigrid or a parallel solver on few compute nodes have not been implemented yet. Using the Q_1 element, the size of the global coarse grid problems is approximately $K := m(2^{L_c}+1)^2$ where m denotes the number of subdomains in the initial grid and L_c the truncation level of the input mesh.[5] L_c is typically chosen so that the resulting problem size K and the solution with UMFPACK is executed faster than setting $L_c = 1$ and suffering from the poor communication to computation ratio of the parallel multigrid scheme on comparatively 'tiny' problem sizes. In practice, on current commodity based architectures, $K = 20\,000$ unknowns is a good heuristic setting. We refer to the thesis by Wobker [242] for detailed numerical experiments.

He also measures the cost of the application of the LU decomposition for the sparse matrices arising in FEAST, which is performed multiple times in the iterative solution process, to range between $\mathcal{O}(K \log K)$ and $\mathcal{O}(K \log\log K)$, which is clearly more efficient than the quadratic runtime for dense matrices. The initial construction of the LU decomposition is performed in the assembly phase of the matrices, and analogously, takes less time than the asymptotically cubic complexity for dense matrices.

The global problem size is approximately $m(2^L+1)^2$ for a refinement level L and m subdomains. In practice, L is often in the range of 8–10, rendering the local problem size orders of magnitude greater than m. Consequently, the serial execution of the global coarse grid problems does not impair the scalability of the SCARC solvers as long as the number of subdomains remains relatively small compared to their refinement level.

Other researchers come to the same conclusion: Rüde et al. pursue an approach similar to ours, using patchwise multigrid smoothers and a decomposition concept termed *hierarchical hybrid grids* [23, 24]. A recent publication [96] reports good weak scalability for a 3D Poisson problem on a regularly refined unit cube domain for up to 9 170 processor cores. They conclude that even for this huge computation with 307 billion degrees of freedom, the coarse grid solver does not become the bottleneck in terms of Amdahl's Law (or rather Gustanfson's Law) in weak scaling scenarios.

---

[5] Here, we assume a scalar equation, for vector-valued problems, this number is multiplied by the number of components. See Section 2.3.4 for details on the treatment of multicomponent equations in FEAST.

2.3.3. Two-Layer ScaRC Solvers

Due to the elliptic character of the underlying equations, even strongly localised effects influence the convergence of the global schemes, as essentially, the number of iterations needed until convergence is determined by the 'worst' local irregularity [130]. This is not an academic problem, these kinds of irregularities in both the meshes and the differential operators are abundant in practical situations. Such local effects and singularities can be resolved by adaptive mesh deformation based on some error estimation procedure, we refer to Grajewski [98] for a discussion of FEAST's adaptivity concepts. To alleviate this problem, the design goal of *two-layer* SCARC as introduced by Kilian [130] is to *hide these local irregularities as much as possible from the outer solver*. This is achieved by modifying the Schwarz smoother of the global multilevel scheme: Instead of the blockwise application of elementary local smoothers (see Section 2.3.1, last paragraph), two-layer SCARC employs full multigrid solvers acting locally on the individual subdomains (*generalised domain decomposition multigrid method*).

The local multigrid solvers only 'see' local (microscopic) anisotropies and are perfectly well suited to treat them: The locality allows for the application of powerful recursive smoothers as in the serial case, no additive blocking is necessary. In particular, the tensor product property of the local meshes in FEAST can be exploited optimally, see Section 2.3.1. Iterating these local solvers until converge to the exact solution on the patch would of course be beneficial for the global multigrid scheme, as the smoothing returns an exact solution for the local subproblem. However, in practice, this is much too expensive in terms of arithmetic operations, so the solver is typically configured to compute an approximate solution only, e. g., gaining one digit. Consequently, all subdomains are 'smoothed equally well' from the point of view of the global, outer multigrid solver. This is not the case when employing elementary smoothers: On a subdomain exhibiting strong local anisotropies, the application of a fixed number of steps of a standard Gauß-Seidel smoother, for instance, reduces the errors much less than on a more regular, 'simple' subdomain, leading to a much more inhomogeneous smoothing of the outer multigrid solver. In practice however, this approach of solving locally leads to potentially significant imbalances in the load distribution among the compute nodes of a cluster executing two-layer SCARC solvers, an issue which has not been addressed yet in FEAST. In the absence of global anisotropies, the two-layer approach thus renders the global problem isotropic even in case of severe local anisotropies, as shown by Kilian [130]. In terms of numerical efficiency, the additive block-Jacobi character of the global SCARC scheme is then entirely alleviated.

The entire approach is highly beneficial in terms of parallel efficiency, as the global ratio of computation and communication is increased significantly [15, 130]: First, all additional work is performed locally, without any communication. Second, the two-layer SCARC scheme in many situations makes the employment of cheap V cycles in the global multigrid feasible. This cycle type performs much less work on small global problem sizes than the more powerful F and W cycles, see Section 2.2.3, and hence, alleviates the low ratio of computation and communication for smaller levels of refinement. Third, as shown by Kilian [130] and revisited by Wobker [242], the two-layer SCARC scheme does indeed reduce the amount of global iterations and hence improves the global convergence rates.

Analogously to the standard case (see previous section), enclosing the global multigrid solver in a Krylov subspace iteration is highly beneficial to compute optimal steplength parameters. Wobker [242] extends this idea and implemented the possibility of using intermediate Krylov solvers also between the global and the local multigrid. Furthermore, he added support to optionally replacing the elementary smoother of the inner multigrid by few steps of such a Krylov scheme, which is in turn preconditioned by the elementary smoother. The benefit in both cases is that the Krylov iterations compute optimal damping parameters for the smoothing results, rendering the entire scheme significantly more robust.

Global Anisotropies

Obviously, the two-layer ScaRC approach only helps indirectly in the presence of *global* anisotropies, as it only 'unburdens' the global solver from having to deal with local irregularities: The global (additive) block-Jacobi character of the ScaRC scheme is not changed in this situation. As a remedy, Becker [15] in his thesis presents *three-layer* ScaRC and *clustering* of subdomains. The former is motivated by the memory hierarchy inside HPC computer systems: Communication between subdomains residing on the same compute node is typically much faster than communication between nodes, a factor of ten is a good estimate for fast Infiniband interconnects, even higher for high-latency ethernet networks. Three-layer ScaRC introduces an intermediate multigrid solver that couples all subdomains (within one MPI process) that reside on one node. *Clustering* of subdomains works similarly: Several subdomains within one MPI process are grouped into one generally unstructured *matrix block*. The grouping is performed so that the decomposition of the entire domain into new, larger subdomains is 'more isotropic'. On the clustered matrix block, strong smoothers like ILU can be applied. Becker performed extensive numerical tests and concludes that the improved numerical behaviour is in most cases outweighed by the increased total time to solution due to the resulting unstructured matrices and the loss of the local tensor product property, and in case of three-layer ScaRC, by the additional amount of arithmetic work. Consequently, we do not use the concepts of clustering or three-layer ScaRC in this thesis. It would however be interesting to evaluate the concept of clustering combined with a standard ScaRC solver as an *alternative* to two-layer ScaRC, as Becker based all his experiments on top of two-layer ScaRC. This is however clearly beyond the scope of this thesis.

Adaptive Coarse Grid Correction

As mentioned previously, the matrices on the individual subdomains are always invertible if set to their *local state*, and consequently, the local systems which are treated by local multigrid solvers are well-defined. By construction, the minimal overlap however leads to a potentially serious problem: The grids on different refinement levels are not nested as illustrated in Figure 2.10, and thus, the local multigrid scheme is nonconforming. This is not an issue for the global scheme by construction of the *minimal overlap*.

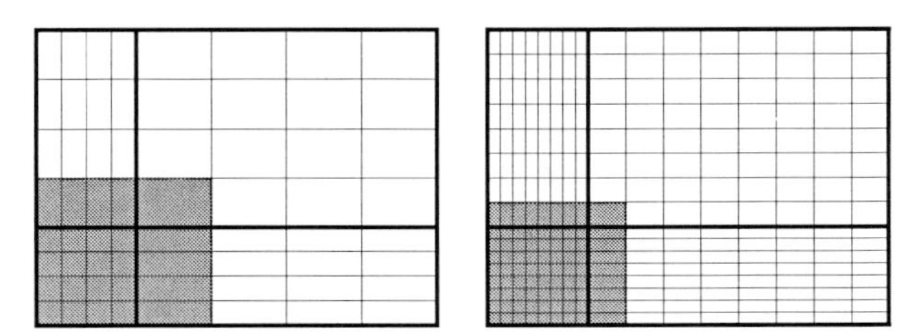

Figure 2.10: Illustration of non-nested grids for the bottom left subdomain. Left: refinement level 2. Right: refinement level 3.

In his thesis, Wobker [242] suggests to apply an *adaptive coarse grid correction* as a remedy, and demonstrates the efficiency by extensive numerical tests. After prolongation, the coarse grid correction $\mathbf{c}$ stemming from a coarser grid 'does not match' the current grid and thus, the current iterate $\mathbf{x}$. The idea now is to damp the correction with a steplength factor α (we omit the indices denoting the current refinement level):

$$\mathbf{x} \leftarrow \mathbf{x} + \alpha \mathbf{c}.$$

This damping parameter is chosen in such a way that the updated vector $\mathbf{x}$ minimises the energy norm

$$(\mathbf{x}-\hat{\mathbf{x}})^{\mathsf{T}}\mathbf{A}(\mathbf{x}-\hat{\mathbf{x}}),$$

$\hat{\mathbf{x}}$ denoting the exact solution. When $\mathbf{A}$ is symmetric this leads to

$$\alpha = \frac{\mathbf{c}^{\mathsf{T}}\mathbf{d}}{\mathbf{c}^{\mathsf{T}}\mathbf{A}\mathbf{c}},$$

where $\mathbf{d} := \mathbf{b} - \mathbf{A}\mathbf{x}$ is the current defect. Apparently, the calculation of the optimal damping factor is comparatively cheap and involves only local data. In this thesis, we apply the adaptive coarse grid correction in all numerical tests.

Summary

The two- and three-layer SCARC schemes and the clustering concept have been proven beneficial in all *numerical* experiments performed so far and all (theoretical) advantages like robustness, convergence etc. have been (numerically) confirmed [15, 130, 242].

However, this only translates to a reduction in total runtime if the additional local work that has to be spent is performed *fast enough*, which in the aforementioned theses is not always the case. This work constitutes an important step forward in this aspect by reducing the time needed for the local solves by one order of magnitude or more.

We conclude our presentation of two-layer SCARC schemes with a general template SCARC solver depicted in Figure 2.11, which is encountered (with suitable modifications) frequently throughout this thesis.

global BiCGStab
preconditioned by
global multilevel (V 1+1)
additively smoothed by
for all Ω_i: **local multigrid**
coarse grid solver: UMFPACK

Figure 2.11: Template two-layer SCARC solver.

2.3.4. Vector-Valued Problems

The guiding idea to treating vector-valued[6] problems with FEAST is to rely on the modular, highly optimised and fully tested core routines for the scalar case in order to formulate robust schemes for a wide range of applications, rather than using the best suited numerical scheme for each application and repeatedly optimising it for new architectures. In particular, the benefit of this approach is that all ongoing and future improvements of FEAST's scalar solvers directly transfer to the solvers for vector-valued systems. Examples include the addition of better multigrid components (e. g., more robust local smoothers such as ADITRIGS), or especially algorithmic and burdensome implementational adaptations tailored to dedicated HPC architectures such as the vector machine NEC SX-8 [20, 45], and hardware co-processors as pursued in this thesis. So to speak, we do not want to *'re-implement the wheel'* every time a new application or a new target architecture is added. The treatment of vector-valued problems has been introduced into FEAST by Wobker

[6]We synonymously use the terms vector-valued, multi-variate and multi-component problems.

[242], and we refer to his thesis for a detailed description. Here, we only illustrate the relevant aspects in the scope of this work.

The observation at the core of the approach is that many vector-valued PDEs, in particular those arising in the application domains of solid mechanics and fluid dynamics, allow for an equation-wise ordering of the unknowns. This rearrangement results, after discretisation, in discrete equation systems consisting of *disjunct* blocks that correspond to scalar subequations.

This special block-structure is exploited in two ways: On the one hand, all standard linear algebra operations on the vector-valued system (e. g., matrix-vector operations, defect computations, dot products) can be implemented as a series of operations for scalar entities, taking advantage of the existing tuned linear algebra components: As the discretisation is performed on the same underlying locally structured grids, the existing routines exploiting the tensor product property can be simply called, avoiding code duplication. On the other hand, the process of solving the vector-valued equation systems can be brought down to the treatment of auxiliary scalar subsystems which can be efficiently solved by scalar SCARC schemes. This procedure is facilitated by the use of special preconditioners that shift most of the overall work to such inner iterations.

SCARC is from now on interpreted as its generalisation to vector-valued problems, which means that the domain decomposition is handled by Schwarz smoothers (see Section 2.3.2) and retains the block-Jacobi character, while the disjunct blocking of the equations is realised by the aforementioned block preconditioner, which can be applied either in a block-Jacobi or in a block-Gauß-Seidel fashion. Consequently, *generalised ScaRC* provides great flexibility in tuning the linear solution schemes to the given problem: The restriction of the global vector-valued system to scalar blocks of equations or to subdomains or simultaneously to both can be performed at any time during the solution process. Concrete examples of solvers are presented in the evaluation of our minimally invasive co-processor integration into FEAST, see Section 6.4.1 (linearised elasticity) and Section 6.5.1 (stationary laminar flow).

Example: Matrix-Vector Multiplication

A matrix-vector multiplication $\mathbf{y} = \mathbf{A}\mathbf{x}$ with the following block structure is a good example to illustrate the treatment of vector-valued problems in FEAST:

$$\begin{pmatrix} \mathbf{y}_1 \\ \mathbf{y}_2 \end{pmatrix} = \begin{pmatrix} \mathbf{A}_{11} & \mathbf{A}_{12} \\ \mathbf{A}_{21} & \mathbf{A}_{22} \end{pmatrix} \begin{pmatrix} \mathbf{x}_1 \\ \mathbf{x}_2 \end{pmatrix}$$

As explained above, the multiplication is performed as a series of operations on the local finite element matrices per subdomain $\bar{\Omega}_i$, denoted by superscript square brackets. The global scalar operators, corresponding to the blocks in the matrix, are treated individually as demonstrated in Figure 2.12.

For $j = 1,2$, do

1. For all $\bar{\Omega}_i$, compute $\mathbf{y}_j^{[i]} = \mathbf{A}_{j1}^{[i]}\mathbf{x}_1^{[i]}$.
2. For all $\bar{\Omega}_i$, compute $\mathbf{y}_j^{[i]} = \mathbf{y}_j^{[i]} + \mathbf{A}_{j2}^{[i]}\mathbf{x}_2^{[i]}$.
3. Communicate entries in $\mathbf{y}_j$ corresponding to the boundaries of neighbouring subdomains.

Figure 2.12: Matrix-Vector multiplication for block-structured systems.

2.3.5. Linearised Elasticity in FEAST

The application FEASTSOLID, built on top of FEAST, provides support for solving problems from solid mechanics. In this thesis, we restrict ourselves to a subset of FEASTSOLID's functionality, namely the fundamental model problem of linearised elasticity and compressible material, i. e., the small deformation of solid bodies under external loads. FEASTSOLID is developed by Hilmar Wobker, and constitutes a large part of his thesis [242].

Theoretical Background

A two-dimensional body is identified with a domain $\bar{\Omega} = \Omega \cup \partial\Omega$ covering it, where Ω denotes the interior and $\Gamma = \partial\Omega$ its boundary. The boundary is split disjunctly into two parts, the Dirichlet part Γ_{D} where displacements are prescribed and the Neumann part Γ_{N} where surface forces can be applied. Furthermore the body can be exposed to volumetric forces such as gravity. In this thesis, we only consider the simple, but nevertheless fundamental, model problem of elastic, compressible material under static loading, assuming small deformations. We use a formulation where the displacements $\boldsymbol{u}(\boldsymbol{x}) = \big(u_1(\boldsymbol{x}), u_2(\boldsymbol{x})\big)^{\mathsf{T}}$ of a material point $\boldsymbol{x} \in \bar{\Omega}$ are the only unknowns in the equation.

The strains are defined by the linearised strain tensor $\varepsilon_{ij} = \frac{1}{2}\left(\frac{\partial u_i}{\partial x_j} + \frac{\partial u_j}{\partial x_i}\right), \quad i,j = 1,2$, describing the linearised kinematic relation between displacements and strains. The material properties are reflected by the constitutive law, which determines a relation between the strains and the stresses. Hooke's law for isotropic elastic material is used, $\boldsymbol{\sigma} = 2\mu\boldsymbol{\varepsilon} + \lambda\,\mathrm{tr}(\boldsymbol{\varepsilon})\boldsymbol{I}$, where $\boldsymbol{\sigma}$ denotes the symmetric stress tensor and μ and λ are the so-called Lamé constants, which are connected to the Young modulus E and the Poisson ratio ν as follows:

$$\mu = \frac{E}{2(1+\nu)}, \qquad \lambda = \frac{E\nu}{(1+\nu)(1-2\nu)} \tag{2.3.1}$$

The basic physical equations for problems of solid mechanics are determined by equilibrium conditions. For a body in equilibrium, the inner forces (stresses) and the outer forces (external loads $\boldsymbol{f}$) are balanced:

$$-\operatorname{div}\boldsymbol{\sigma} = \boldsymbol{f}, \qquad \boldsymbol{x} \in \Omega.$$

Using Hooke's law to replace the stress tensor, the problem of linearised elasticity can be expressed in terms of the following elliptic boundary value problem, called the *Lamé equation*:

$$-2\mu\operatorname{div}\boldsymbol{\varepsilon}(\boldsymbol{u}) - \lambda\,\mathbf{grad}\operatorname{div}\boldsymbol{u} = \boldsymbol{f}, \qquad \boldsymbol{x} \in \Omega \tag{2.3.2a}$$

$$\boldsymbol{u} = \boldsymbol{g}, \qquad \boldsymbol{x} \in \Gamma_{\mathrm{D}} \tag{2.3.2b}$$

$$\boldsymbol{\sigma}(\boldsymbol{u}) \cdot \boldsymbol{n} = \boldsymbol{t}, \qquad \boldsymbol{x} \in \Gamma_{\mathrm{N}} \tag{2.3.2c}$$

Here, $\boldsymbol{g}$ are prescribed displacements on Γ_{D}, and $\boldsymbol{t}$ are given surface forces on Γ_{N} with outer normal $\boldsymbol{n}$. For details on the elasticity problem, see for example the textbook by Braess [34].

Solution Strategy

In order to solve vector-valued linearised elasticity problems with the general strategy outlined in Section 2.3.4, the equations are reordered according to the unknowns $\boldsymbol{u} = (u_1, u_2)^{\mathsf{T}}$ representing the displacements in x and y-direction, respectively. This technique is called *separate displacement ordering* [9]. Rearranging the left hand side of Equation (2.3.2a) accordingly yields:

$$-\begin{pmatrix} (2\mu+\lambda)\partial_{xx} + \mu\partial_{yy} & (\mu+\lambda)\partial_{xy} \\ (\mu+\lambda)\partial_{yx} & \mu\partial_{xx} + (2\mu+\lambda)\partial_{yy} \end{pmatrix} \begin{pmatrix} u_1 \\ u_2 \end{pmatrix} = \begin{pmatrix} f_1 \\ f_2 \end{pmatrix} \tag{2.3.3}$$

The domain $\bar{\Omega}$ is partitioned into a collection of several subdomains $\bar{\Omega}_i$ as explained in Section 2.3.1, each of which is refined to a logical tensor product structure. The weak formulation of Equation (2.3.3) is considered to apply a finite element discretisation with conforming bilinear elements of the Q_1 space, the only element supported in FEAST at the time of writing. The resulting linear equation system can be written as $\mathbf{Ku} = \mathbf{f}$. Corresponding to representation (2.3.3) of the continuous equation, the discrete system has the following block structure,

$$\begin{pmatrix} \mathbf{K}_{11} & \mathbf{K}_{12} \\ \mathbf{K}_{21} & \mathbf{K}_{22} \end{pmatrix} \begin{pmatrix} \mathbf{u}_1 \\ \mathbf{u}_2 \end{pmatrix} = \begin{pmatrix} \mathbf{f}_1 \\ \mathbf{f}_2 \end{pmatrix}, \tag{2.3.4}$$

where $\mathbf{f} = (\mathbf{f}_1, \mathbf{f}_2)^\mathsf{T}$ is the vector of external loads and $\mathbf{u} = (\mathbf{u}_1, \mathbf{u}_2)^\mathsf{T}$ the (unknown) coefficient vector of the finite element solution. The matrices $\mathbf{K}_{11}$ and $\mathbf{K}_{22}$ of this block-structured system correspond to scalar elliptic operators (cf. Equation (2.3.3)), hence the generalised SCARC solvers are applicable.

Note, that the scalar elliptic operators appearing in Equation (2.3.3) are *anisotropic*. The degree of anisotropy a_{op} depends on the material parameters (see Equation (2.3.1)) and is given by

$$a_{\text{op}} = \frac{2\mu + \lambda}{\mu} = \frac{2 - 2\nu}{1 - 2\nu}. \tag{2.3.5}$$

Solver Outline

We illustrate the core step of the solution process with a basic iteration scheme, a preconditioned defect correction method:

$$\mathbf{u}^{(k+1)} = \mathbf{u}^{(k)} + \omega \tilde{\mathbf{K}}_{\text{B}}^{-1} (\mathbf{f} - \mathbf{K}\mathbf{u}^{(k)}) \tag{2.3.6}$$

This iteration scheme acts on the global system (2.3.4) and thus couples the two sets of unknowns $\mathbf{u}_1$ and $\mathbf{u}_2$. The *block-preconditioner* $\tilde{\mathbf{K}}_{\text{B}}$ explicitly exploits the block structure of the matrix $\mathbf{K}$. We use a *block-Gauß-Seidel* preconditioner $\tilde{\mathbf{K}}_{\text{BGS}}$ in this thesis, so that each iteration of the global defect correction scheme consists of the following three steps:

1. Compute the global defect (cf. Section 2.3.4):

$$\begin{pmatrix} \mathbf{d}_1 \\ \mathbf{d}_2 \end{pmatrix} = \begin{pmatrix} \mathbf{f}_1 \\ \mathbf{f}_2 \end{pmatrix} - \begin{pmatrix} \mathbf{K}_{11} & \mathbf{K}_{12} \\ \mathbf{K}_{21} & \mathbf{K}_{22} \end{pmatrix} \begin{pmatrix} \mathbf{u}_1^{(k)} \\ \mathbf{u}_2^{(k)} \end{pmatrix}$$

2. Apply the block-preconditioner

$$\tilde{\mathbf{K}}_{\text{BGS}} := \begin{pmatrix} \mathbf{K}_{11} & \mathbf{0} \\ \mathbf{K}_{21} & \mathbf{K}_{22} \end{pmatrix}$$

   by approximately solving the system $\tilde{\mathbf{K}}_{\text{BGS}} \mathbf{c} = \mathbf{d}$. This is performed by two scalar solves per subdomain and one global scalar matrix-vector multiplication:

   (a) For each subdomain $\bar{\Omega}_i$, solve $\mathbf{K}_{11}^{[i]} \mathbf{c}_1^{[i]} = \mathbf{d}_1^{[i]}$.

   (b) Update right hand side: $\mathbf{d}_2 = \mathbf{d}_2 - \mathbf{K}_{21} \mathbf{c}_1$.

   (c) For each subdomain $\bar{\Omega}_i$, solve $\mathbf{K}_{22}^{[i]} \mathbf{c}_2^{[i]} = \mathbf{d}_2^{[i]}$.

3. Update the global solution with the (eventually damped) correction vector:

$$\mathbf{u}^{(k+1)} = \mathbf{u}^{(k)} + \omega \mathbf{c}$$

In our numerical and performance tests in Section 6.4, this defect correction scheme is replaced by a (multi-component) multigrid iteration, constituting the generalisation to vector-valued problems of the template solver depicted in Figure 2.11. The procedure, in particular the application of the block-preconditioner, is however identical to the basic iteration we use here for illustrative purposes.

2.3.6. Stokes and Navier-Stokes in FEAST

The application FEASTFLOW, built on top of FEAST, provides support for solving problems from fluid dynamics. It is developed by Sven Buijssen and constitutes a large part of his thesis [44]. In this thesis, we only use a subset of FEASTFLOW's feature set, and restrict ourselves to model problems from stationary laminar flow. We refer to Buijssen's thesis for all details we intentionally omit in the following. The textbook by Donea and Huerta [61] is also recommended for a graduate-level presentation of the theoretical background of the model.

Theoretical Background

The *Navier-Stokes* equations are derived under certain assumptions and simplifications from the general conservation laws for mass, momentum and energy in continuum mechanics. They govern the flow of incompressible, isothermal, homogeneous and isotropic Newtonian fluids and gases. The *kinematic viscosity* ν, the ratio between inertial and viscous forces, is assumed to be constant. It is inversely proportional to the *Reynolds number* of the fluid; more precisely, the relation is $Re = UL/\nu$ where U is the characteristic fluid velocity and L the characteristic length. Low Reynolds numbers correspond to highly viscous fluids, small spatial dimensions or low flow speeds. The stationary Navier-Stokes equations read in their dimensionless form:

$$\begin{aligned} -\frac{1}{Re}\Delta \boldsymbol{u} + (\boldsymbol{u}\cdot\mathbf{grad})\boldsymbol{u} + \mathbf{grad}\ p &= \boldsymbol{f} \text{ in } \Omega \\ \operatorname{div} \boldsymbol{u} &= 0 \text{ in } \Omega \\ \boldsymbol{u} &= \boldsymbol{u}_D \text{ on } \Gamma_{\mathrm{D}} \\ -\frac{1}{Re}(\boldsymbol{n}\cdot\mathbf{grad})\boldsymbol{u} + p\boldsymbol{n} &= \boldsymbol{t} \text{ on } \Gamma_{\mathrm{N}} \end{aligned} \tag{2.3.7}$$

Here, $\boldsymbol{u}$ denotes velocity, p pressure, $\boldsymbol{f}$ body forces, $\Omega \subseteq \mathbb{R}^2$ the domain with outer unit normal vector $\boldsymbol{n}$, and Γ_{D} and Γ_{N} the boundary parts with, respectively, Dirichlet and natural boundary conditions (i. e., inflow, outflow and adhesion/slip conditions). A common simplification are the *Stokes* equations for very low Reynolds numbers. In this case, the viscous term dominates and the nonlinear transport term $(\boldsymbol{u}\cdot\mathbf{grad})\boldsymbol{u}$ can be omitted, leading to a linear problem.

Discretisation

The weak formulation of equation system (2.3.7) is considered to apply a finite element discretisation with conforming bilinear elements of the Q_1 space for both $\boldsymbol{u}$ and p. This Galerkin approach exhibits instabilities which stem from dominating convection and from the violation of the discrete LBB-condition (also known as inf-sup condition) [40]. For stability on arbitrary meshes, pressure-stabilisation (PSPG) and streamline-upwind stabilisation (SUPG) is applied choosing the mesh-dependent parameters in accordance with Apel et al. [6]. The resulting discrete nonlinear equation system reads

$$\left[\begin{pmatrix} \mathbf{A}_{11} & \mathbf{0} & \mathbf{B}_1 \\ \mathbf{0} & \mathbf{A}_{22} & \mathbf{B}_2 \\ \mathbf{B}_1^{\mathsf{T}} & \mathbf{B}_2^{\mathsf{T}} & \mathbf{0} \end{pmatrix} + \begin{pmatrix} \mathbf{C}_{11} & \mathbf{C}_{12} & \mathbf{C}_{13} \\ \mathbf{C}_{21} & \mathbf{C}_{22} & \mathbf{C}_{23} \\ \mathbf{C}_{31} & \mathbf{C}_{32} & \mathbf{C}_{33} \end{pmatrix}\right] \begin{pmatrix} \mathbf{u}_1 \\ \mathbf{u}_2 \\ \mathbf{p} \end{pmatrix} = \begin{pmatrix} \mathbf{f}_1 \\ \mathbf{f}_2 \\ \mathbf{g} \end{pmatrix} + \begin{pmatrix} \mathbf{c}_1 \\ \mathbf{c}_2 \\ \mathbf{c}_3 \end{pmatrix},$$

with

$$\mathbf{A}_{11} := \frac{1}{Re}\mathbf{L}_{11} + \mathbf{N}_{11}(\mathbf{u})$$
$$\mathbf{A}_{22} := \frac{1}{Re}\mathbf{L}_{22} + \mathbf{N}_{22}(\mathbf{u}),$$

in short $\mathbf{Kv} = \mathbf{h}$ where the matrices $\mathbf{L}_{ii}$ correspond to the Laplacian operator and $\mathbf{N}_{ii}(\mathbf{u})$ to the convection operator. $\mathbf{B}$ and $\mathbf{B}^\top$ are discrete counterparts of the gradient and divergence operators while $\mathbf{C}_{ij}$ stem from the discretisation of the PSPG and SUPG stabilisation terms. For the case of an isotropic mesh, we notice the following: $\mathbf{C}_{33}$ is identical to a discretisation of the pressure Poisson operator, scaled with the mesh size αh^2 where α is a fixed stabilisation parameter.

Solution Algorithm

The nonlinear problem is reduced to a sequence of linear problems by applying a fixed point defect correction method,

$$\mathbf{v}^{(k+1)} = \mathbf{v}^{(k)} + \omega\tilde{\mathbf{K}}_{\mathrm{B}}^{-1}(\mathbf{h} - \mathbf{K}\mathbf{v}^{(k)}) \qquad k = 1,\ldots,$$

where the application of $\tilde{\mathbf{K}}_{\mathrm{B}}$ can be identified with the solution of linearised subproblems with the nonlinear residual as right hand side and using the solution $\mathbf{u}$ from the previous nonlinear iteration for assembly of the $\mathbf{N}_{ii}(\mathbf{u})$ part of $\mathbf{K}$. These still vector-valued linearised subproblems are subsequently tackled with help of a pressure Schur complement approach. We illustrate it with the following basic iteration, but prefer a Krylov subspace solver such as BiCGStab for increased numerical efficiency in the tests in Section 6.5:

$$\begin{pmatrix}\mathbf{u}^{(n+1)}\\ \mathbf{p}^{(n+1)}\end{pmatrix} = \begin{pmatrix}\mathbf{u}^{(n)}\\ \mathbf{p}^{(n)}\end{pmatrix} + \mathbf{K}_{\mathrm{S}}^{-1}\left[\begin{pmatrix}\mathbf{f}\\ \mathbf{g}\end{pmatrix} - \begin{pmatrix}\mathbf{A} & \mathbf{B}\\ \mathbf{B}^\top & \mathbf{C}\end{pmatrix}\begin{pmatrix}\mathbf{u}^{(n)}\\ \mathbf{p}^{(n)}\end{pmatrix}\right] \tag{2.3.8}$$

$\mathbf{A}$ is a block-structured matrix consisting of the linearised matrices $\mathbf{A}_{ij}$, $\mathbf{B}^\top$ is defined as $(\mathbf{B}_1^\top, \mathbf{B}_2^\top)$ and the vectors $\mathbf{u}^{(n)}$ and $\mathbf{f}$ as the iterates of the solution $(\mathbf{u}_1, \mathbf{u}_2)^\top$ and right hand side $(\mathbf{f}_1, \mathbf{f}_2)^\top$, respectively. The preconditioner $\mathbf{K}_{\mathrm{S}}$ is defined as the lower block triangular matrix

$$\mathbf{K}_{\mathrm{S}} := \begin{pmatrix}\mathbf{A} & 0\\ \mathbf{B}^\top & -\mathbf{S}\end{pmatrix}$$

involving the pressure Schur complement matrix

$$\mathbf{S} := \mathbf{B}^\top\mathbf{A}^{-1}\mathbf{B} - \mathbf{C}.$$

It can be shown that the square of the iteration matrix of the preconditioned system

$$\hat{\mathbf{K}} := \mathbf{I} - \mathbf{K}_{\mathrm{S}}^{-1}\begin{pmatrix}\mathbf{A} & \mathbf{B}\\ \mathbf{B}^\top & \mathbf{C}\end{pmatrix}$$

vanishes [166], which is equivalent to the associated Krylov space, $\mathrm{span}\{\mathbf{r}, \hat{\mathbf{K}}\mathbf{r}, \hat{\mathbf{K}}^2\mathbf{r}, \hat{\mathbf{K}}^3\mathbf{r}, \ldots\}$, having dimension 2. This implies that—with exact arithmetics—any Krylov subspace solver would terminate in at most two iterations with the solution to the linear system arising in system (2.3.8), using the preconditioner $\mathbf{K}_{\mathrm{S}}$. Few iterations with a 'good' approximate $\tilde{\mathbf{K}}_{\mathrm{S}}$ of $\mathbf{K}_{\mathrm{S}}$ hence suffice to solve system (2.3.8).

In summary, the basic iteration (2.3.8) entails the following steps:

1. Compute the global defect

$$(\mathbf{d}_1, \mathbf{d}_2, \mathbf{d}_3)^\top = \begin{pmatrix}\mathbf{f}\\ \mathbf{g}\end{pmatrix} - \begin{pmatrix}\mathbf{A} & \mathbf{B}\\ \mathbf{B}^\top & \mathbf{C}\end{pmatrix}\begin{pmatrix}\mathbf{u}^{(n)}\\ \mathbf{p}^{(n)}\end{pmatrix}$$

2. Apply the block preconditioner $\tilde{\mathbf{K}}_S$ by approximately solving

$$\mathbf{A}(\mathbf{c}_1,\mathbf{c}_2)^\top = (\mathbf{d}_1,\mathbf{d}_2)^\top \tag{2.3.9}$$

and

$$\mathbf{S}\mathbf{c}_3 = -\mathbf{d}_3 + \mathbf{B}_1^\top\mathbf{c}_1 + \mathbf{B}_2^\top\mathbf{c}_2 \tag{2.3.10}$$

3. Update the global solution with the (damped) correction vector:

$$(\mathbf{u}^{(n+1)},\mathbf{p}^{(n+1)})^\top = (\mathbf{u}^{(n)},\mathbf{p}^{(n)})^\top + \omega(\mathbf{c}_1,\mathbf{c}_2,\mathbf{c}_3)^\top$$

The second step is the most expensive one and requires closer examination. Note that the diagonal block matrices $\mathbf{A}_{ii}$ correspond to scalar operators stemming from the momentum equations, so FEAST's tuned scalar and multivariate solvers can be applied. To treat the block $\mathbf{S}$ efficiently, an appropriate preconditioner for the pressure Schur complement is required. It has been shown that the pressure mass matrix $\mathbf{M}_p$ is a good preconditioner for the diffusive part of $\mathbf{S}$ [225]. The use of a lumped mass matrix $\mathbf{M}_p^L$ reduces solving equation (2.3.10) to scaling the right hand side with the inverse of this diagonal matrix. As the convective part is neglected, this choice of the preconditioner is only favourable for stationary Navier-Stokes problems at low Reynolds numbers.

3

A Short History of GPU Computing

Graphics processing units (GPUs) are primarily designed for one particular class of applications, *rasterisation and depth-buffering based interactive computer graphics*. One could argue that this is no longer the case, as certain features are being added to the processors that are not needed in graphics workloads, a consequence of GPUs transforming into a viable general-purpose (data-parallel) computing resource. However, to help understand why and how GPUs and the field of *GPU computing* (synonymously: *GPGPU, general-purpose computation on graphics processing units*) which this thesis contributes to, have matured in the way they did and what makes GPUs (fundamentally) different to commodity CPUs, we are convinced that a historical perspective is necessary. At the core of this chapter, we try to answer the question 'Why are GPUs fast, and what can we do with them?', and present all necessary background, aspects of computer architecture, paradigm changes in the underlying programming model, and bibliographies of important results and publications as we proceed.

Section 3.1 is concerned with the 'legacy GPGPU' approach to program graphics hardware through graphics APIs, while Section 3.2 covers 'modern GPGPU' through dedicated compute APIs such as NVIDIA CUDA or AMD Stream. In Section 3.3 we discuss GPUs in terms of parallel computing taxonomies and parallel programming models. The chapter concludes with an outlook on future programming APIs in Section 3.4, future graphics architectures and the convergence of CPU and GPU designs.

3.1. Legacy GPGPU: Graphics APIs

This section is structured as follows: Section 3.1.1 introduces the *graphics pipeline* as a hardware *and* software abstraction concept. The evolution of GPUs towards programmability is presented in Section 3.1.2, this development culminated into what we refer to as the first wave of GPU computing in the years 2003–2006. Section 3.1.3 and 3.1.4 discuss important features of these *DirectX 9 class GPUs* and the underlying programming model. In Section 3.1.5 we present a bibliography of publications from these years that are relevant in the scope of this thesis. Section 3.1.6 is devoted to the limitations of single precision. Finally, Section 3.1.7 describes NVIDIA's GeForce 6 architecture as an exemplary hardware design for the early years of GPGPU.

3.1.1. The Graphics Pipeline

The rendering of photo-realistic, high-resolution images at interactive frame rates has extremely high computational requirements:

- At the time of writing, the 1080p full high definition standard defines a screen resolution of 1920×1080, slightly more than two million pixels.

- In complex applications, each pixel requires thousands or more operations for geometry processing, screen-space mapping, lighting and shadowing, material simulation, texturing, and advanced volumetric effects.
- In particular in the domain of action-oriented computer games, enthusiasts debate vigorously at which frame rate an animation is perceived as continuous. This is of course dependent on the intensity of the rendered scene, almost still images require much lower frame rates as, e. g., rendering of explosions. Most games try to maintain a frame rate of at least 30–60 frames per second.

This means that interactive photo-realistic computer graphics can often require up to 10^{11} operations per second, and GPUs must deliver this enormous amount of compute power. Looking at graphics workloads in more detail, three important observations can be made:

- Graphics workloads are inherently massively parallel. In fact, the huge performance demands can only be satisfied by exploiting this parallelism to extreme scales, at least compared to other (single-chip) processors. Basically, the GPU architecture is centred around a large number of fine-grained parallel processors.
- The bandwidth demands of graphics tasks, in particular multi-texturing with advanced, anisotropic filtering, are insatiable. Furthermore, the memory subsystem must be able to serve many concurrent requests.
- Roughly speaking, the human visual system operates on the millisecond scale. In contrast, a modern processor executes at a clock rate of more than one Gigahertz, so single operations complete in the order of nanoseconds. This gap of six orders of magnitude implies that the latency of individual operations is less important than overall throughput.

All commodity GPUs (and the APIs—application programming interfaces—used to program them, see below) are organised in a similar way, the so-called *graphics pipeline*. This concept was first introduced by Silicon Graphics Inc. (SGI) in 1992 with the first version of the OpenGL standard, even though at that time, not all features were implemented in hardware. As an abstraction of the actual implementation, the pipeline divides the computation (i. e., the rendering of an image) into several disjunct *stages* which are explained below. The pipeline is feed-forward, which naturally leads to *task parallelism* between the stages. Within each stage, *data parallelism* is trivially abundant, as all primitives are treated separately. From a hardware perspective, the pipeline concept removes the necessity of expensive control logic to counteract typical hazards induced by the parallelism such as read-after-write, write-after-read, and synchronisation, deadlocks and other race conditions. To maximise throughput over latency, the pipeline is very deep, with thousands of primitives in flight at a time. In a CPU, any given operation may take on the order of 20 cycles between entering and leaving the processing pipeline (assuming a level-1 cache hit for data); on the GPU, in contrast, operations may take thousands of cycles to finish. In summary, the implementation of the graphics pipeline in hardware allows to dedicate a much larger percentage of the available transistors to actual computation rather than to control logic, at least compared to commodity CPU designs. In the following, the abstraction of the graphics pipeline and its hardware implementation in the form of GPUs are used synonymously; two concrete hardware realisations are discussed in Section 3.1.7 and Section 3.2.2.

CPUs are dealing with memory bandwidth and latency limitations by using ever-larger hierarchies of caches. The working set sizes of graphics applications have grown approximately as fast as transistor density. Therefore, it is prohibitive to implement a large enough caching hierarchy on the GPU chip that delivers a reasonably high cache hit rate and maintains coherency. GPUs do have caches, but they are comparatively small and optimised for spatial locality, as this is the relevant case in texture filtering operations. More importantly, the memory subsystem is designed to maximise streaming bandwidth (and hence, throughput) by latency tolerance, page-locality, minimisation of read-write direction changes and even lossless compression.

Finally, another important aspect is that the market volume of interactive computer games amounts to billions of dollars per year, creating enough critical mass and market pressure to drive rapid hardware evolution, in terms of both absolute performance and broadening feature set (*economies of scale*).

A Canonical Graphics Pipeline

Figure 3.1 depicts a simplified, canonical pipeline. In detail, the stages perform the following tasks:

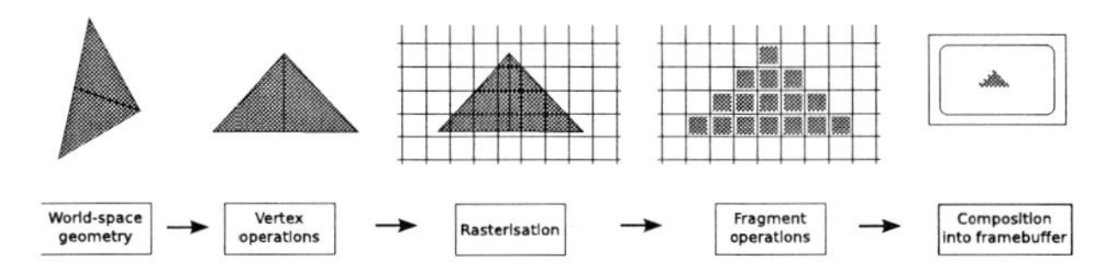

Figure 3.1: Conceptual illustration of the graphics pipeline.

Vertex operations and primitive assembly: The input of the graphics pipeline is a list of geometric primitives in a three-dimensional world coordinate system. The individual primitives are formed from vertices, which are transformed individually into screen space. Additionally, each vertex is *shaded* by computing its interaction with the lights in the scene. After shading, all geometric primitives are transformed into triangles, the fundamental primitive supported natively in hardware. As all vertices are treated independently of each other, parallelism in this stage is substantial.

Rasterisation: The *rasteriser* processes triangles independently and generates a *fragment* (a 'proto-pixel') at each pixel covered in the output image. As triangles may overlap in screen space, different triangles may generate different fragments at the same pixel location. All per-vertex attributes such as colours, texture coordinates and normals, are interpolated over the covered region of the screen, and associated with the fragments.

Fragment operations: At this stage, the final colour of each fragment is computed from the interpolated vertex attributes, again independently for each fragment. Additionally, values from texture maps (global memory buffers containing images mapped to surfaces) can be included. This stage is typically the computationally most demanding stage.

Composition: In this final stage, all fragments' colours corresponding to one pixel are blended with the colour already present in the framebuffer, taking into account their depth information (*z-buffer*) for occlusion or transparency. The output of this stage is one colour value for each pixel, which is written into the framebuffer from where it is read by the display device.

It is important to note that the fundamental data type of the pipeline is a four-tuple, both for vertices (homogeneous 3D coordinates) and fragments (RGBA colour).

Graphics APIs

The hardware is not exposed directly to the programmer, in fact, most details of the hardware realisation of the graphics pipeline are proprietary and largely secret. Instead, well-defined APIs offer a set of data containers and functions to map operations and data to the hardware. These

APIs are typically implemented by the vendors, as a set of libraries and header files interacting with the low-level device driver.

DirectX (more specifically, Direct3D as a subset of DirectX) and OpenGL are the two dominant APIs to program graphics hardware. DirectX is restricted to Microsoft Windows (and Microsoft's gaming consoles such as the XBox), while OpenGL has been implemented for, e. g., Windows, Linux, MacOS and Solaris. Both APIs are defined by consortia in which hardware and software vendors collaborate closely. The DirectX specification is headed by Microsoft, whereas the open Khronos group leads the development of OpenGL, more specifically, the OpenGL ARB (architecture review board) as part of the Khronos consortium. The pipeline concept has been the integral component of both APIs since their initial revisions, in 1995 and 1992 respectively.

As both APIs map to the same hardware, there is usually a one-to-one correspondence; typically, no (major) features exist that are only exposed through one of the APIs. The fundamental design difference between the two APIs is how new hardware features are exposed. The goal of DirectX is to specify a set of features that hardware must implement for a longer product cycle, typically three years. It is therefore convenient to identify a certain class of GPUs by the highest DirectX version it supports, one speaks for instance of *DirectX 10 class hardware*. As the hardware development cycles are usually shorter, OpenGL includes the concept of *extensions* to expose new or experimental features faster. Not all extensions are supported on all hardware, and software developers have to check at runtime if a given extension is supported. Only after extensions are supported by a wide range of GPUs, they are considered for inclusion in the OpenGL core, and consequently, the OpenGL version numbers are incremented at a slower rate than for DirectX.

In the domain of computer games, DirectX is almost exclusively used, and with Linux increasing its market shares, is not even restricted to Microsoft Windows alone anymore: This is the goal of the ongoing *wine* project.[1] However, in academia and for 'professional' applications, OpenGL is often favoured.

3.1.2. Evolution Towards Programmability

We begin our short survey of the evolution towards programmability of GPUs for both graphics and non-graphics applications in the late 1990s. An article by Blythe [29] covers the earlier development.

The first generations of dedicated commodity 3D accelerators were configurable but not programmable. For instance, fragment operations like multi-texturing and filtering were supported quite early in hardware, but the programmer could only configure how different textures were combined from a fixed set of possibilities such as linear interpolation. This hardware was too restricted for general-purpose computations, with the exception of the 1990 SIGGRAPH paper by Lengyel et al. [146] on using the rasteriser for robot motion planning which was ahead of its time in retrospective.

Until the late 1990s, the number of GPU manufacturers consolidated to very few: NVIDIA and ATI dominate high-performance graphics and are continuously competing, with neither of them being able to take the lead permanently. VIA and Intel also produce graphics chips integrated in the motherboards' chipsets; their performance (and feature set) is often not sufficient for GPU computing, and we ignore these processors in this thesis. The landscape might change in the near future, as Intel has announced the *Larrabee* graphics processors (see Section 3.4).

[1] `http://www.winehq.org/`

Pre-DirectX 8 GPUs

NVIDIA's GeForce 256, introduced in late 1999, was the first accelerator that implemented the entire pipeline in hardware by introducing vertex processing capabilites, then called 'hardware T&L' (transform and lighting). Anecdotally, the term *GPU* to distinguish graphics processors from CPUs was coined during the launch of this card. It was also the first to add programmability to the fragment stage with its so-called *register combiners* which allowed limited combination of textures and interpolated colours to compute final fragment colours.

At this time already, researchers began to explore the use of graphics hardware for non-graphics operations and applications. We refer to the PhD theses by Harris [107] and Krüger [139] for details and references. In the scope of this thesis, we note that—to the best of our knowledge—Rumpf and Strzodka [188] were the first to map components of a finite element solver for a PDE-based application to graphics API calls and hence to the hardware: They consider the numerical solution of nonlinear diffusion problems which appear in denoising, edge enhancement and related (multiscale) image processing tasks. For instance, they map linear combinations and matrix-vector multiplications (for fixed stencils) to blending, function evaluations to lookup tables in texture memory, index shifts to coordinate transformations, and reduction operations such as norms and dot products to histogram computations with a small number of buckets. Their paper also discusses the limitations of the available fixed precision number formats in much detail.

DirectX 8 – The Beginning

In late 2000, *shader model 1* as part of the DirectX 8 standard introduced an assembly-level language for vertex and fragment processing, and the first GPUs implementing it were ATI's Radeon 8500 and NVIDIA's GeForce 3 [148]. Programmable shading marked a significant deviation from the *fixed function pipeline*, enabling developers to write code without worrying about setting pipeline states. Vendors added a shader compiler to their device drivers that mapped shader program assembly code to actual hardware instructions (ISA). Capabilities were limited, in particular in terms of branching and looping. Additional restrictions existed in the number of textures that could be sampled, the length of compiled programs and the number of static and dynamic instructions that each shader could perform. Nonetheless, a multitude of effects became possible that could not be achieved previously, or only with cumbersome and slow multipass rendering.

This hardware generation inspired more researchers to explore the suitability and applicability for general-purpose computations. Mark Harris' PhD thesis contains an almost complete bibliography of related research as of 2003 [107]. Here, we point out the work by Harris et al. [111] as an example, who implemented a physically accurate PDE simulation of boiling water on GeForce 3 and 4 hardware, and Strzodka [208] on applying emulation techniques (see Section 4.2) to extend the available fixed precision formats.

In 2002, Mark Harris coined the term *GPGPU* ('general-purpose computation on graphics processing units') and one year later founded GPGPU.ORG[2], a community website dedicated to cataloguing the use of GPUs for general-purpose computation, and to providing a central resource and discussion forum for GPGPU researchers and developers. Since 2005, the author of this thesis has contributed significantly to this site, both as moderator on the discussion forums and as editor of the developer pages, e. g., by writing tutorial code.

3.1.3. DirectX 9 GPUs – The First Wave of GPGPU 2003–2006

Common to all early work on using GPUs for general-purpose computations was the lack of floating point precision, see above for examples and references. Fixed precision was also not sufficient

[2] `http://www.gpgpu.org`

for the rendering of many photo-realistic effects in real time, for instance HDR (high dynamic range) rendering, volumetric effects such as soft, accurate shadows, fog, clouds, fire and smoke, and natural effects such as rendering of water surfaces, see the textbook edited by Fernando [72] for details. In late 2002, DirectX 9 introduced *shader model 2*, which lifted many restrictions and added, among other things, hardware support for reading from and—importantly—rendering to floating point textures. However, the standard did not specify the exact number format (and hence, resulting precision) to be supported, see below for details.

Year	Model	Chip	DX version	VS	FS	TEX	Bandwidth
Jan. 2001	GeForce 3	NV20	8.0	1	4	8	7.4 GB/s
Oct. 2001	Radeon 8500	R200	8.1	2	4	8	8.8 GB/s
Dec. 2001	GeForce 4 TI 4600	NV25	8.1	2	4	8	10.4 GB/s
Aug. 2002	Radeon 9700 PRO	R300	9.0	4	8	8	19.8 GB/s
Jan. 2003	GeForce FX 5800 Ultra	NV30	9.0	2	4	8	16.0 GB/s
Apr. 2004	GeForce 6800 Ultra	NV40	9.0c	6	16	16	35.0 GB/s
May 2004	Radeon X800 XT PE	R420	9.0b	6	16	16	35.8 GB/s
Jun. 2005	GeForce 7800 GTX	G70	9.0c	8	24	24	38.4 GB/s
Oct. 2005	Radeon X1800 XT	R520	9.0c	8	16	16	48.0 GB/s

Table 3.1: Hardware features of NVIDIA and ATI GPUs 2001–2005, given for the top-end models at launch time of the respective generation (Abbreviations: VS–vertex shaders, FS–fragment shaders, TEX–texture units).

ATI's Radeon 9700 (which actually became available before the standard due to marketing reasons on Microsoft's side) and NVIDIA's GeForce FX were the first GPUs implementing shader model 2.0. Until 2006, each vendor released three DirectX 9 class GPU series, NVIDIA's GeForce FX, 6 and 7 series, and the Radeon 9, X and X1K series boards by ATI, who were acquired by AMD in the process (October 2006). The later models supported shader model 3.0 which did not mandate a tremendous amount of new features (in the scope of this thesis) but lifted many restrictions on the static and dynamic length of shader programs. In terms of task parallelism, nothing was changed in those days: The pipeline was implemented with programmable vertex and fragment stages, and in fixed-function logic otherwise. However, data parallelism capabilities within the individual stages of the graphics pipeline, and consequently also the bandwidth to off-chip memory[3], was significantly increased. Table 3.1 summarises the evolution of NVIDIA and ATI GPUs until 2006, both in terms of parallelism and other characteristics such as memory bandwidth which is the decisive performance factor for our applications. Section 3.1.7 presents NVIDIA's GeForce 6 architecture in more detail.

High level shading languages, replacing the assembly languages of DirectX 8 and the corresponding OpenGL extensions, significantly improved programmability of DirectX 9 class GPUs. HLSL and GLSL [186] are integral components of DirectX and OpenGL, respectively. NVIDIA's Cg [154] was developed alongside HLSL, and can be used platform-independently. The choice of the shading language is, with the apparent exception of HLSL, completely left to the programmer, all languages are semantically and syntactically almost identical.

2004 also marked the replacement of the AGP technology with the PCIe bus, which is significantly faster. As the connection between the GPU board and the host memory can be a severe bottleneck, the doubling of transfer rates from 2 GB/s to 4 GB/s (bidirectional 8 GB/s) was warmly received by developers.

[3]Of course, increasing the number of compute resources does not automatically increase memory bandwidth, in fact, GPU architects managed to scale the memory subsystem with the rest of the chip by increasing the width of the memory bus and partitioning the memory to enable parallel access.

Floating Point Support and Render-to-Texture

NVIDIA's DirectX 9 GPUs have supported a 32 bit single precision (s23e8) floating point format (almost IEEE-754 [122] conforming) right from the start. Due to performance reasons, computation was not entirely IEEE compliant, see the work by Hillesland and Lastra [116] and Daumas et al. [56] for detailed error measurements. However, these slight deviations from the IEEE standard were not considered a severe drawback. These GPUs also supported a 'half precision' 16 bit (s10e5) format natively in hardware, which allowed faster processing and reduced bandwidth requirements. At first, ATI only supported a 24 bit non-IEEE format, which was later improved to a quasi-IEEE single precision format with similar error bounds as for the NVIDIA models by the R500 architecture.

The framebuffer however remained (and still remains) at a fixed precision format of at most 8 bits per colour channel (RGBA with 32 bit colour depth), not surprisingly because display devices scan this memory buffer directly. The second important innovation in the evolution of GPGPU was thus hardware support for so-called *render-to-texture (RTT)*. This feature exposes a feedback loop in the graphics pipeline via multipass rendering without loss of precision. The output of a rendering pass can be directed to a texture, and read again in a subsequent pass. In our early implementations of FEM multigrid components in graphics hardware, we made excessive use of this feature, see Section 5.2.3 for implementational details.

The emanation of RTT is a good example of how hardware features are subsequently exposed via OpenGL extensions: The first implementation of RTT through the so-called `pBuffer`-extensions was very heavyweight, as it was based on the underlying window system (WGL on Windows, X11 on Linux). Context switches, i. e., changing the output texture, were expensive, and on Linux, RTT was not available and had to be emulated by 'copy-to-texture'. Our first GPU codes in late 2004 used this technique, and the learning curve was very steep as virtually no documentation and only one tutorial code existed. In mid 2005, the OpenGL ARB released the extension `EXT_framebuffer_object` (FBO), which allowed lightweight, cross-platform RTT and a lot more flexibility and performance. Yet still, vendor-dependent extensions defined the actual number format and dimensions of floating point textures that could be rendered to.

3.1.4. Legacy GPGPU Programming Model

In legacy GPGPU, the hardware has only been exposed via graphics APIs. In the following, we present the underlying programming model of abstracting the graphics pipeline as a *data-parallel* computation engine, and describe key concepts of mapping general-purpose computations to graphics terminology and thus sequences of graphics API and shader program calls. The book chapters by Harris [109], Lefohn et al. [144], Harris and Buck [110] and Woolley [244] provide more information, beyond the focus on numerical linear algebra in this thesis. An alternative abstraction is the *stream programming model*, see Section 3.1.5 for a brief discussion and further references. Section 5.2 describes our implementation of numerical linear algebra kernels and multigrid solver components.

Overview

The fundamental data structure on CPUs is a one-dimensional array, stored contiguously in memory. When programming GPUs through graphics APIs, the basic data structure is a two-dimensional memory buffer, called *texture*. Consequently, indices (*texture coordinates*) are tuples. In OpenGL, texel centres are addressed, and we use a variant with unnormalised coordinate ranges: The texture coordinates of the first texel are always $(0.5, 0.5)$, and the coordinates $(0.75, 0.75)$ map to the same texel. More precisely, with the coordinate rounding mode we employ, all coordinates of the interval form $([i, i+1[, [j, j+1[)$ for fixed integers i, j map to the same texel.

The first step in any GPGPU program is to allocate texture memory in a floating point format for the input and output data, and download the data onto the device. The programmer then configures the graphics pipeline as follows for a simple computation that performs the entire work in the fragment stage:

1. The output texture is attached to a framebuffer object for render-to-texture.
2. The viewport is set to a plain 1:1 mapping between world coordinates and screen coordinates by means of a simple orthogonal projection, so that the input geometry covers the entire output texture.
3. Textures holding input data are bound to texture image units, and texture filtering is disabled for each unit.
4. Array indices for input data are mapped to texture coordinates corresponding to the corners of rectangular regions in the input textures.
5. A viewport-sized quadrilateral is rendered to trigger the computation.
6. The vertex processor performs the (fixed-function) orthogonal projection; and the rasteriser generates a fragment corresponding to each item in the output texture.
7. The fragment processor executes a computational kernel (a shader program) for each fragment and thus for each item in the output texture independently. This kernel gathers input data from the input textures using the interpolated texture coordinates, and finally stores the result at the position in the output texture it corresponds to.
8. The output texture can either be read back to the host, or used as input for subsequent passes.

The entire computation is thus performed in a fully data-parallel, SPMD (single program multiple data) and possibly also SIMD (single instruction multiple data) fashion. The input and output textures must be disjunct, and there is no communication between threads (instances of the fragment program) within the granularity of the entire rendering pass.

In some kernels, it is necessary to output more than one texel. *Multiple render targets (MRT)* provide this functionality for (hardware- and vendor-dependent) up to four or eight different textures, which must all have the same format.

Kernels: Fragment Programs

We consider a simple elementwise addition of two long vectors to illustrate how the actual computation is performed. On the CPU, this operation is naturally implemented in a for-loop over all elements. It is obvious that this operation is trivially data-parallel, the computation for each element is performed independently of all the others.

As explained in Section 3.1.3 on page 51, the computational workhorse of pre-DirectX 10 GPUs is the fragment stage. Therefore, the loop body is converted to a *fragment program* to perform computations. It is executed in parallel on the fragment processors as explained below. At runtime, each fragment program 'knows' its index or, more precisely, its fragment coordinates. With the one-to-one mapping set up as above, these coordinates correspond to the respective texels in the currently bound output texture. In other words, each fragment writes exactly one element of the output texture (in the output domain). Each fragment program may read data from additional input textures, which is called a *texture lookup*. The number of different input textures is limited and hardware-dependent. The fragment program also has access to scalar parameters which are passed in just as regular parameters, as well as a number of predefined registers for interpolants (see below). Relative indexing into input textures based on the fragment's own coordinates is possible, e. g., to read local neighbourhoods.

```
1 #extension GL_ARB_texture_rectangle : enable
2 uniform sampler2DRect textureX;
3 uniform sampler2DRect textureY;
4 void main()
5 {
6   float x = texture2DRect(textureX, gl_TexCoord[0].st).x;
7   float y = texture2DRect(textureY, gl_TexCoord[0].st).x;
8   gl_FragColor.x = x + y;
9 }
```

Listing 3.1: Example GLSL fragment program to add two vectors.

A fragment program to add two vectors stored in textures is depicted in Listing 3.1. After declaring the texture samplers, the values are read from the two textures at the index passed in through the first `gl_TexCoord` special-purpose register, which in our implementation always holds the fragment's own position. The computed result is written to the special purpose register `gl_FragColor`. Note that the output register does not contain indices at all.

Rendering Passes: Executing Computations

If the viewport is configured with the one-to-one mapping introduced above, then the rendering of a quadrilateral that covers the entire viewport[4] traverses through the graphics pipeline as follows:

1. The vertex processor converts the four vertices from world space to screen space. The transformed vertices are passed, along with their coordinates, to the rasteriser.

2. The rasteriser generates *exactly one fragment* for each element in the output texture by interpolating the vertex coordinates across the geometry. These fragments are streamed to the fragment processors.

3. The fragment processors execute the currently active fragment program as explained in the previous paragraph independently for each fragment. The order in which individual fragments are assigned to the available fragment processors is undefined, and the fragments cannot communicate.

4. The resulting values (fragment colours) are written to the currently bound output texture (render target), at each fragment's predefined location.

To execute computations on a subset of the output domain, smaller quadrilaterals, line segments or even points can be drawn.

Coming back to the analogy of a for-loop on a CPU, steps 1,2 and 4 correspond to the loop header (they set up and launch the computation), and step 3 corresponds to the loop body (it performs the actual computation). In short: The rendering of a screen-sized quadrilateral results in the operation defined by the currently active fragment program being performed for all elements in the output texture simultaneously.

[4]This is also often referred to as a 'screen-sized quad', i. e., a quadrilateral with the four vertices $(0,0)$, $(0,M)$, (M,M) and $(M,0)$.

Vertex Processor: Lookup Pattern Generator

As explained in Section 3.1.1, fragments have, beyond their colour, different attributes which are interpolated by the rasteriser based on the corresponding vertex attributes. In the described model of the graphics pipeline as a computational engine, the vertex processor currently only performs the coordinate transformation for the one-to-one mapping between rendered geometry and output texture coordinates. Together, the vertex and rasterisation stages are much more capable and can in fact be configured to generate and interpolate arbitrary data across the output domain. In our case, numerical linear algebra, we use the *texture coordinate interpolants* to generate lookup patterns which are uniform across the output domain, but may vary for different input textures, see Section 5.2.3. The interpolated values are available to each fragment program in special-purpose registers. This is an example to *balance the workload between pipeline stages*, an important performance tuning technique. Consider for instance a fragment program that computes the average value of each element and its surrounding eight neighbours in a large, say $1\,000 \times 1\,000$, input texture. Instead of performing the index arithmetic (computing $(i+1,j),(i-1,j),\ldots$ for each fragment (i,j) one million times in each instance of the fragment program, it is performed four times, for each vertex, and the result is interpolated by the rasteriser. This is quite effective, as the vertex and rasterisation stages are typically underutilised anyway in the computations in the scope of this thesis.

Listing 3.2 depicts a minimal vertex program that applies the one-to-one mapping and writes out the corresponding texture coordinates to be interpolated by the rasteriser.

```
1 void main()
2 {
3   gl_Position = ftransform();
4   gl_TexCoord[0] = gl_Vertex;
5 }
```

Listing 3.2: Example 'passthrough' GLSL vertex program that generates texture coordinates for the one-to-one mapping.

3.1.5. Bibliography of Relevant Publications

At first, implementations got broken on a regular basis for instance with new driver releases, and programming only became less cumbersome once FBO support got settled (cf. Section 3.1.3). A retrospective quote by Michael Wolfe, Compiler Engineer with The Portland Group (PGI), summarises the general attitude in these early days quite well:[5]

> *This is truly heroic programming. It's like using a chain saw to carve blocks of ice: in the right hands, it can produce something beautiful, but one wrong mistake and all you have is ice cubes (or worse).*

Despite all confusing details both in terms of hardware features and software support that made programming cumbersome and error-prone in 2003–2006, and admittedly prevented the immediate, broad emanation of *GPU computing* in retrospective, interest among researchers was raised to a critical level. A lot of groundbreaking research was published. Here, we briefly summarise the most important results that inspired the pursuit of this thesis in late 2004, and we refer to the survey articles by Owens et al. [174, 175] for more details, in particular for other application domains.

Harris et al. [112] simulate realistic volumetric clouds by solving the three-dimensional Euler equations, coupled with thermodynamic conservation laws, buoyant forces and water phase

[5] `http://www.hpcwire.com/features/Compilers_and_More_Programming_GPUs_Today.html`

transitions. Their discretisation is based on the popular 'stable fluids' approach by Stam [202], which is unconditionally stable and hence allows for large timesteps[6], a prerequisite for interactive, *visually accurate* simulations (the authors use the term *visually-realistic*). By using Jacobi and Gauß-Seidel iterations in both the conventional and a sophisticated, vectorised formulation tailored to the specifics of the underlying GeForce FX GPUs, the approach is still fast enough to include self-shadowing and light scattering within the simulated clouds. In a subsequent publication, Harris [108] uses similar techniques to solve the Navier-Stokes equations.

Krüger and Westermann [140] present a general framework for operations from numerical linear algebra on GPUs, and discuss many implementational techniques and optimal data structures. They compose the Gauß-Seidel and the conjugate gradient schemes from functions provided by their library, and solve, again using a discretisation based on Stam's approach, the wave equation and the Navier-Stokes equations in two space dimensions on regular, cartesian grids.

Bolz et al. [30] and Goodnight et al. [92] were the first to realise a simple V cycle multigrid in graphics hardware. Their techniques exploit the regular grids and the resulting use of a matrix stencil to a great extent, holding all data required for a discretisation with 1025^2 mesh points inside the limited device memory. Hence, their work is quite different from our GPU-accelerated multigrid solvers (cf. Chapter 5). As before, the application is the simulation of fluid flow using Stam's approach and a very specific Jukowski transformation, respectively.

As mentioned previously, Strzodka and his collaborators were the first to use finite element discretisations, and they continue to pay special attention to accuracy issues in their work on DirectX 9 GPUs [209, 212, 213]. Their later work includes multigrid solvers for applications in image registration and segmentation, again based on regular cartesian grids in the pixel plane.

A noteworthy publication is the work of the Visualization Lab of Stony Brook University [67]. They describe, for the first time, how an existing cluster (and an associated MPI-based distributed memory application) can be improved significantly by adding GPUs not for visualisation, but for computation. To the best of our knowledge, they were—at that time—the only group being able to target realistic problem sizes, including very preliminary work on scalability and load balancing of GPU-accelerated cluster computations. Their application is a Lattice-Boltzmann solver for fluid flow in three space dimensions, using 15 million cells. As usual in the Lattice-Boltzmann world, single precision suffices.

Horn [119] was the first to implement the *scan primitive* (see Section 5.4.4) on GPUs in $\mathcal{O}(N \log N)$ work complexity, i. e., his implementation is not work-efficient compared to the serial case, and applied it to the problem of non-uniform stream compaction, the removal of certain elements from a data stream based on a masking operation. In 2006, Greß et al. [100] and Sengupta et al. [196] presented the first $\mathcal{O}(N)$ scan implementations.

Lefohn et al. [145]—though targeting pure graphics applications like adaptive high-resolution shadow maps and octree-based painting—pursue groundbreaking research on irregular, adaptive data structures. By separating data structures from application code, they are able to re-use high-level building blocks such as stacks, quadtrees and octrees in many fundamentally different contexts, significantly simplifying the implementation in terms of programming and debugging effort. Lefohn's PhD thesis presents the approach in full detail [143].

Kass et al. [124] implement an efficient solver for tridiagonal systems via cyclic reduction. Their application domain are depth-of-field effects in computer graphics, in particular computer-generated feature film. The tridiagonal solver is at the core of the alternating direction implicit scheme they propose to approximately solve the heat equation to create depth-of-field blur effects interactively.

For the sake of completeness, we conclude with references to dense LU decomposition [76], searching and sorting [94] and a general memory model for GPUs and associated techniques to

[6]The other way round, this naturally means that the simulations are the less accurate the more interactive they are.

maximise performance in bandwidth-bound computations [95].

The work by Sheaffer et al. [198, 199] is different from the publications summarised above, as they do not present particular applications, but instead address reliability of graphics processors. Exponential semiconductor device scaling has lead to feature sizes that are so small that they are vulnerable to, e. g., cosmic or terrestial radiation and minimal voltage deviations, leading to an increasing transient soft error probability of 8 % per year. In graphics workloads, a single pixel error often remains unnoticed, whereas for general-purpose computations, reliability of the hardware is much more important. In their papers, they suggest a redundancy and recovery mechanism that—at moderately increased transistor count and performance penalty—improves reliability dramatically in a way completely transparent to computations and graphics applications.

Programming Abstractions and High Level Languages

To improve programmability and accessibility, and in particular to hide all graphics-related concepts and terminology from the programmer, early high level languages and libraries soon surfaced. All these efforts started out as pure research projects, and exposed the GPU as a data-parallel, *data-stream based* computing engine. The two most important ones are *Brook for GPUs* and *Sh*, which succeeded in gathering a larger userbase.

Brook for GPUs, also known as BrookGPU, was developed at Stanford University by Buck et al. [43]. It is a compiler and runtime implementation for graphics hardware of the Brook stream programming language. In fact, the graphics pipeline can be interpreted as a specialisation of the general *stream architecture* and programming paradigm, see the work by Dally et al. [55] and Khailany et al. [127]. BrookGPU is implemented as an extension to ANSI C. It provides backends for both OpenGL and DirectX and supports NVIDIA and ATI GPUs on Windows and Linux. At the time of writing, BrookGPU is no longer actively supported by the original developers. Instead, AMD has used it as the basis of *Brook+*, the high level compiler and data-parallel language included in AMD Stream (see Section 3.2.3).

Sh is a metaprogramming language for programmable GPUs and was developed at the University of Waterloo by McCool et al. The first ideas were published already in 2002 [157], and an exhaustive paper appeared at SIGGRAPH 2004 [156]. Sh is implemented as a library on top of C++, and supports NVIDIA and ATI GPUs on Windows and Linux. At the time of writing, it is no longer actively supported by its developers. Instead, it has been commercialised into RapidMind Inc.[7], and significantly expanded to support not only GPUs, but also the Cell processor and multicore CPUs.

A similar commercial product was offered by PeakStream Inc., called the *PeakStream Platform for Many-Core Computing*, targeting both GPUs and CPUs. This company was founded by a former head of NVIDIA's GPU architecture group, who was responsible for early GeForce products and the graphics chip in the first XBox console. PeakStream was acquired by Google in 2007, and their software is no longer available.

Other noteworthy contributions are *Scout*, a GPU programming language for large-scale scientific computations [158, 159] and *Accelerator* from Microsoft Research [218], a linear algebra library that integrates seamlessly into C# and uses just-in-time compilation to fragment shaders. As one example of research in the compiler community targeting auto-parallelisation for graphics processors, we reference *CGiS* [152].

[7] `http://www.rapidmind.net`

3.1.6. Restrictions of Single Precision

Common to all early work on GPU computing as listed in Section 3.1.5 is that *visual accuracy* was sufficient. Owens et al. state in their 2007 survey of the field [175]:

> *Finally, while the recent increase in precision to 32 bit floating point has enabled a host of GPGPU applications, 64 bit double precision arithmetic remains a promise on the horizon. The lack of double precision hampers or prevents GPUs from being applicable to many very large-scale computational science problems.*

In fact, the lack of double precision floating point arithmetic was first regarded as a 'knock-out criterion' in most of the scientific computing community. Combined with the rather 'hacky', ad-hoc approach, the emerging field of GPU computing was not taken seriously at first.

However, using double precision for the entire computation is often not necessary, careful analysis shows that computational precision and result accuracy are related in a highly nonlinear way (see Chapter 4 for details). We accepted this challenge and in our first work on GPUs, evaluated the applicability of mixed precision schemes to GPUs: By letting the CPU execute few, high-precision correction steps and the GPU the bulk of a linear solver, we were able to achieve exactly the same accuracy despite performing up to 95 % of the arithmetic work in single precision. The resulting hybrid CPU-GPU solver executed on average 3–5 times faster than its conventional CPU counterpart, subsequent work with native double precision available on the device improved the speedup to a factor of 20–40 (cf. Section 5.7 and Section 5.8.8). We presented these initial results at a small conference in Germany [82]. To the best of our knowledge, this conference paper was the first to achieve results of high accuracy on GPUs.

3.1.7. Example Architecture: NVIDIA GeForce 6

NVIDIA's GeForce 6 chip, codenamed NV40, was released in April 2004. It comprises 222 million transistors, the core clock is 400 MHz, and the memory is clocked at 550 MHz double data rate for an effective memory clock of 1.1 GHz. Figure 3.2 depicts a block diagram, the data flow through the pipeline is from top to bottom. The whole architecture is scalable by design to different price-performance regimes by varying the number of vertex processors (VP), fragment processors (FP) and pixel blending units (PB). At launch time, two different variants of the GeForce 6800 were available, the 'vanilla' 6800 with 5 VPs and 12 FPs and PBs each, and the 6800 Ultra with 6, 16 and 16 respectively.

The vertex processor follows a flexible MIMD (multiple instructions multiple data) design. Each processor's data path comprises a 4-wide vector multiply-add (MAD) unit and a scalar special function unit for transcendentals (we ignore the instruction unit for the moment), as well as a texture unit. All units support quasi-IEEE 32 bit floating point arithmetic. It is also equipped with constant and temporary registers, and can execute up to 512 static (65 536 dynamic) instructions. Branching and looping are supported efficiently. The processor can issue instructions to both the vector and scalar data paths at every clock cycle.

Primitive assembly and the rasteriser are implemented in fixed-function logic. The setup stage performs view-frustum culling to remove primitives outside the viewport. The rasteriser traverses geometric primitives in a DRAM-page friendly order, and interpolates all per-vertex attributes across each triangle. Besides position, colour, depth and fog, up to ten generic 4-wide vector attributes are supported at full 32 bit precision per entry, which are typically used for (but not restricted to) texture coordinates. Interpolated attributes are passed to the fragment processors in special-purpose registers.

Each fragment processor consists of two 4-wide vector units, a special function unit, and a texture unit. The texture unit can additionally perform filtering (linear, bilinear, trilinear and

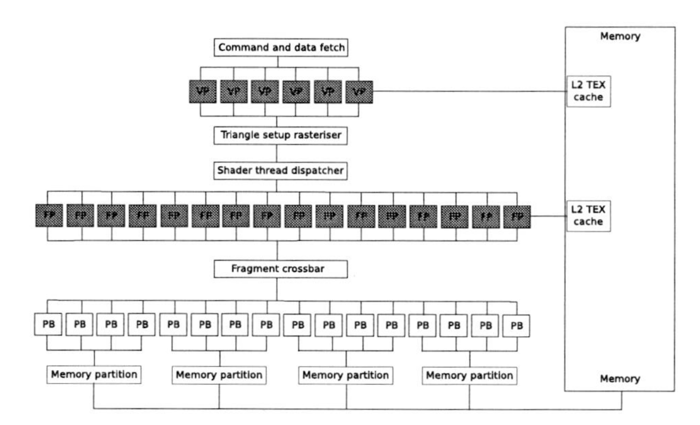

Figure 3.2: Block diagram of the NVIDIA GeForce 6 GPU.

anisotropic), but only in 16 bit floating point per colour channel, which makes filtering inappropriate for our applications. Each fragment processor has enough registers to keep hundreds of threads in flight simultaneously. The latency of a texture fetch is on the order of hundreds of clock cycles, and by switching to another thread as soon as a thread stalls for a memory transaction, this latency can be hidden very effectively, minimising the amount of idle clock cycles. This is a realisation of a core idea in graphics processor design to maximise overall throughput instead of minimising the latency of an individual operation. The thread management is very lightweight, and entirely implemented in silicon, fundamentally different to heavyweight threads on commodity CPU designs. The fragment processor works on quads (2×2 pixels) at a time to allow for efficient computation of derivatives for texture filtering. Due to assigning many threads to one fragment processor (called a thread group) and the realisation of the instruction unit, the computation has SIMD characteristics (single instruction multiple data). The exact size of it is kept secret but microbenchmarks indicate groups of 32×32 or 64×64 pixels.[8]

Each processor has a small level-1 texture cache, and in addition, there is a larger second level cache shared among all processors. These caches are optimised for streaming throughput with many hits and misses in flight, but local reuse of data. Phrased differently, data reuse is maximised for spatial locality in two dimensions. The reason for this is texture filtering: Texture maps are 2D images, and filtering is a strongly localised operation taking small neighbourhoods of pixels into account.

Without texture fetches, each processor can perform 12 floating point operations per clock cycle (not counting filtering operations in the texture unit), or four if the first vector unit issues a texture fetch. The floating point ALUs also have superscalar characteristics: The components of each 4-wide vector can be treated either by a 4-vector single instruction, or by 3/1 or 2/2 dual-issue with different instructions, e. g., for the RGB and alpha channels. The maximum program length for the fragment processor is 65 536 static and dynamic instructions. Although branching and looping only have small overhead at the granularity of the SIMD group, branch divergence can significantly degrade performance and should be minimised.

The fragment processor supports output to four 4-wide render targets, for a total of 16 single precision floating point values at every clock cycle.

All fragments pass through a fragment crossbar so that every memory location can be reached.

[8]GPUBench Utility (Stanford University Graphics Lab): `http://graphics.stanford.edu/projects/gpubench`

The pixel blending units are realised as fixed-function, performing depth and stencil testing as well as blending, which is supported only for 16 bit floating point values and therefore not a viable feature for our applications. Additionally, multisampling and supersampling are supported for fixed-point formats. Each pixel pipeline connects to a specific memory partition, each partition has its distinct 64-pin connection to DRAM, for a total memory bus width of 256 bit.

We refer to the articles by Montrym and Moreton [164] and Kilgariff and Fernando [129] for more detailed information on this architecture.

3.2. NVIDIA CUDA and AMD Stream

This section briefly outlines the changes introduced to the graphics pipeline with the DirectX 10 specification. In particular, it explains why graphics hardware transitioned to a unified shader architecture, with the same physical processing units being able to execute instructions for each programmable stage of the pipeline. This change in hardware design, combined with the vendors recognising the potential (and the potential future market revenues) of GPUs as a pure computing device, lead to the introduction of languages and APIs for *(data-parallel) GPU computing* without the overhead of a graphics-centric programming model. In fact, the hardware and API designers worked in very close cooperation, and some features have even been added to the hardware that are not accessible via traditional graphics APIs. Both NVIDIA and AMD also launched brands in 2007 targeting the high performance computing community, which can anecdotally be described as 'GPUs without display connectors'.

Section 3.2.1 outlines the reasons that lead to unified architectures. NVIDIA CUDA and AMD Stream are presented in Section 3.2.2 and 3.2.3, describing both the underlying architecture and the programming model. Section 3.2.4 extends the bibliography from Section 3.1.5 to the time from 2006–2009. Finally, Section 3.2.5 concludes with a discussion of double precision support.

3.2.1. DirectX 10 and Unified Architectures

Historically, the functionality of the two programmable stages of the graphics pipeline has evolved in different ways, see Section 3.1. For vertex processing, low latency operations and high-precision floating point arithmetic are more important, while fragment processors are mostly optimised for high-throughput texture filtering with less restrictive precision requirements. The higher complexity of per-vertex operations is further underlined by the fact that vertex shading became programmable first and has later been realised in a fully MIMD way, while the fragment processors operate in a more throughput-oriented SIMD fashion, see Section 3.1.7. In contrast, using GPUs for general-purpose computations has been mostly realised in the fragment stage because of its high-bandwidth connection to memory, as sketched in Section 3.1.4. Nonetheless, the feature set of the two processors has been converging with each new generation [28, 105].

Typical graphics workloads require much more processing power in the fragment stage than in the vertex stage. Early programmable GPUs reflected this by providing two to three times as many fragment as vertex processors, see Table 3.1 on page 52. However, not all scenes are 'typical', and the tradeoff is always a compromise for the most common situations. Such a *static load balancing* gives way to many inefficiencies due to extreme variations in workload distribution. For instance, rendering many large triangles leaves the vertex stage mostly idle while the fragment processors are busy. The opposite is true for geometry-intensive scenes with many small triangles.

DirectX 10 (*shader model 4.0*) was officially approved in 2006 [28] and adds a third programmable stage to the graphics pipeline, the so-called *geometry shader*.[9] This stage, located between the vertex- and rasterisation stage, amplifies input primitives by emitting additional lines, points or triangles (used for instance in tessellation or instancing). Closely coupled with it is the *stream output* feature, that allows to feed the output of the geometry shader back to the start of the pipeline without going through the fragment stage as with the older, non-standardised *render-to-vertex-array* approaches. Additionally, vertex processors are now required to support access to texture memory, a feature called *vertex texture fetch* that previously was not mandatory.

Designing three separate processors thus became more and more complex, demanding and expensive both in terms of die area and development cost. As the required functionality of the

[9] The specification of the DirectX standard is created in close cooperation between hardware designers, application developers and API architects, and is therefore always a tradeoff between cost and complexity. See Section 3.1.1 on page 49 for details.

programmable stages converged, GPU architects reconsidered the strict task-parallel pipeline in favour of a *unified shader architecture*. DirectX 10 defines a single 'common core' virtual machine as the basis for realisations of each of the programmable stages in hardware, in fact making a unified architecture mandatory for compliant hardware implementations. The unified architecture enables dynamic load balancing of resources between the stages, which can be implemented directly in hardware in a very efficient way.

To be historically correct, we should mention that the first unified graphics processor, Xenos, was developed by ATI for the XBox360 gaming console already in 2005 [5].

Standardisation of Important Features

The DX10 standard made a lot of features relevant for general-purpose computations mandatory, features that previously were vendor-dependent and often only exposed via OpenGL extensions. In the scope of this thesis, the following ones are most important:

Render-to-texture and arbitrary non-power-of-two texture dimensions are finally standardised. Hardware implementations must support 32 bit single precision floating point storage and arithmetic (FP32) throughout the entire pipeline (including RTT), considerably close to IEEE-754 [122]: Basic arithmetic operations (addition, subtraction, multiplication) are accurate to 1 ulp,[10] rather than 0.5 ulp required in IEEE-754. Division and square root are accurate to 2 ulp. Denormalised numbers are flushed to zero, and floating point specials (not-a-number, infinity) are fully implemented. Implementations are also required to support 32 bit integer storage and arithmetic. All shader programs must support true dynamic control flow. In addition, the minimum number of render targets, instruction slots per shader program and registers is increased, as well as the maximum texture size.

Due to the transition towards a unified shader architecture, all features are required to be implemented in hardware, except for the costly FP32 texture filtering and multisample antialiasing. Furthermore, all traditional fixed-function capability that is expressible in terms of programmable constructs has been eliminated from the pipeline and the core API. This includes vertex transform and lighting, point sprites, fog, and alpha testing. The consequence is obvious: Hardware implementations can dedicate more of the available transistors to the unified processing units, simplifying the design by removing opportunities to offload special tasks to dedicated fixed function units; and making a larger fraction of the on-chip computing resources accessible and available to general-purpose computing.

We refer to Blythe [28] for more details and a description of the exact requirements and changes introduced with the DirectX 10 specification.

Unified Graphics and Compute

Legacy GPU computing took the form of mapping algorithms to the individual stages of the graphics pipeline via graphics APIs like OpenGL or DirectX. Undisputedly, the standardisation of features in DirectX 10 marked a very important step forward towards writing cross-platform, vendor-independent GPGPU applications. However, something entirely different happened that tremendously picked up the pace for GPU computing, and resulted in the entire research area gaining a lot of momentum.

As the programmable parts of the pipeline are responsible for more and more computation within the graphics pipeline abstraction, the architecture of GPUs has migrated from a strict pipelined task-parallel architecture to one that is built around a single unified massively multithreaded, data-parallel programmable unit. In computer architecture, this class of designs is

[10] *Units in last place* are an error measure for floating point operations, see the IEEE-754 standard or the survey article by Goldberg [91] for details.

known as a *processor array*. It was not uncommon in legacy GPGPU programming that the entire task parallelism of the hardware pipeline had to be ignored, and only the data parallelism within the fragment stage could be exploited (cf. Section 3.1.4), leaving much computational horsepower of the chips potentially unused. The new view on the hardware enables a much cleaner programming model, and both AMD and NVIDIA recognised this and exposed the processor array on their chips directly for data-parallel computation through special non-graphics APIs. Inspired by the early GPGPU work and the possibility to gain market shares in the high performance computing domain, these APIs were developed in close cooperation of hardware and API architects. In fact, some features were even added to the hardware that are only exposed through these compute-oriented APIs and not through graphics APIs at all.

AMD and NVIDIA followed different approaches, which are outlined in the next two sections.

3.2.2. NVIDIA CUDA

This section explains CUDA, NVIDIA's approach to GPU computing. CUDA was developed simultaneously with the GeForce 8 architecture (internal code name 'Tesla architecture') which is explained below, and publicly announced in November 2006. Nowadays, the term CUDA is used as the name of both the underlying parallel computing architecture and the (data-parallel) programming model.

The CUDA architecture includes a device-independent assembly language (PTX, short for *parallel thread execution*) and compilation technology that is the basis on which multiple parallel language and API interfaces are built for CUDA-capable NVIDIA GPUs. In the scope of this thesis, C for CUDA is the most relevant one. Other interfaces include OpenCL and DirectX compute shaders, see Section 3.4. The assembly language PTX is exposed to allow, e. g., middleware interfaces. C for CUDA uses the standard C language with extensions, and exposes hardware features that are not available explicitly through OpenGL or Direct3D.

At the time of launch, NVIDIA spelled out the acronym as *compute unified device architecture*, but has since transitioned to using it as a fixed term as explained above. Early publications refer to the architecture as *Tesla* and to the programming model as CUDA, which can be confusing: Tesla also is the internal code name of the architecture of all three generations of CUDA-capable GPUs released to date, but is now almost exclusively used as the brand name for the product line targeting the high performance computing domain, see the end of this section.

The GeForce 8 Architecture

Figure 3.3 depicts a functional block diagram of the GeForce 8800 GTX (chip name G80), the first CUDA-capable GPU at the time of launch. The GeForce 8 also is the first GPU compliant with the DirectX 10 specification.

The design is built around a scalable processor array (SPA) of *stream processor* 'cores' (ALUs, also called *thread processors*, abbreviated SP), organised as *streaming multiprocessors* (SM) or *cooperative thread arrays* (CTA) of eight SPs each, which in turn are grouped in pairs into independent processing units called *texture processor clusters* (TPC). For the remainder of this thesis, we ignore the additional level of the TPC and just consider the chip at the granularity of the multiprocessors, because the TPC clustering is a pure graphics-motivated design choice.[11] The GeForce 8800 GTX comprises 16 multiprocessors for a total of 128 thread processors. By varying the number of SMs per chip, different price-performance regimes can be targeted. In contrast to previous generation hardware, these processors are scalar, as it became increasingly difficult to automatically vectorise instructions for the traditional 4-tuple RGBA data format.

[11] The TPC corresponds to the quad-pixel processing of previous designs, see Section 3.1.7.

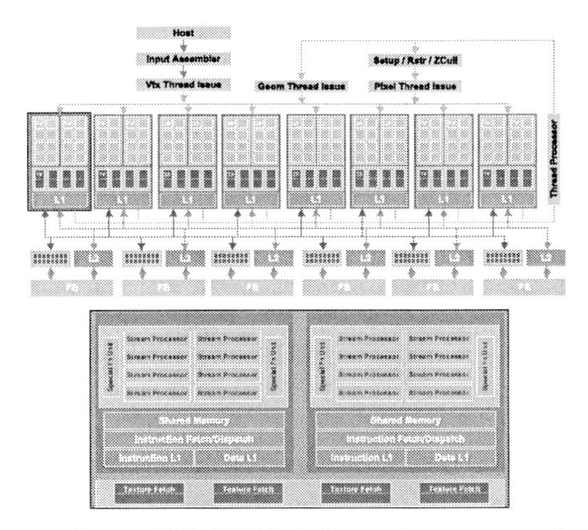

Figure 3.3: NVIDIA GeForce 8800 GTX block diagram. Image courtesy Owens et al. [176].

At the highest level, the SPA performs all computations, and shader programs from the programmable stages are mapped to it using dynamic load balancing in hardware. The memory system is also designed in a scalable way, with external, off-chip DRAM control and composition processors (ROP – raster operation processors) performing colour and depth frame buffer operations like antialiasing and blending directly on memory streams to maximise performance. A powerful interconnection network (realised via a crossbar switch) carries computed pixel values from the SPA to the ROPs, and also routes (texture) memory requests to the SPA, using on-chip level-2 caches. As in previous designs, these caches are optimised for streaming throughput and strongly localised data reuse, see Section 3.1.7 on page 59.

All fixed-function hardware is grouped around the SPA. The data flow for a typical rendering task and thus, the mapping of the graphics pipeline to this processor, is as follows: The input assembler collects per-vertex operations and a dedicated unit distributes them to the multiprocessors in the SPA, which execute vertex and geometry shader programs. Results are written into on-chip buffers, and passed to the Setup-Raster-ZCull unit, in short the rasteriser, which continues to be realised as fixed-function hardware for performance reasons. Rastered fragments are routed through the SPA analogously, before being sent over the interconnection network to the ROPs and to off-chip memory. The SPA accepts and processes work for multiple logical streams simultaneously, to allow for dynamic load balancing. A dedicated unit called *compute work distribution* dispatches blocks of work accordingly. Three different clock domains control the chip, the reference design of G80-based graphics boards prescribes the following values: Most fixed-function and scheduling hardware uses the core clock of 575 MHz, the SPA runs at 1350 MHz, and the GDDR3 memory is clocked at an effective 1.8 GHz (900 MHz double data rate). The chip is fabricated in a 90 nm process and comprises almost 700 million transistors, a significant increase compared to 220 million for the GeForce 6800 Ultra, which is only two generations (2.5 years) older.

The streaming multiprocessor is at the core a unified graphics and compute processor. Each SM (see Figure 3.3, bottom) comprises eight streaming processor cores (ALUs), two special function units, a multithreaded instruction fetch and issue unit, disjunct data and instruction level-1 caches, a read-only constant cache and 16 kB shared 'scratchpad' memory allowing arbitrary read and write operations. Each ALU comprises scalar floating point multiply-add as well as integer and logic operations, whereas the special function units provide transcendental (trigonometric, square root, logarithm and exponentiation) functions as well as four additional scalar multipliers used for attribute interpolation.

To dynamically balance the shifting vertex, geometry, pixel and compute thread workloads, each multiprocessor is hardware multithreaded, and able to manage and execute up to 768 concurrent threads with zero scheduling overhead. The total number of threads concurrently executing on a GeForce 8800 GTX is thus 12 288. Each SM thread has its own execution state and can execute its own independent code path. However, for performance reasons, the chip designers implemented a *single instruction multiple thread (SIMT)* execution model, creating, managing and executing threads in groups of 32 called warps.[12] Every instruction issue time, the scheduler selects a warp that is ready to execute and issues the next instruction to the active threads of the warp. Instructions are issued to all threads in the same warp simultaneously (the warp executes a common instruction at a time), so there is only one instruction unit per multiprocessor. Full efficiency is realised when all 32 threads of a warp agree on their execution path, as it is commonly known in other SIMD architectures. If threads of a warp diverge at a data-dependent conditional branch, the warp serially executes each branch path taken, disabling threads that are not on that path, and when all paths complete, the threads converge back to the same execution path. Branch divergence occurs only within a warp; different warps execute independently regardless of whether they are executing common or disjointed code paths. Different interpretations of the hardware in terms of a taxonomy for parallel computing are discussed in more detail in Section 3.3. The SIMT programming model is a superset of the well-known SIMD model which performs instructions simultaneously on (short) data vectors, e. g., 4-tuples in SSE on commodity CPUs.

As with previous generation GPUs, hardware multithreading with zero-overhead scheduling is exploited to hide the latency of off-chip memory accesses, which can easily reach more than 1 000 clock cycles. This approach again maximises throughput over latency, in particular for memory-bound workloads.

To increase generality for compute tasks, the memory system supports arbitrary load-store instructions in contrast to the graphics-oriented texture filtering model and precomputed, fixed output addresses as in previous hardware. In particular, scattered output is possible, meaning that individual threads can write their output to arbitrary, dynamically computed memory locations. However, in case different threads write to the same memory location within the granularity of a compute kernel (see the next paragraph), it is only guaranteed that one arbitrarily selected value will be written, the writes are not serialised and there is no explicit protection from read-after-write (RAW) or write-after-read (WAR) hazards. The important consequence of this feature is that input and output regions may now overlap in contrast to previous hardware generations, enabling in-situ computations. The drawback is that the programmer is responsible for avoiding all hazards, but at least for our applications, this is only a minor issue and can in practice be accomplished easily.

We describe the memory hierarchy from the bottom up: The *constant memory* is shared between multiprocessors, implemented in the form of a register file with 8192 entries with a typical latency of 2–4 clock cycles. Constant memory is cached, but the cache is not coherent to save logic and thus, constant memory is read-only as the name implies. The *shared memory* per multiprocessor is implemented in 16 DRAM banks, reaching a similarly low latency as long as certain restrictions are met for the location each thread within a warp accesses, see the programming guide [171] and Section 5.3.1 for details. Multiprocessors can only communicate data via off-chip DRAM. The bus width is 384 pins, arranged in six independent partitions for a maximum theoretical bandwidth of 86.4 GB/s, more than a factor of two faster compared to the launch model of the previous generation (see Table 3.1 on page 52). This bandwidth is however only achievable if re-

[12]NVIDIA's GPU architects showed good sense of humour, as the term warp originates from weaving, which they identified as the 'first parallel thread technology'.

quests from several threads can be *coalesced* into a single, greater memory transaction, to exploit DRAM burst reads and writes. The hardware performs this coalescing only if strict rules for data size and warp-relative addresses are adhered to, see Section 5.3.1 for details and examples. Here we already note that non-coalesced memory requests can perform one order of magnitude slower, in other words, a memory-friendly data access pattern respecting locality has greater impact on performance than on commodity CPUs. This is due to the throughput-oriented design of the GPU, while CPU memory systems with their deep hierarchy of caches minimise latencies.

For more details on the architecture, we refer to an article by Lindholm et al. [149], on which most of the material in the previous paragraphs is based.

Programming Model for Compute Tasks, C for CUDA

The key idea of the CUDA programming model is to expose the scalable processor array and the on- and off-chip memory spaces directly to the programmer, ignoring all fixed function components. This approach is legitimate from a transistor-efficiency point of view as the SPA constitutes the bulk of the chip's processing power. In the following, we only very briefly summarise C for CUDA, and refer to Section 5.3.2 and Section 5.4 for more details on implementing multigrid solvers for sparse linear systems. The book GPU Gems 3 [168] and an article by Nickolls et al. [169] provide much more information; and for a more hands-on introduction, the programming guide [171] and the slides from various conference tutorials given by experts from both NVIDIA and academia are recommended.[13] Furthermore, an article by Boyd [33] presents an excellent overview of data-parallel programming techniques on GPUs in general.

C for CUDA comprises both an extension of standard C as well as a supporting runtime API. Instead of writing compute kernels in a shader language like GLSL, the programmer uses C for CUDA. A set of additional keywords exists to explicitly specify the location within the memory hierarchy in which variables used in the kernel code are stored, for instance, the `__constant__` qualifier marks a variable to be stored in fast, read-only constant memory. For the actual kernel code, essentially all arithmetic and flow control instructions of standard C are supported. It is not possible to perform operations like dynamically allocating device memory on the GPU from within a kernel function, and other actions typically in the operating system's responsibility.

Each kernel instance corresponds to exactly one device thread. The programmer organises the parallel execution by specifying a so-called *grid of thread blocks* which corresponds to the underlying hardware architecture: Each thread block is mapped to a multiprocessor, and all thread blocks are executed independently. The thread blocks thus constitute virtualised multiprocessors. Consequently, communication between thread blocks within the granularity of the kernel launch is not possible (except for slow atomic integer operations directly on memory which we omit here for the sake of brevity), but there is always an implicit barrier between kernel invocations: This means that all memory transactions performed by one kernel are guaranteed to have completed when the next kernel launches, and programmers can thus synchronise data between kernels via global memory (there is no warranty that the contents of the constant and shared memory spaces are preserved between kernels, so at the end of each kernel function, the result must be stored in global memory anyway). Programmers must not make any explicit or implicit assumption on the order of execution of the thread blocks, or even on the order of threads within each block, which are contiguously split into warps as explained in the previous paragraph. The threads within each block may however communicate via the 16 kB shared memory on the multiprocessor, using a lightweight barrier function to synchronise the warps of the block. Our high-performance implementation of a sparse matrix-vector multiplication (cf. Section 5.3.2 on page 129) illustrates how the performance of memory-bound kernels can be significantly improved by staging memory

[13] `http://gpgpu.org/developer#conference-tutorials`

accesses through shared memory, and by interpreting shared memory as a cache with a user-controlled replacement policy.

Restrictions apply to the maximum number of threads per block (currently 512) and the dimensions of the grid. As several thread blocks are resident on one multiprocessor at the same time (it supports up to 768 concurrently active threads, or 24 warps), shared memory and the registers are split disjunctly, and thus, not all configurations of partitioning the computation into thread blocks may work, referred to as 'multiprocessor occupancy'. Running very register-heavy kernels in large blocks may result in only one block being active per multiprocessor. In this case, there are potentially not enough threads active concurrently, and memory latency cannot be adequately hidden. The size of the thread blocks should always be a multiple of the warp size (currently 32), and the programs should be written in such a way that the threads within a warp follow the same execution path, as otherwise, the warp is serialised and both sides of branches are executed for all threads in the warp. The threads within each thread block should also exhibit memory access patterns that allow the coalescing of requests into larger memory transactions.

In order to parameterise code with the dimensions of the grid and the number of threads per block, and to be able to compute memory addresses of offsets in input- and output arrays, the current configuration is available via special keywords that are mapped to reserved input registers for each thread. In other words, each thread can look up its block number, the number of threads per block and its offset within the block. This also allows to mask certain threads from execution.

The so-called 'launch configuration', the partitioning of the problem into a grid of thread blocks, is realised via a minimal extension of the C language on the host side. The kernel is called just like any other procedure, passing input and output arguments as pointers to device memory and using a special notation to pass the configuration.

The *CUDA runtime API* provides all necessary routines to allocate memory on the device, to copy data to and from the device, and to query device parameters such as the number of multiprocessors, the limits of the launch configuration, the available memory etc.

The CUDA tool chain includes `nvcc`, the CUDA compiler driver. All non-CUDA related parts of a given piece of code are passed on to the system default compiler, `gcc` on Linux systems. `nvcc` can be configured to output a binary object file that can be linked into larger applications, or raw PTX assembly, or standard C code that can be compiled with any other compiler by linking to the appropriate CUDA libraries. In our implementations, we decided to separate the CUDA kernel code from the rest of the application in small compilation units that contain only the kernel and some wrapper code to launch it. These files are compiled with `nvcc` into object files that are added to the entire application during linking. This approach allows to use for instance the Intel compiler suite, which is known to generate faster executables than the GNU compiler suite on all of the (Linux) machines we use, for the CPU part of our applications.

Since the initial version, NVIDIA has continuously added features to CUDA both in terms of hardware (see Section 3.2.5) and software. Backward and forward compatibility is realised by assigning each new GPU model a so-called *compute capability* that can be queried using the runtime API. The programming guide [171] documents improvements in software like the exposure of overlapping computation with PCIe transfers, a feature called *streams* (cf. Section 5.3.1).

Dedicated High Performance Computing Product Line

Tesla is NVIDIA's product line targeting the high performance computing domain, not to be confused with the internal code name for the architecture underlying the G80 chip and its two successors. All hardware in this brand is based on consumer-level products, with few but important modifications: These GPUs do not have display connectors; and the on-board memory is significantly increased up to 4 GB per GPU for the latest models to enable calculations on much larger datasets. These products are subject to much more rigorous testing than the GPUs intended for the mass market, and to increase reliability and stability, their memory clock is reduced.

NVIDIA provides three different solutions, all based on the same chip: The *GPU computing processor* is a single GPU in the same PCIe form factor as a regular graphics card. The *deskside supercomputer* is an external chassis with proprietary PCIe-based connector cables comprising two GPUs, which is discontinued at the time of writing and has been replaced by the *personal supercomputer*, a multi-GPU workstation. Finally, the *GPU computing server* is a 1U rack-mounted chassis housing four GPUs. It uses a proprietary connector that combines two GPUs into one PCIe slot, and separate power and cooling, and is designed to enable dense commodity-based GPU-accelerated clusters. Despite being offered in separate chassis, GPUs continue to be co-processors in the traditional sense, and a standard CPU is always needed to control them.

To be technically correct, Tesla is NVIDIA's third brand of GPUs, the second one which has been available before is Quadro, used in production and engineering, for instance in CAD workstations. These GPUs also undergo much more rigorous testing than their consumer-level counterparts, and the corresponding display driver is certified to work with established software in the field.

CUBLAS and CUFFT

The CUDA toolkit includes optimised implementations of the BLAS collection and FFTs on CUDA-capable GPUs, which require only minor changes to existing codes to benefit from GPU acceleration. However, the full feature set of corresponding CPU implementations is not available yet at the time of writing. Nonetheless, the release marks an important step forward towards employing the GPU also by the average user, unwilling to learn C for CUDA and get accustomed to the unfamiliar programming model.

3.2.3. AMD CTM and Stream

As the implementations for this thesis target NVIDIA hardware, AMD's approach is presented in less detail, focusing on the relevant information for the historical argument in this chapter.

The Radeon R600 Architecture

Figure 3.4 depicts a block diagram of AMD's first DirectX 10 compliant GPU, the Radeon HD 2900XT, released in May 2007, six months after NVIDIA's G80 architecture. The chip is codenamed R600 and follows a VLIW (very large instruction word) design.

This GPU contains 320 scalar stream processing units arranged into four SIMD arrays of 80 units each. For each group of four ALUs, there is a fifth, more complex unit responsible for transcendentals, together forming a 5-way superscalar shader processor or functional unit. Each functional unit has its own branch execution unit for control flow and conditionals, as well as a number of general-purpose registers. All functional units are connected to an on-chip kilobit ring-bus memory controller, which connects to off-chip DRAM via eight independent 64 bit channels for a total of 512 pins, resulting in a peak memory bandwidth of 105.6 GB/s. This means that out of the 2140 pins that connect the chip with the rest of the graphics board, roughly 25 % are responsible just to move data. These pins are distributed along the edges of the 20 by 21 mm die, an engineering tour de force besides the fact that the whole chip comprises 700 million transistors. The memory subsystem supports arbitrarily scattered writes of computed results. We refer to the presentation by Mike Mentor of AMD given at the 2007 Hot Chips 19 conference[14] for more details, as well as to other resources on the web.[15]

[14] `http://www.hotchips.org/archives/hc19/`

[15] See for example `http://hothardware.com/Articles/ATI-Radeon-HD-2900-XT--R600-Has-Arrived`

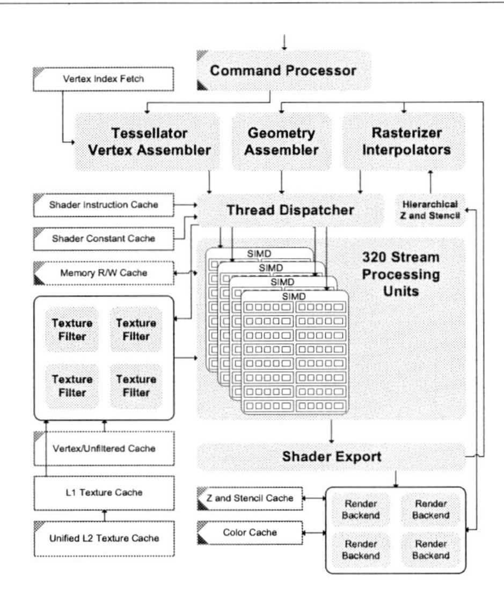

Figure 3.4: AMD Radeon HD 2900XT block diagram: Gray ovals indicate logic units and red-bordered rectangles indicate memory units. Green triangles at the top left of functional units denote units that read from memory, and blue triangles at the bottom left symbolise writes to memory. Image courtesy Owens et al. [176].

Programming Model

AMD took a different approach to exposing the processor array of their GPUs for general-purpose computation. At SIGGRAPH 2006, Peercy et al. [180] presented CTM, short for *close to the metal*. CTM provides a very low-level hardware abstraction layer (HAL) to access the stream processors directly: Computation is performed by loading an ELF binary of offline-compiled raw ISA assembly into a dedicated command buffer, binding input and output arrays to the stream processors, and defining a domain over the outputs on which to execute the binary. No graphics-specific features are available through CTM.

Furthermore, AMD's tool chain includes the *compute abstraction layer* (CAL), which adds higher level constructs and follows the *stream computing* paradigm (cf. the end of Section 3.1.5). CAL also includes compilation support to generate ISA binaries from GLSL or HLSL shader programs. AMD also used the BrookGPU language and compiler from Stanford University (cf. as above) and extended it with scatter support, DirectX 10 features like integer support and a backend that compiles directly to CTM, known as *Brook+*. In the past years, AMD proceeded to brand their approach as *Stream*, and discontinued to use the term CTM.

The latest versions of Stream also include a GPU-accelerated (subset of) ACML, AMD's core math library of optimised implementations of, e. g., BLAS and FFT. AMD also works closely with third-party middleware companies, for instance, the aforementioned RapidMind software platform maps directly to HAL for optimal performance on AMD GPUs. Recently, AMD Stream has been continuously expanded and we refer to the programming guide for details.[16]

[16] `http://www.amd.com/stream`

Dedicated High Performance Computing Product Line

AMD brands their HPC product line as *FireStream*, previously also as the AMD Stream processor. The name stems from AMD's professional series GPUs labeled FireGL. Similar to NVIDIA's GPU computing processor, these are single-board solutions with increased but downclocked memory. AMD does not currently offer dedicated dense server solutions.

3.2.4. Bibliography of Relevant Publications

Estimates vary on the number of published papers in the field since AMD Stream and NVIDIA CUDA pushed the second wave of GPU computing, we believe 750 is a realistic number as of the time of writing. Success has been reported in many different application areas, and for problems that have previously not been feasible due to hardware limitations preventing an efficient implementation. This is also reflected by the increasing amount of dedicated GPU computing workshops and mini-symposia at application-specific conferences, e. g., in fluid dynamics or astrophysics. In the following, the most relevant publications in the scope of this thesis are briefly summarised, the selection is unavoidably subjectively biased. The survey articles by Owens et al. [176] and Garland et al. [77] are highly recommended.

Buatois et al. [41, 42] were the first to implement the SPMV kernel efficiently by applying blocking techniques, relying on the observation that in many practical situations, nonzero entries are not randomly distributed, but typically grouped (eventually after the application of a reordering technique) into small blocks which can be treated as dense. The resulting data layout is known as *block-CSR*, see for instance the textbook by Barrett et al. [13]. Bell and Garland [21, 22] and Baskaran and Bordawekar [14] improve their work and analyse different data storage techniques beyond the standard CSR format. Reported speedups over tuned CPU implementations reach an order of magnitude, but depend on the sparsity pattern of the matrices.

Multigrid solvers and multilevel methods in general have been refined by many researchers, including the author of this thesis and his co-workers (see Chapter 5 and Chapter 6). Kazhdan and Hoppe [125] are able to realise a full V cycle in only two kernels, specifically tailored to image processing in the gradient domain, with applications in tone mapping and stitching of gigapixel images. Due to the size of their input images, they have to use out-of-core techniques, which are particularly challenging from both a numerical point of view (higher-order finite elements are required to gain enough accuracy in only two passes) and because of the bandwidth bottleneck incurred by the comparatively narrow PCIe interface. Feng and Li [71] use a GPU-accelerated multigrid solver for power grid analysis, and Molemaker et al. [161] and Cohen and Molemaker [49] use multigrid to solve pressure Poisson problems in CFD simulations for feature film.

Another noteworthy publication using finite element methods on GPUs is the work by Komatitsch et al. [135], who implemented a high order spectral element method on graphics hardware, used to simulate the propagation of seismic waves and shocks from earthquakes through a three-dimensional model of the earth. They are able to significantly outperform an established and tuned CPU code. We emphasise that the authors pay special attention to analysing the impact of single precision and conclude that single precision gives exactly the same results for reasonably large problem sizes as double precision on the CPU. The main reason for this is that their explicit approach does not require the solution of linear systems (condition number), and that their physical fields do not vary greatly. The author of this thesis has been honoured to be able to contribute to extending the implementation to GPU clusters in later work [136].

Incompressible fluid flow has been implemented on a single GPU by Tölke and Krafczyk [222], using the Lattice-Boltzmann method to achieve a speedup of two orders of magnitude over a singlecore CPU code. Thibault and Senocak [220] report one order of magnitude speedup for a structured grid finite difference solver using a projection method. They compare a multithreaded

CPU code with a GPU code on a four-GPU Tesla S1070 blade. Unfortunately, they restrict themselves to using simple Jacobi iterations for the pressure Poisson solver. Incompressible flow governed by the Euler equations has received more attention, all of the following are based on finite difference discretisations and explicit timestepping schemes: Brandvik and Pullan [38] use a single GPU for common Euler benchmarks in two space dimensions. Elsen et al. [65] present simulations on complex geometries and use sophisticated numerical techniques like a two-grid relaxation, in three space dimensions. Corrigan et al. [52] focus on unstructured grid techniques, like the former work, on a single GPU. Phillips et al. [182] map a multi-block compressible Euler code to an eight node cluster with dual-GPU boards. Finally, Micikevicius [160] presents efficient techniques to map three-dimensional stencil computations (e. g., finite differences on regular grids) to the CUDA architecture, and techniques to overlap computation and communication in multi-GPU setups. All these papers report at least one order of magnitude speedup.

Cohen and Molemaker [49] present OpenCurrent, a framework that has been designed from scratch to facilitate fluid simulations on graphics hardware. In addition to their already mentioned full re-implementation, Phillips et al. [182] also describe an approach to integrate GPUs into their Fortran solver MBFLO. These two publications are closest in spirit to the methodology we present and evaluate in Chapter 6.

Common to the all these publications on fluid and gas flow is that in most cases, single precision is sufficient. This is not surprising for the standard Lattice Boltzmann method, and Tölke and Krafczyk [222] present detailed comparisons of drag and lift coefficients with reference solutions of their benchmark problem. With the exception of Cohen and Molemaker [49], the accuracy is not addressed beyond simply stating that the results agree with the ones computed by their CPU counterparts.

The scan primitive is used in parallel programming to solve recurrences and operations in which each output requires global knowledge of the inputs. Sengupta et al. [197] were able to implement the scan primitive very efficiently on the CUDA architecture, and they demonstrate significant speedups on the same GPU over previous implementations. More importantly, they introduce the *segmented scan* to the GPU, a parallel primitive needed for several complex operations. A book chapter provides a more practically-oriented explanation [113], and they make their implementation available in the form of `CUDPP`, the *CUDA Data Parallel Primitives Library*.[17] Using their tuned scan framework, they demonstrate novel GPU implementations of quicksort and sparse matrix-vector multiply, and analyse the performance of the scan primitives, several sort algorithms that use the scan primitives, and a visually-accurate shallow-water fluid simulation using the scan framework for the underlying tridiagonal matrix solver. A more recent paper by Satish et al. [192] significantly improves the presented sorting algorithms, and Dotsenko et al. [62] present algorithmic and implementational improvements for the scan and segmented scan primitives by balancing work- and step-efficient variants.

As more and more GPU clusters are being deployed, GPU acceleration of large-scale applications is becoming increasingly interesting feasible. The Euler solver by Phillips et al. [182] has been mentioned already. Other noteworthy contributions include Stone, Phillips et al. [183, 205] and Anderson et al. [4], who worked on GPU acceleration of molecular dynamics applications. The November 2008 edition of the TOP500 list of supercomputers[18] has been the first to contain a GPU-accelerated cluster, the Tsubame machine at the Tokyo Institute of Technology. Fatica [70] explains the underlying implementation of the Linpack benchmark in C for CUDA. A year later, the Tianhe-1 system installed at the National Super Computer Center in Tianjin (China) debuted at position #5.[19] It has a hybrid design with Intel Xeon processors and AMD GPUs used as accel-

[17] `http://gpgpu.org/developer/cudpp`
[18] `http://top500.org/lists/2008/11`, entry #29
[19] `http://top500.org/lists/2009/11`, entry #5

erators. Finally, Kindratenko et al. [132] address various issues they solved related to building and managing large-scale GPU clusters, such as health monitoring, data security and privacy, resource allocation and management, and job scheduling.

3.2.5. Double Precision Floating Point Support

The first GPU to support double precision floating point arithmetic was AMD's FireStream 9170, the professional variant of the Radeon HD 3800. It is based on the RV670 chip generation, the second generation of DirectX 10 compliant hardware by AMD. It was released in November 2007. NVIDIA's GeForce GTX 280 (chip name Tesla 10, or GT200) became available seven month later, in June 2008. As before, we concentrate on NVIDIA's chip, as this is the hardware we have used for the experiments in this thesis.

NVIDIA Tesla 10 Architecture

NVIDIA's approach is the addition of a dedicated double precision unit to each multiprocessor, while leaving the number of thread processors and special function units (see Section 3.2.2) unchanged. Each of the now ten thread processing clusters (TPC) contains three multiprocessors in the GTX 280 (8 TPCs with 2 MPs each on previous-generation hardware), for a total of 30 multiprocessors, 30 double precision units and 240 single precision units per chip, which almost doubles the raw amount of shader resources compared to the G80 design. The memory interface now comprises eight instead of six raster operation units, each connected via a dedicated 64-pin connection to its memory partition for a total bus width of 512 bit. The launch model, the GeForce GTX 280, exhibits a peak memory bandwidth of 142 GB/s, which marks a 4.5-fold improvement over the first GeForce 6 model in four years and four hardware generations, compared to roughly a factor of two for CPUs in that time spam. The core clock of the GTX 280 is 600 MHz, the shaders operate at 1.3 GHz, and the memory is clocked at 1.1 GHz for an effective double-data-rate of 2.2 GHz. One additional improvement of the GT200 chip that is worth mentioning is the simplification of the memory coalescing rules, enabling faster performance for many compute kernels. The GT200 also pushed the limits of chip manufacturing, comprising 1.4 billion transistors covering 576 mm^2 die surface area built in a 65 nm process.

As the design implies, the peak double precision performance is roughly one eighth of the single precision performance. For bandwidth-limited applications like the ones presented in this thesis, this is however not an issue as long as enough blocks can be scheduled onto each multiprocessor to hide stalls due to latency of off-chip memory accesses. We present results on this hardware generation in Section 5.3.3 and in Sections 5.5–5.8 where we also discuss the impact of native double precision support on our mixed precision solvers.

3.3. GPUs and Parallel Programming Models

In the previous sections, we have presented the emerging field of GPU computing in its historical evolution, describing hardware features and corresponding programming models simultaneously. There are of course many other approaches, and in the following, the most important ones are briefly outlined, 'picking up loose trails' as we proceed. Here, GPUs which we consider to be *(emerging) many-core processors* are also compared to other multicore and multithreaded designs such as the STI Cell BE processor, Sun's Niagara2, an Intel Xeon quadcore processor, and a NEC SX-8 CPU as a representative of a dedicated supercomputing vector architecture.

3.3.1. GPUs as Parallel Random-Access Machines

The parallel random-access machine (PRAM) is a popular model for designing and analysing parallel algorithms, see for example the textbook by Grama et al. [99]. The model comprises three different classes:

Exclusive Read Exclusive Write (EREW) Multiple processors must not simultaneously access the same memory address neither for reading nor for writing.

Concurrent Read Exclusive Write (CREW) Any set of processors may read the same memory address, but only exclusive, single-processor memory writes are permitted.

Concurrent Read Concurrent Write (CRCW) Any set of processors may concurrently read and write memory addresses. The resolution of concurrent writes is further classified as:

- **Common** All concurrent writes must be the same value for the value to be actually written.
- **Arbitrary** The write value is arbitrarily selected from the processors concurrently writing to the same memory address.
- **Priority** Writes are prioritised statically or dynamically, and the value from the processor with the highest priority is stored in memory.
- **Combining** The final value is composited from all values via an associative and commutative operator.

In the following, the characterisation in the thesis by Lefohn [143] is extended to the latest hardware generations.

Early Programmable GPUs

Early programmable GPUs up to those compliant with the DirectX 9 specification belong to the class of *concurrent read exclusive write (CREW)* PRAMs. Computational kernels (i. e., shader programs) may simultaneously read from arbitrary locations in texture memory, but may only write to a single, a priori known address. The main limitation of this class is that so-called scatter operations are not permitted, meaning operations that write to dynamically computed memory addresses. It is possible to emulate scatter by rendering individual points, which in theory corresponds to a *concurrent read concurrent write (CRCW)* PRAM, subclass *priority writes*. Due to the inefficiency of the approach, this is with some noteworthy exceptions not practically relevant in general.

DirectX 10 Class GPUs

These newer GPUs belong to the class of *concurrent read concurrent write (CRCW)* PRAMs, and to the subclass of *arbitrary writes*: Scattered writes to dynamically computed addresses are supported, but when several threads write to the same memory location concurrently, there is no guarantee which thread's value will actually be written, see Section 3.2.2 for details. With atomic memory operations, one can avoid real concurrency, so that the order is still undetermined but at least all values contribute, which is important for associative operations. Atomic memory operations are however very slow on current hardware and are not supported for floating point data. Consequently, the correctness and determinism of any given algorithm is in the programmer's responsibility.

3.3.2. GPUs and Vector SIMD Architectures

SIMD (single instruction multiple data, [74]) is a very efficient way to increase the power- and area-efficiency of a processor core, as it amortises the complexity of decoding an instruction stream and the cost of control structures across multiple ALUs. Section 3.1.7, 3.2.2 and Section 3.2.3 present examples of concrete GPU architectures, and discuss the impact on performance of exploiting the SIMD characteristics of the designs. For instance, the SIMT model (single instruction multiple thread) at the base of NVIDIA CUDA implies that optimal throughput can only be achieved when all threads within a SIMD block (called 'warp') follow the same execution path: Divergent branches are effectively serialised, resulting in reduced performance.

The actual hardware implementation of SIMD control flow divergence treatment varies from architecture to architecture, but in general, divergence always means that fewer ALUs do useful work. On a chip with SIMD width w, the worst case behaviour is $1/w$ of the chip's peak floating point performance.

The integration of (short-vector) SIMD units into chip designs is also considered as a very efficient countermeasure for the ILP (instruction level parallelism) wall problem: In general, current up to 6-way superscalar CPU designs struggle hard to extract enough independency from the instruction stream to achieve more than 1.5 instructions per clock cycle [106]. The obvious drawback is that either the programmer or the compiler have to vectorise the code, because the hardware no longer succeeds in exploiting instruction level parallelism.

As mentioned previously, the ability of multithreaded architectures to hide thread stalls is another very important aspect, in our context, stalls due to memory access latency. A metric that captures this effect is incorporated in the comparison below.

Comparison of SIMD Characteristics

Table 3.2 compares the SIMD characteristics of GPUs with that of other architectures, the Cell BE processor, an Intel Xeon quadcore processor (Core2 microarchitecture, Harpertown), Sun's hardware-multithreaded Niagara2 chip and a NEC SX-8 CPU, a supercomputer with a dedicated vector architecture. This discussion expands the characterisation given by Fatahalian and Houston [68]. All data is given for single precision floating point operations only. For the Cell BE processor [181], the PPU cores are ignored and only the SPEs are counted, and analogously on the SX-8 CPU, the scalar unit is omitted from the argument. Similarly, on the Harpertown CPU, only the short-vector units (SSE) are counted and the sequential FPU is ignored. On GPUs, double precision and special function units are not taken into account. Also, only recent models are discussed, as the details of earlier graphics processors are not disclosed by the vendors. The upcoming Larrabee chip (see Section 3.4.2) is included for the sake of completeness, it is not publicly released yet and information about it is sparse [194].

The values of interest in Table 3.2 are the number of cores per chip, the number of ALUs per core, the SIMD width and a metric called 'max occupancy', which is defined as the maximum ratio of hardware-managed thread execution contexts to simultaneously executable threads, rather than an absolute count of hardware-managed execution contexts. This ratio is a measure of the ability

Type	Processor	Cores/Chip	ALUs/Core	SIMD Width	Max Occupancy
	Intel Core2Quad	4	4-8	4	1
CPU	STI Cell BE	8	4	4	1
	Sun UltraSPARC T2	8	1	1	4
	AMD Radeon HD 4870	10	80	64	25
GPU	NVIDIA GeForce GTX 280	30	8	32	128
	Intel Larrabee	> 16	1	16	> 4
HPC	NEC SX-8	1	1	512	1

Table 3.2: Comparison of SIMD characteristics. Note that the Xeon CPU only has one physical SSE unit per core (4 ALUs), but can perform more than one operation on it in a superscalar way. Table adapted and extended from Fatahalian and Houston [68].

of an architecture to automatically hide thread stalls via context switches in massive hardware multithreading.

The numbers clearly indicate the benefits of the specialisation of GPUs as an architecture for data-parallel computations, by maximising SIMD width and latency hiding simultaneously. Section 3.4.3 discusses implications of these trends.

3.3.3. Other Programming Models

Implicitly Parallel Programming Models

Hwu et al. [120] argue for an *implicitly parallel programming model* for many-core microprocessors. Their pragmatic rationale is that the growing complexity and parallelism of processor designs should not be fully exposed to the programmer any more, who cannot be dealing with, e. g., all the cache sizes and microarchitecture details. C for CUDA is in their opinion a step in the right direction for the paradigm change in the programming model, as the parallelism is only exposed where it needs to be. However, they are aware that a higher level of abstraction is not feasible yet, because compilers, performance analysis tools and the entire programming tool chain is simply not able at this time to automatise the parallelisation progress.

As a proof of concept, Stratton et al. [206] implemented a compiler that maps C for CUDA onto commodity multicore CPUs, which was later extended to FPGAs by Papakonstantinou et al. [177]. The RapidMind Platform (see Section 3.1.5) and OpenCL (see next section) are also good examples of programming environments for multiple parallel heterogeneous architectures.

Message Passing between Distributed Memories

Stuart and Owens [214] pursue an entirely different avenue of research. They demonstrate how a subset of the MPI message passing paradigm for distributed memory architectures can be realised within the constraints of a data-parallel architecture, using current GPUs as an example. For a number of test cases with different communication patterns, the authors report the overhead to be tolerable, always within 5 %. Also very promising is the CUDASA extension to CUDA, which extends the concept of mapping kernel execution to a grid of blocks of threads to similar grids and blocks to execute entire tasks and jobs. Tasks in their nomenclature are abstracted in a bus layer, allowing to efficiently distribute a batch of kernels across multiple GPUs within one compute node. Similarly, a network layer abstracts an entire GPU-accelerated cluster. CUDASA stands for *compute unified device and systems architecture* and is developed by Strengert et al. [165, 207].

3.4. Future APIs and Architectures

Speculation about future graphics and compute hardware in general is always dangerous. This section nonetheless summarises a few directions and publications which support that the contributions of this thesis in the field of GPU computing will remain valid independent of near-future hardware development.

3.4.1. OpenCL and DirectX 11

In December 2008, the Khronos group announced the ratification of OpenCL, the *Open Compute Language*.[20] The leading hardware, software and middleware companies worked together to create the specification [128] in a remarkably short amount of time. Similar to OpenGL, OpenCL is a royalty-free open industry standard. OpenCL targets general-purpose parallel programming of heterogeneous systems, and aims at providing a uniform programming environment for software developers to write efficient, portable code for high-performance compute servers and desktop computer systems using a diverse mix of multicore CPUs, GPUs and Cell-type architectures. At the time of writing, the first non-beta implementations are becoming available. The most important aspects of OpenCL (version 1.0) are:

- OpenCL supports both data- and task-parallel compute models with the explicit aim to use all heterogeneous resources in a system.
- The design goal is to provide a low-level 'ecosystem foundation', without middleware or convenience functions. Vendors and third-party developers are free to provide wrappers and tools.
- Double precision floating point arithmetic is optional.
- OpenCL implementations must be portable across devices, in a vendor-independent way. Vendor-specific functionality (such as double precision support) is available through extensions similar to OpenGL.
- The software stack has three layers: Compute kernels are written in a language based on C99 with some extensions (*OpenCL C for Compute Kernels*), this language is syntactically and semantically very similar to C for CUDA. A platform layer API abstracts the hardware and supports context creation, memory allocation etc.; and a runtime API provides support to launch kernels and dynamically manage scheduling and resources.
- The memory model specifies several address spaces for data, private to so-called work-items and work-groups, as well as constant and global memory. This terminology resembles the threads, blocks and shared memory per block in CUDA.

We refer to the OpenCL web page and the specification for more details.

DirectX 11

The next version of the DirectX standard, DirectX 11, specifies so-called compute shaders. They are explicitly designed for general-purpose yet graphics-related computations such as game physics. NVIDIA already announced that compute shaders will be supported on all CUDA-enabled GPUs. It will be interesting which features in terms of program control, memory spaces, communication between compute threads etc., are made mandatory in the specification.

[20] `http://www.khronos.org/opencl`

3.4.2. Intel Larrabee and Future Graphics Architectures

The *Larrabee* microprocessor is Intel's foray into the domain of discrete graphics cards. So far, the company has focused on graphics functionality integrated in mainboard chipsets, and remarkably, owns a graphics chips market share of almost 50 %.[21] Intel presented the architecture at the SIGGRAPH conference in August 2008 [194], but in late 2009 they announced that the first generation of this chip design will not result in a commercial product.

Larrabee uses multiple in-order x86 CPU cores (based on the original Pentium I design), which have been augmented with a 16-wide vector unit and a few fixed function logic blocks for graphics-specific tasks. Each core is connected to a high-speed on-chip interconnection network and its local subset of the level-2 cache. The subsets maintain coherency. Intel claims that the design is much more powerful than current GPU 'cores', as Larrabee's cores are x86-based and support, for instance, subroutines and better memory page handling. In particular, functionality like blending and rasterisation that is performed in fixed function logic on current GPUs, is realised entirely in software on Larrabee.

Performance data is not yet public, so it remains to be seen if this promising design is competitive as a general purpose parallel computing architecture.

Stanford GRAMPS

Researchers in the graphics and architecture groups at Stanford University are actively developing GRAMPS [215], a programming model for graphics pipelines encompassing heterogeneous many-core commodity resources. One specific aim of GRAMPS is to support both graphics and non-graphics workloads. The execution model contains both fixed-function and application-programmable stages that exchange data via so-called execution queues. The interesting design choice of the approach is that the number, type and connectivity of these processing stages is defined in software, permitting arbitrary pipelines or even processing graphs. One can argue that this continues the idea of the DirectX 10 standard, which removed all fixed-function functionality that is expressible in terms of programmable constructs from the pipeline, see Section 3.2.1 on page 62 and the article by Blythe [28]. The authors have implemented GRAMPS simulators, one similar to current GPUs, and one similar to multicore CPUs. Their evaluation results of three different pipelines (DirectX, raytracing and a hybridisation of the two) on the two emulators are promising.

Future Graphics Architectures

In his ACM Queue article, Mark [153] outlines trends and requirements of future graphics architectures, arguing from a historical perspective and the way GPUs have evolved over the past years. He examines the limitations of the traditional depth-buffered, rasterisation based pipeline, in particular for raytracing image synthesis; and collects supportive arguments for general-purpose graphics hardware from both a graphics and a non-graphics point of view. In his opinion, the hardware pipeline is likely to be replaced with a software pipeline, built on top of massively (hardware-) multithreaded task- and data-parallel cores. Peak performance is only obtained if SIMD execution is exploited due to the ILP wall (instruction-level parallelism): Even the best superscalar CPUs today struggle to average more than 1.5 instructions per cycle. His prediction of the memory spaces is that the trend of current GPUs continues: Data locality will become increasingly important, an example is the shared memory approach in CUDA and OpenCL to enable communication within a batch of compute threads.

[21] Jon Peddie Research, Report for Q1-2009

3.4.3. CPU and GPU Convergence

The tendency of graphics hardware to become increasingly general up to a point from where on architects incorporate dedicated fixed-function logic for certain functionality to improve performance further has existed for a long time, Myer and Sutherland [167] describe this already in 1968 as the 'wheel of reincarnation'. This classic citation has been used in the past in the context of the emerging field of GPU computing. However, we believe that now, CPU and GPU designs are indeed converging. Many researchers, in particular system architects, come to the same conclusion, see for example Bakhoda et al. [12], Che et al. [48], Yeh et al. [246] and Hwu et al. [121]. The most important supportive arguments of this claim are:

- GPU capabilities have become more and more general.
- The extrapolation of future graphics workloads implies a fully programmable software pipeline.
- CPU designs are becoming increasingly parallel.
- All microprocessor designs aiming at performance are facing the same fundamental physical problems such as heat, power dissipation and leaking voltage.
- GPUs demonstrate a successful approach to task- and data parallelism on the same chip, at extreme scales compared to current commodity CPU single-chip multicore designs.
- GPUs demonstrate a feasible computer architecture approach to maximise performance per Watt, which has become a very important metric ('green computing').

On the other hand, the convergence of architectures imposes many challenges:

- The programming model for massively hardware-multithreaded architectures is still an open research question, in particular in the compiler community. NVIDIA's C for CUDA and upcoming OpenCL implementations are however mature enough for real-world deployment.
- Raw parallelism is not the only point: The SIMD width of future hardware is likely to increase. On the other hand, experience with dedicated vector computers has told us that vectorisation is highly nontrivial for many diverse tasks, and is currently mainly left in the responsibility of the programmer rather than the compiler.
- Bandwidth to off-chip memory on current multicore systems scales with the number of sockets per compute node and not with the number of cores per physical processor. On CPUs, this implies NUMA (non uniform memory access) effects, as not all cores can access locations in main memory at the same speed. The increasing number of cores per socket worsens this problem. Despite this drawback, the actual external bandwidth of recent CPU and memory designs has increased remarkably with dual and triple channel DRAM technology. On the other hand, GPUs incorporate scalable memory systems, which are already approaching a hard limit: There can be only a fixed number of pins to carry the data.
- Cell-type processors (broadly speaking, the Larrabee design fits into this category) and conventional CPU designs can drive operating systems, at the cost of increased hardware complexity. These designs typically exhibit a much narrower connection to off-chip memory than specialised processors such as GPUs, which can be considered as a legacy design drawback because of the standards of mainboards: The northbride-southbridge architecture connects much more clients than just the floating point units with the off-chip memory.

- With discrete co-processors such as current GPUs, architects have much more freedom. However, the open question is the interconnect between such boards and the rest of the compute nodes: From the very beginning, the AGP and PCIe buses imposed a severe bandwidth bottleneck. Current PCIe2.0 x16 technology combined with asynchroneity between data transfers and independent computation starts to provide a decent connection to the host, but at some point during a (large-scale) computation, data has to be sent across a network. One possible solution might be to equip discrete boards with Ethernet or Infiniband technology. NVIDIA engineers for instance are beginning to think publicly about such features.

4

Mixed and Emulated Precision Methods

Emulated and mixed precision techniques have been known since the 1960. However, little attention has been paid to hardware efficiency aspects and to performance improvements that can be achieved using these techniques. This chapter is structured as follows: Section 4.1 motivates the use of mixed and emulated precision methods. Here, we also briefly revisit floating point arithmetic and present a number of instructive examples highlighting common pitfalls in floating point computations. Section 4.2 explains techniques to emulate high precision formats using only (native) low precision data and operations, and thus techniques that are useful when no native high precision format is available. Section 4.3 presents mixed precision methods from a historical and numerical perspective. Here, we also introduce our mixed precision iterative refinement solver tailored to the solution of large sparse linear systems of equations. Mixed and emulated precision techniques are discussed from a hardware efficiency and hardware design point of view in Section 4.4. Finally, we analyse the precision requirements of the accurate solution of Poisson problems (prototypical for all applications considered in this thesis) in much detail in Section 4.5.

All applications in later chapters of this thesis use mixed or emulated precision methods in one concrete realisation or another, and we refer to Chapters 5 and 6 for detailed numerical tests. With the exception of Section 4.5, this chapter is in most parts a revised and updated version of a survey paper previously published by the author of this thesis together with Robert Strzodka and Stefan Turek [85].

4.1. Motivation

The idea behind mixed precision methods is to perform a *large* part of the computation in low precision and only a small part in high precision without sacrificing the accuracy of the high precision result. Historically, *mixed precision iterative refinement* methods have been developed to increase the accuracy of a result computed with a linear or nonlinear solver, see Section 4.3. In this thesis, we extend the original work and motivate, employ and analyse mixed precision schemes for sparse linear systems also in terms of performance improvements and hardware efficiency. Therefore, we make no assumption on the characteristics of high and low precision in this chapter, except that unless otherwise noted, we assume that the size (bit length) of the low precision format is half the size of the high precision format to simplify the presentation. Low and high precision floating point units can reside on the same chip, with a potentially significant difference in execution speed, for instance on the Cell BE processor, GPUs from the latest hardware generation, or in the SIMD SSE units on a conventional CPU. High precision may not be available on every chip in a given system, for instance in a hybrid compute node comprising CPUs and (older) GPUs.

If sufficient precision is not available at all in a native floating point format, an *emulation* of a high precision format is required. It is even possible in some cases that an emulated format improves performance even if a native high precision format is available, in particular when the low precision format is much faster than the high precision one, as for instance on current GPUs and the first generation Cell processor.

4.1.1. Examples

After briefly revisiting floating point number representations to fix our notation, the first example recapitulates representation, roundoff and cancellation errors in floating point arithmetic. The second, very instructive example highlights the unintuitive relation between computational precision and result accuracy.

Floating Point Number Representations and Notation

The IEEE-754 standard [122] defines floating point numbers in the the following format: The first bit s is used to represent the sign. The following bits encode the exponent with base two, in a biased form to be able to store it in an unsigned format. Some special exponents are reserved. For instance, with eight bits and the corresponding bias $2^8 - 1$, the exponent range $[-126, 127]$ is represented. The remaining bits in a floating point number encode the mantissa (significant), normalised so that the leading binary digit is 1 and does not need to be stored explicitly anymore. We use the following shorthand notation: $sMeE$ denotes a floating point number with M mantissa bits and E exponent bits. The standard IEEE-754 single and double precision floating point number formats are thus denoted as $s23e8$ and $s52e11$ respectively.

For an excellent introduction to floating point number systems and arithmetic in general, we refer to Goldberg [91].

Roundoff and Cancellation Errors

Floating point systems are finite, and hence, not all real values in $\mathbb{R}$ can be represented exactly. This is obvious for irrational numbers like π, but it is also true for rational numbers like $1/9$. Whenever such a number is represented in a floating point format, a *representation error* occurs. Consider the following floating point numbers, which are all exactly representable in the standard single precision s23e8 format:

$$a = 1 \qquad b = 0.00000004 \qquad c = 0.9998 \qquad d = 1.0002$$

Additive and multiplicative *roundoff errors* occur when computing—in s23e8—the terms $a+b$ and $c \cdot d$ respectively, because even if the input data for a computation is exact in the finite floating point system, the result data is not necessarily exactly representable:[1]

$$1.00000004 = 1 + 0.00000004 =_{\text{fl}} 1 \qquad 0.99999996 = 0.9998 \cdot 1.0002 =_{\text{fl}} 1$$

Taking either the expressions $a+b$ or $c \cdot d$, subtracting 1 and multiplying by 10^8 introduces *cancellation*:

$$4 = (1.00000004 - 1) \cdot 10^8 =_{\text{fl}} 0 \qquad -4 = (0.99999996 - 1) \cdot 10^8 =_{\text{fl}} 0$$

Cancellation thus promotes the small roundoff error of 0.00000004 to an absolute error of 4 and a relative error of order one. The order of operations is crucial:

$$\begin{aligned} (1 + 0.00000004 - 1) \cdot 10^8 &=_{\text{fl}} 0 \\ (1 - 1 + 0.00000004) \cdot 10^8 &=_{\text{fl}} 4 \end{aligned}$$

[1] The notation $x = y =_{\text{fl}} z$ abbreviates that x is the analytic evaluation of the expression y and z the evaluation in floating point, respectively.

Consequently, floating point arithmetic is not associative. This is very important, and very difficult to account for in practice, as the numerical values of input data are typically not known a priori: Compilers simply do not have enough knowledge of the data to rearrange computations in order to minimise cancellation effects, and representation and roundoff errors cannot be avoided by definition.

The situation is further complicated in practice as many compilers by default map floating point arithmetic to the 80 bit wide extended precision generic floating point unit (FPU) of current x86 and x86_64 CPU designs when compiling without optimisation, as one typically does in the implementation and debugging phases of program development. As a consequence, cancellation effects might not always be visible during debugging.

Rump's Famous Example

Rump [187] presents a famous example that contradicts the intuitive understanding of the relation between computational precision and result accuracy. He considers the evaluation of the polynomial

$$f = 333.75b^6 + a^2(11a^2b^2 - b^6 - 121b^4 - 2) + 5.5b^8 + \frac{a}{2b}$$

for $a = 77617.0$ and $b = 33096.0$. At first sight, this polynomial seems relatively harmless. All input data is exactly representable in 32 bit floating point, so the only errors that occur in the computation are roundoff and, more importantly, cancellation. The exponentiations are realised by successive multiplications to focus on the basic arithmetic alone. The results on Rump's machine, an IBM S/370 main frame, are:

single precision (s23e8)	$f = 1.172603...$
double precision (s52e11)	$f = 1.1726039400531...$
quad precision (s112e15)	$f = 1.172603950053178...$

It seems that increasing the computational precision verifies more and more digits, the result becomes increasingly reliable. However, not even the sign is correct, as the true result, computed with interval arithmetic and guaranteed to be correct in the first 15 decimal places, is:

$$f = -0.827396059946821...$$

Loh and Walster [150] point out that this example is not reproducible in IEEE-754 arithmetic due to different rounding modes. They suggest a small modification of the original expression that exhibits the same behaviour using standard IEEE-754 arithmetic with the round-to-nearest mode. The rearranged polynomial reads:

$$f = (333.75 - a^2)b^6 + a^2(11a^2b^2 - 121b^4 - 2) + 5.5b^8 + \frac{a}{2b} \tag{4.1.1}$$

It is evaluated for the same values of a and b to yield the same results as Rump's original example. The values of a and b satisfy the relation $a^2 = 5.5b^2 + 1$. Substituting into Equation (4.1.1), expanding the expression and cancelling out common terms reduces the polynomial to

$$f = \frac{a}{2b} - 2$$

and the exact result can be computed without severe cancellation effects. A more transparent form of Equation (4.1.1) is thus

$$f = 5.5b^8 - 2 - 5.5b^8 + \frac{a}{2b}$$

and cancellation is apparent to cause the severe errors: The eighth power of a five-digit number requires 40 decimal places after the exponents have been aligned with the one-digit number to perform the subtraction, resulting in cancellation of the one-digit number.

This example is admittedly very artificial. Nonetheless, it clearly demonstrates that increasing the precision in a computation does not necessarily increase the accuracy of a computed result. Also, as pointed out in the previous paragraph, the effects of cancellation in a given floating point computation cannot be determined a priori, simply because the input data is unknown. To some extent, one can rely on compilers to eliminate common subexpressions, but as soon as computations become too complex or the input data causes cancellation errors for basic arithmetic operations, these problems can in general not be avoided. In fact, they are inherent to floating point systems as such.

4.1.2. Benefits of Emulated and Mixed Precision Schemes

The simple idea of utilising different precision formats in the same algorithm has surprisingly many beneficial properties, which we discuss in the following.

Accuracy

For many problems and application domains, a mixed precision method can obtain exactly the same final accuracy compared to performing the entire computation exclusively in the corresponding high precision format. For certain algorithms, like the solution of linear systems of equations, the resulting savings are particularly large: It is often possible to perform more than 90 % of all operations in the low precision format, without affecting the final accuracy.

Obviously, mixed precision methods are not applicable in all cases. If a physical field has a sufficient dynamic range or if a system matrix is too ill-conditioned (i. e., the condition number is close to the inverse of the machine epsilon of the low precision format), the low precision format may not be able to represent it accurately enough to capture the (extreme) range of scales. However, we demonstrate that mixed precision methods are always applicable and beneficial for the application domains examined in this thesis (cf. Section 5.6 and Section 5.7). In particular, mixed precision methods yield the correct solution when the use of a low precision format alone breaks down.

Memory

Using low precision rather than high precision allows to store twice as many variables in the same amount of memory, the data block size doubles. Analogously, twice the amount of data can be transferred in the same amount of time over a bus with a fixed width, the effective bandwidth and thus the computational intensity doubles. This applies to all storage levels in the memory hierarchy, from disk and network down to the caches and registers. In view of the memory wall problem, performing the majority of the computation in a low precision format is thus very beneficial. On the other hand, we need to store all data in both the high and low precision format, which increases the memory footprint in off-chip memory.

Computation

The number of transistors required to implement a hardware multiplier grows quadratically with the size of the operands (bit length). Thus, hardware designers can use the same number of transistors required for one high precision multiplier to implement four low precision multipliers. The factor is not exactly four, as there is some overhead associated with each unit, and for floating point numbers the ratio of the exponent to the mantissa bits has additional impact. Also, in hardwired architectures this quadratic advantage cannot be exploited so easily since the data paths are fixed. Therefore, hardwired *dual-mode* FPUs, like the SSE units in current CPUs, have only a linear and not quadratic efficiency, offering twice and not four times as many operations in low than in high precision. Reconfigurable architectures allow direct control over the data paths and hence immediately benefit from the quadratic savings. As a direct consequence of the above, the number of operations required in the emulation of a high precision format with low precision arithmetic grows quadratically with the quotient of the high and low precision number format sizes. For floating point operations even more operations are required, because of the special treatment of the exponent. Finally, doubling the bit length of the operands only results in a linear increase of the size of hardware adders.

4.2. High Precision Emulation

This section summarises the *emulation* of high precision formats, using only native low precision values and arithmetic. Section 4.2.1 explains the differences between exact and redundant emulation, Section 4.2.2 presents algorithms to perform arithmetic in the *double-single* format of unevaluated sums of low-precision values. We discuss the tradeoff between throughput and precision in Section 4.2.3. Section 4.2.4 concludes with related work on GPUs and an outlook on extending the methodology presented in the following to even higher precision.

4.2.1. Exact and Redundant Emulation

For unsigned integers the emulation of higher precision arithmetic with lower precision units is straightforward. An addition of two m bit operands generates a $m+1$ bit, and a multiplication a $2m$ bit result. Integer units typically provide a carry bit that stores the additional bit of the sum, and subsequent operations can include this carry bit in the addition, such that an arbitrary long chain of m bit words can be processed to emulate a km bit format. In case of the multiplication, the hardware either provides the entire $2m$ bit result of a $m \times m$ bit multiplication and thus the product of two km bit operands results from the addition of $\frac{1}{2}k(k+1)$ mixed $m \times m$ bit products; or the multiplier delivers only a m bit result, then each m bit word must be split into two $m/2$ bit words, yielding four times as many mixed products to process.

The emulation of higher precision floating point numbers follows the same lines, but there is an additional difficulty as the bits belonging to the exponent must be treated differently. So while an emulated km bit integer has exactly the same precision as a native km bit integer (*exact emulation*), floating point emulations with the same property are typically very expensive in software due to the treatment of rounding modes and denormalisation etc. Instead, a partially redundant number representation with less precision can be chosen in favour of efficiency (*redundant emulation*). This technique has been introduced by Wolfe [243], Kahan [123], Møller [163], Knuth [133], Dekker [58] and other researchers in the 1960s. In the literature, it is inconsistently referred to as *double-single*, *native-pair arithmetic* or *double-length floating point*. We follow the first nomenclature.

If the hardware natively provides for example an s23e8 format, a higher precision format can be emulated by spreading the mantissa over two native variables and using the exponents to align the two mantissas. In other words, values in the emulated format are represented as unevaluated sums of native values, i. e., a high order term and a low order correction term (*multi-component format*). The resulting format is able to represent 46 bit of significant and is thus approximately equivalent to an s46e8 format. On the other hand, the effective range of the exponent is slightly reduced. In this respect the double-single format is even less precise than an exact s46e8 format and is clearly inferior to the corresponding native high precision s52e11 format that has the same bit length: $52/46 = 1.13$, $2^{11}/2^{8} = 8$. However, the two separate exponents allow to represent some numbers, e. g., $1+2^{-100}$, that are rounded off even in the s52e11 format. So the comparison to an s46e8 format is only approximate.

For the sake of completeness, we also mention *multi-digit formats*. These techniques do not store values as unevaluated sums with separate exponents, but as sequences of digits coupled with a single exponent. This format is slower as it does not exploit native low precision operations. It is widely used in symbolic in contrast to numeric computations.

4.2.2. Double-Single Operations

Arithmetic operations on multi-component low precision floating point values require a careful treatment of overflows and underflows between the high and low order parts, using only low precision arithmetic operations. In this section, we present the algorithms for addition and multiplication as concrete examples. We refer to Bailey et al. [11] for reference implementations of division and transcendental functions. The following algorithms are adapted from their `DSFUN90` package which is freely available and provides valuable information as well as implementational details.

The addition of two double-single values `(c.hi,c.lo) = (a.hi,a.lo) + (b.hi,b.lo)` is straightforward and given in Figure 4.1. The high order parts are added, then the low order parts are added including the error from the high order addition. Finally the overflow from the low order component is included in the high order part. In total, emulated arithmetic for addition requires 11 native operations.

1. Compute high order sum and error:
```
t1 = a.hi + b.hi
e = t1 - a.hi
```
2. Compute low order term, including error and overflows:
```
t2 = ((b.hi - e) + (a.hi - (t1 - e))) + a.lo + b.lo
```
3. Normalise to get final result:
```
c.hi = t1 + t2
c.lo = t2 - (c.hi - t1)
```

Figure 4.1: Double-single addition.

Multiplication is slightly more complicated and the algorithm is given in Figure 4.2. The emulation is significantly cheaper on systems that provide a fused multiply-add ($y = a + b \cdot c$). In a fused multiply-add operation, rounding and normalisation in the underlying floating point system are not performed until the final addition, i. e., the product $b \cdot c$ is not rounded before the accumulation to a. If no fused multiply-add is available, the low and high order products during the multiplication have to be split separately using a suitable multiplicand, e. g., $2^{13}+1$ in the case of s23e8 single precision. 18 native arithmetic operations are required for an emulated precision multiplication, and 32 if no fused multiply-add is available.

4.2.3. Normalisation

Both the addition and the multiplication algorithms perform a normalisation step at the end of the computation. In an effort to reduce the instruction count in the emulation one can skip this step, thus trading performance for precision. Before normalisation, the lower term `t2` in the addition of two numbers a, b has an exponent of at most

$$\text{ExponentOf}(\max(\texttt{a.hi},\texttt{b.hi})) - 23 + 2,$$

assuming an s23e8 native format. If `a.hi` and `b.hi` have the same sign, then `t1` has an exponent of ExponentOf(max(`a.hi`,`b.hi`)) or one higher, so the mantissas are already almost aligned, and one could skip the normalisation. In the worst case this leads to the loss of two bits of significant in a subsequent operation. However, if `a.hi` and `b.hi` have different signs, then the normalisation step is much more important, as the exponent of `t1` can greatly vary. In the worst case `t1` evaluates to zero and the normalisation step is crucial to transfer the mantissa from the lower to the upper term. Multiplications are generally safe, since the high order term `a.hi * b.hi` dominates the result and thus its exponent.

The general effect of fewer normalisations is an increase of the redundancy of the format, because this allows for a double-single c with

$$\text{ExponentOf}(\texttt{c.lo}) > \text{ExponentOf}(\texttt{c.hi}) - 23.$$

1. Compute initial high order approximation and error:
   - If a fused multiply-add is available:
     ```
     c11 = a.hi * b.hi
     c21 = a.hi * b.hi - c11
     ```
   - If no fused multiply-add is available:
     ```
     cona = a.hi * 8193.0
     conb = b.hi * 8193.0
     a1 = cona - (cona - a.hi)
     b1 = conb - (conb - b.hi)
     a2 = a.hi - a1
     b2 = b.hi - b1
     c11 = a.hi * b.hi
     c21 = (((a1 * b1 - c11) + a1 * b2) + a2 * b1) + a2 * b2
     ```
2. Compute high order word of mixed term:
   ```
   c2 = a.hi * b.lo + a.lo * b.hi
   ```
3. Compute (c11, c21) + c2 using Knuth's trick, including low order product:
   ```
   t1 = c11 + c2
   e = t1 - c11
   t2 = ((c2 - e) + (c11 - (t1 - e))) + c21 + a.lo * b.lo
   ```
4. Normalise to get final result:
   ```
   c.hi = t1 + t2
   c.lo = t2 - (c.hi - t1)
   ```

Figure 4.2: Double-single multiplication.

This is just a specific example of the general tradeoff between the latency of operations and the redundancy of the operand representation in arithmetic hardware circuits.

4.2.4. Extensions to Higher Precision and Applicability on GPUs

With native double (s52e11) or extended precision (s64e15) support in current CPUs, emulation techniques are most often used to construct higher precision formats, e. g., by combining two native double precision values into a *quad precision* or even arbitrary precision format [114, 147, 184, 200]. These approaches are applicable if the underlying floating point format satisfies certain minimal prerequisites, e. g., faithful rounding and the existence of guard bits, and we refer to the references for details. In some cases it is possible to trade branches for a few additional native operations, see for example Shewchuk [200] for a rigorous discussion.

As explained in Section 3.2.5 on page 73, native double precision floating point support on GPUs recently became available in late 2007 (AMD products) and June 2008 (NVIDIA products). In parallel to our work [85], researchers have studied the applicability of emulated precision techniques on graphics hardware.

Da Graça and Defour [53] analyse the applicability of double-single emulation techniques and conclude that because of faithful rounding and guard bits, the emulated precision arithmetic is well suited for GPUs. Similar work on complex and quad-precision numbers has been presented by Thall [219], and a residue based number representation with variable precision is discussed by Payne and Hitz [178].

In contrast to emulation, Dale et al. [54] suggest to equip GPUs with dual-mode floating point units, realised through small-scale reconfigurability. Their approach improves double precision performance over emulation but lowers the transistor efficiency for single precision arithmetic, so the decision depends on how often double precision arithmetic is indispensable for the GPU. The development cycle of GPUs is typically three years and the latest chip generations (released

2008) implement double precision arithmetic with dedicated units. Thus, at the time of writing it is unclear if the suggested approach will be available in future hardware.

Even though native double precision arithmetic is available on modern GPUs, its performance is approximately an order of magnitude slower than in single precision. This is similar to the first generation of the Cell BE processor [240]. Therefore, in some cases the intermediate use of emulated precision operations can be beneficial for performance. One specific example is the use of Kahan's summation algorithm [123] for the accumulation step in the computation of dot products, but we do not pursue this topic further in this thesis.

4.3. Mixed Precision Iterative Refinement

Section 4.3.1 provides a concise overview of mixed precision methods, summarising related work. We present and analyse our modified mixed precision scheme specifically tailored for large, sparse linear systems in Section 4.3.2.

4.3.1. Background and Related Work

Iterative refinement methods have been known for more than 100 years already. They gained rapid interest with the arrival of computer systems in the 1940s and 1950s. Wilkinson et al. [32, 155, 239] combined the approach with accumulated inner products as a mechanism to assess and increase the accuracy of computed results for linear system solvers, and provided a solid theoretical foundation of these methods. Shortly afterwards, Moler [162] extended the analysis from fixed point to floating point arithmetic.

The core idea is to use the residual of a computed solution as a right hand side to solve an auxiliary correction problem. This process is iterated, and it is essential that the residuals are computed with higher than working precision (see below). The algorithm in its original form to solve a linear system $\mathbf{Ax} = \mathbf{b}$ for a square $N \times N$ matrix $\mathbf{A}$ is shown in Figure 4.3.

1. Initialise:
   $k = 0, \quad \mathbf{x}^{(0)} = \mathbf{0}$
2. Compute residual in *high precision*: $\mathbf{d}^{(k)} = \mathbf{b} - \mathbf{Ax}^{(k)}$
3. Solve in *low precision*:
   $\mathbf{Ac}^{(k)} = \mathbf{d}^{(k)}$
4. Update in *high precision*:
   $\mathbf{x}^{(k+1)} = \mathbf{x}^{(k)} + \mathbf{c}^{(k)}$
5. Check for convergence, otherwise set $k = k + 1$ and continue with step 2.

Figure 4.3: Mixed precision iterative refinement scheme in its original form.

Wilkinson and his colleagues showed that the computed solution $\mathbf{x}^{(k^\star)}$ after convergence in $k^\star$ iterations is the exact solution of $(\mathbf{A} + \delta\mathbf{A})\mathbf{x} = \mathbf{b}$ where $\delta\mathbf{A}$ is dependent on $\mathbf{b}$ and uniformly bounded. The upper bound m of $\|\delta\mathbf{A}\|_2$ depends on the details of the floating point system used, in particular on the intermediate rounding modes of accumulations. If $m\|\mathbf{A}^{-1}\|_2 \geq 1$, then $\mathbf{A}$ is *'too ill-conditioned'* [155] for the precision of the computation. In this case $\mathbf{A} + \delta\mathbf{A}$ could be singular in the given floating point system and no solution can be obtained. On the other hand, if $m\|\mathbf{A}^{-1}\|_2 = 2^{-p}$ for some $p > 1$, then the successive iterates $\mathbf{x}^{(k)}$ of the refinement process satisfy ($\hat{\mathbf{x}}$ denoting the exact solution)

$$\|\hat{\mathbf{x}} - \mathbf{x}^{(k+1)}\|_2 \leq \frac{2^{-p}}{1 - 2^{-p}} \|\hat{\mathbf{x}} - \mathbf{x}^{(k)}\|_2 \tag{4.3.1}$$

and the refinement procedure converges to working accuracy. Wilkinson and his co-workers also state that usually, the statistical distribution of rounding errors will ensure faster convergence.

We note that the procedure in itself is very general and can in principle be used with any solver, especially iterative ones for large sparse matrices. In their original work, Wilkinson et al. suggest the use of an LU decomposition of the system matrix to solve the auxiliary systems for the varying right hand sides.

Related Work

A number of research efforts have been devoted to improve and extend this method, and in the following, we briefly explain selected approaches that are relevant in the context of this thesis. For a more comprehensive overview, see Demmel et al. [60] and Zielke and Drygalla [248] and the references therein.

Turner and Walker [235] analyse the applicability of the iterative solver GMRES(m) in the context of iterative refinement methods and conclude that they can achieve the same accuracy as a reference double precision solver with roughly the same arithmetic work. The original GMRES(m) solver already involves restarts and is therefore particularly well suited for the interleaving of low precision computations with few high precision corrections. The numerical examples in their publication are based on discretised linear elliptic PDEs and the scheme is implemented in FISHPACK and achieves noteworthy speedup factors.

Geddes and Zheng [78] propose an iterative refinement method based on a linear Newton iteration. They demonstrate their method for a wide range of different problems such as least-squares fits of overdetermined systems and the solution of multivariate nonlinear polynomial equations. For each specialised adaptation of their algorithm, they provide a detailed cost analysis and thorough numerical tests. This work, in particular, demonstrates that the algorithmic mixed precision optimisation can be successfully applied to various problems.

Langou et al. [142] evaluate iterative refinement techniques for dense matrices on a wide variety of modern CPUs such as the Opteron, Itanium, Cray, PowerPC and Cell processors. The speedup is achieved by executing the most expensive operation, namely the LU decomposition, in lower precision. This decomposition can be computed much faster in single than in double precision and dominates the solution time with $\mathcal{O}(N^3)$ arithmetic work. Once the LU decomposition is available, few iterations of the iterative refinement scheme (each requiring only $\mathcal{O}(N^2)$ work) suffice to solve the problem. Adapting results from Stewart [203] the authors prove a convergence condition with an upper bound for the number of required iterations, which depends on the low and high precision bit lengths and the condition number.

More general error analysis of fixed and mixed precision iterative refinement techniques is provided by Stewart [203] and Higham [115].

4.3.2. Mixed Precision Iterative Refinement for Large, Sparse Systems

In the following, we describe a mixed precision iterative refinement algorithm tailored to the solution of large sparse linear equation systems stemming from finite element discretisations. In particular, it is not feasible to apply an LU decomposition or similar direct methods in this context, so the resulting scheme comprises two iterative solvers.

The core idea of the algorithm is to split the solution process into a computationally intensive but less precise inner iteration and a computationally simple but precise outer correction loop. The loops are coupled by a scaling heuristics to enlarge the exponent range locally and consequently the dynamic range of the scheme as a whole. To solve the defect problem, an arbitrary iterative solver executing in low precision can be employed. In our implementation and numerical tests in Section 5.6 and Section 5.7, we make use of conjugate gradient and multigrid solvers.

Another view on this scheme is that the the inner solver is a black-box *preconditioner* ensuring, e. g., the reduction of residuals by a prescribed number of digits for an outer iteration of Richardson-type. Of course, other outer solvers can be employed, for example a multigrid or a BiCGStab iteration. We analyse performance and accuracy of such advanced solvers in Chapter 6.

Let $\varepsilon_{\text{inner}}$ and $\varepsilon_{\text{outer}}$ denote the (relative or absolute) stopping criteria for the inner and outer solver and $\mathbf{Ax} = \mathbf{b}$ the linear equation system to be solved in high precision. Furthermore, α and d^0 are two scalars used for scaling and convergence control. Defect and iteration vectors are denoted by **d** and **c**, respectively. Subscript `low` and `high` indicate the precision of the data vectors. The conversion between the two precision formats is intentionally left abstract, the details depend on the target hardware: The conversion typically involves duplicating the values into a new array with different precision, that might be stored in a different memory address space. In any case, no special operations are performed apart from a regular number format conversion. The template form of the algorithm is given in Figure 4.4.

Input: System matrix $\mathbf{A}_{\text{high}}$, right hand side $\boldsymbol{b}_{\text{high}}$,
convergence parameters $\varepsilon_{\text{inner}}$ and $\varepsilon_{\text{outer}}$
Output: Solution $\mathbf{x}_{\text{high}}$

1. Set initial values and calculate initial defect:
   $\alpha_{\text{high}} = 1.0$
   $\mathbf{x}_{\text{high}} =$ initial guess
   $\mathbf{d}_{\text{high}} = \mathbf{b}_{\text{high}} - \mathbf{A}_{\text{high}}\mathbf{x}_{\text{high}}$
   $d^0_{\text{high}} = \|\mathbf{d}_{\text{high}}\|$
2. Set initial values for inner solver and convert data to low precision:
   $\mathbf{A}_{\text{low}} = \mathbf{A}_{\text{high}}$
   $\mathbf{d}_{\text{low}} = \mathbf{d}_{\text{high}}$
   $\mathbf{c}_{\text{low}} = \mathbf{0}$
   $d^0_{\text{low}} = \|\mathbf{d}_{\text{low}} - \mathbf{A}_{\text{low}}\mathbf{c}_{\text{low}}\|$
3. Iterate inner solver until
   $\|\mathbf{d}_{\text{low}} - \mathbf{A}_{\text{low}}\mathbf{c}_{\text{low}}\| < \varepsilon_{\text{inner}} d^0_{\text{low}}$
4. Update outer solution:
   $\mathbf{x}_{\text{high}} = \mathbf{x}_{\text{high}} + \alpha_{\text{high}}\mathbf{c}_{\text{low}}$
5. Calculate defect and its norm in high precision:
   $\mathbf{d}_{\text{high}} = \mathbf{b}_{\text{high}} - \mathbf{A}_{\text{high}}\mathbf{x}_{\text{high}}$
   $\alpha_{\text{high}} = \|\mathbf{d}_{\text{high}}\|$
6. Check for convergence
   $\alpha_{\text{high}} < \varepsilon_{\text{outer}} d^0_{\text{high}}$
   otherwise scale defect
   $\mathbf{d}_{\text{high}} = \frac{1}{\alpha_{\text{high}}}\mathbf{d}_{\text{high}}$
   and continue with step 2.

Figure 4.4: Mixed precision iterative refinement solver for large, sparse linear systems of equations.

The correctness and convergence of the algorithm given in Figure 4.4 follows trivially, assuming that the same conditions hold as for the more general algorithm in Section 4.3.1: The additional scaling heuristics do not change the argument of the convergence proof, because the scaling changes all relevant quantities by the same amount. More importantly, the convergence condition in Equation 4.3.1 holds even for inexact solutions of the auxiliary problems by an iterative method, as long as the residuals are reduced in the respective norm. Obviously, convergence is not guaranteed if the inner iterative solver does not converge, which is analogous to the inner system being 'not too ill-conditioned'.

In high precision, the only arithmetic work that is performed is a matrix-vector multiplication, a vector update and a norm calculation in each outer iteration. The matrix is assembled once, and this step must be executed in high precision or otherwise, the high precision correction loop does not gain accuracy at all. Consequently, there is no additional cost other than the conversion of the matrix to low precision, which can be performed in an initialisation phase. The additional memory requirements depend on the actual hardware, in most common cases, the memory footprint of the method increases by the quotient of the high and low precision bit length, because all data has to be stored twice, once in high and once in low precision. Otherwise, the bandwidth savings from executing the solver in low precision cannot be exploited. Like every mixed precision scheme, the approach is inapplicable if the defect problem cannot be represented in low precision at all.

As scaling heuristics, we tested both normalising the defect vector (as shown in Figure 4.4) and shifting the exponent by an appropriate power of two, and both work equally well. Experiments revealed that the scaling heuristics is particularly useful when using very aggressive compiler settings (or intrinsics on the GPU[2]) that disable IEEE conformance and thus trade accuracy for speed. In these cases, the unscaled variant of the solver is in some tests prone to diverging or producing NaNs. Based on these experiments, we always use the scaling heuristics in our tests, as its additional computational costs are minimal and can fully be incorporated in other operations, thus not increasing the bandwidth requirements at all.

All stopping criteria for iterative schemes can be used, in particular prescribing a threshold for the (relative) reduction of the norm of the residuals by a fixed number of digits, and executing the inner solver for a fixed number of iterations. The latter has the disadvantage that eventually too many inner iterations, especially in the last step, or too frequent update steps are performed. Which of these drawbacks is more critical depends on the actual hardware, in many cases an update step can be considered costly compared to one iteration of the inner solver. This is particularly relevant for co-processor architectures where the data bus between the different chips is comparatively narrow. Section 5.7 presents numerical experiments in this regard.

Another way to view this scheme is a preconditioned high precision defect correction or Richardson iteration. The preconditioner executes in lower precision than the solver loop. Typically, Richardson defect correction iterations have to be damped. In our experience, this is not necessary; to date we have not found configurations which required damping to ensure convergence.

We finally note that the iterative refinement approach can be cascaded: Instead of solving a given problem expensively in high precision, the solution process is split up recursively into a series of defect equations which are solved in successively decreasing precision. This way, most of the arithmetic work is performed in cheap low precision floating point arithmetic, with the resulting beneficial reduction in bandwidth requirements. The feasibility on this cascade depends on the available hardware accelerated precisions, and the minimal precision necessary to gain relative accuracy reasonably fast on a defect problem of given condition. One concrete example on modern CPUs is the solution of problems that require quadruple precision, which is not currently available natively in hardware. Instead of employing an emulated *quad-double* format for the entire computation, only a few defect calculations need to be performed expensively when employing a mixed precision iterative refinement scheme.

The mixed precision approach can also be formulated for arbitrary convergent iterative schemes, which Robert Strzodka and the author of this thesis briefly examined in 2006 [210].

[2]One example in C for CUDA is the `__sqrt()` intrinsic that is significantly less accurate than the library function `sqrt()`.

4.4. Hardware Efficiency on FPGAs

This section continues the discussion of hardware efficiency started in the introduction of this chapter.

4.4.1. Field Programmable Gate Arrays

Field programmable gate arrays (FPGAs) allow to configure any hardware design and run it as though such a chip had been fabricated. This is very useful for testing future chips, but is also widely exploited to map problem-specific algorithms into a hardware configuration, generating a processor designed specifically for the application in mind. FPGAs are available in a wide variety of possibilities, ranging from PCIe-based boards (equipped with a small microprocessor and additional off-chip memory) to configurations with direct access to a wide system bus, like the FPGA-enhanced Cray XD1 and SGI Altix supercomputers.

Hardwired processors operate on fixed-width formats, and thus, they usually have dedicated single or double precision floating point units. They also have fixed data paths and synchronised scheduling. In contrast, FPGAs have fully configurable memory and logic elements, data paths and clock generators. Since FPGAs consequently allow the configuration of arbitrary number formats, it is possible to cascade the mixed precision approach and use several different precision formats within the FPGA. In particular, it is easy to generate fused operations that apply rounding only to the final result of a composed operation, e. g., for a more accurate accumulation of s23e8 data an s23e8 + s36e11 $\rightarrow$ s36e11 accumulator can be configured. In addition, there are many options to trade latency, throughput and area in operations. For instance, given a bit length n of the operands, an $n \times n$ bit adder can have $\mathcal{O}(1)$ area and $\mathcal{O}(1/n)$ throughput or $\mathcal{O}(n)$ area and $\mathcal{O}(1)$ throughput. For more background and additional information we refer to general surveys of reconfigurable hardware by Bondalapati and Prasanna [31] and Compton and Hauck [51].

4.4.2. Mixed Precision Solvers on FPGAs

During the work on our survey article on mixed, emulated and native precision computations [85], Robert Strzodka and the author of this thesis have also investigated the aforementioned benefits of FPGAs. Our results have been presented at the FCCM 2006 conference [211]. We refer to the original publication for details and only summarise the most important ideas: Using the Poisson problem as a typical yet fundamental model problem, we have shown that most intermediate operations can indeed be computed with single precision or even smaller floating point formats and only very few operations—less than 2 % for the best configuration we established—must be performed in double precision to obtain the same accuracy as a full double precision solver. Consequently, an FPGA can be configured with many parallel single rather than few resource hungry double precision units. One key idea to achieve this has been a fully pipelined version of the conjugate gradient scheme which we developed, see Section 5.7.2. We have evaluated this solver with different iterative refinement schemes, stopping criteria and precision combinations in a wide range of numerical tests. All our experiments have been performed in software using the tool chain from Xilinx due to the lack of actual hardware with sufficient resources for the interesting large problem sizes. This work and the aforementioned research by Langou et al. [142] have later been extended by Sun et al. [216] and in a diploma thesis supervised by the author of this thesis [217].

4.5. Precision and Accuracy Requirements of the Poisson Problem

Throughout this chapter, we have motivated mixed precision methods rather from a 'hardware-oriented' point of view. This section establishes the need for high precision from a numerical perspective, and we demonstrate that single precision is not accurate enough in the context of this thesis. While technically, one counterexample would be sufficient, we gather evidence for a wider range of problems to gain some insight—from a practical perspective—why higher precision is needed to compute accurate results. Therefore, our study is numerical, and a thorough and rigorous theoretical analysis is not the aim of the following. All experiments are based on the Poisson problem, which is a prototypical yet fundamental test problem for all applications considered in this thesis: In the linearised elasticity application (see Section 2.3.5 and Section 6.4), the displacements in horizontal and vertical direction correspond to scalar elliptic second-order PDEs; and similarly for the velocity components in the fluid flow application (see Section 2.3.6 and Section 6.5).

There are potentially many factors that cause single precision to fail. In Section 4.5.1, the influence of the condition number of the linear system stemming from the discretisation is examined. Section 4.5.2 presents an experiment aiming at the crossover point at which floating point errors (accumulation of truncation and cancellation errors in the course of a computation) dominate other errors, most importantly, the discretisation error. Finally, Section 4.5.3 studies the influence of reduced precision on the convergence behaviour of iterative solvers from a practical point of view. We conclude with a summary in Section 4.5.4.

Test Procedure

Unless noted otherwise, all tests are performed on a uniformly refined unitsquare domain, and bilinear quadrilateral conforming finite elements (Q_1) are used for the discretisation. A refinement level L corresponds to $(2^L)^2$ elements and thus $N = (2^L + 1)^2$ degrees of freedom, which are enumerated in a rowwise fashion, yielding a band matrix structure. The tests thus correspond to a single subdomain in FEAST, see Section 2.3. Homogeneous Dirichlet boundary conditions are prescribed along the whole boundary. We consider several analytical functions and their Laplacians to construct Poisson problems with known exact solutions, cf. Section 2.1.3. These problems thus use different right hand side data, but the matrix always remains the same. We measure the relative L_2 error of the computed solution against the analytical reference solution, as explained in Section 2.1.3. A computed solution is considered accurate if the L_2 errors are reduced by a factor of four (proportionally to h^2) per refinement step up to $L = 10$, the finest discretisation we apply for these tests.

The L_2 error is computed in double precision; and to minimise errors in the input data, the matrix and right hand side are always assembled in double precision and casted to single precision as needed. Special care is taken to ensure that the data formats and computations use exactly the specified precision, in an IEEE-754 compliant way: Appropriate compiler flags ensure that intermediate results are always written to memory instead of being kept in potentially higher-precision registers, and optimisations for speed are disabled if they break conformance.

Three different solvers are used for the experiments. Conjugate gradients (as a representative of Krylov subspace schemes) and multigrid, both in single and double precision respectively, are taken from the `libcoproc` library by the author of this thesis, see Section 5.1.2 on page 110. As the accuracy of these iterative solvers depends on a (user-specified) stopping criterion, additionally a direct sparse LU decomposition solver is employed. UMFPACK [57], which is used by FEAST (see Section 2.3.2), unfortunately supports double precision only, so we implemented an interface to SuperLU [59] instead. We experimented with SuperLU's parameters, but found that the default settings (in particular, the AMD column ordering) minimises the fill-in and thus the memory requirements of the computed LU decomposition. For the largest problem size, the employed configuration allocates roughly 6 GB of memory, the 'natural ordering' option requested more than 16 GB of memory, which is the maximum available to us. In short, SuperLU is used as a 'black-box' solver with default settings.

4.5.1. Influence of the Condition Number

For symmetric and positive definite matrices with largest and smallest eigenvalues $\lambda_{\max}$ and $\lambda_{\min}$, the *spectral condition number* is defined as $\text{cond}_2(\mathbf{A}) = \lambda_{\max}/\lambda_{\min}$. An important result from numerical error analysis for a given linear system $\mathbf{Ax} = \mathbf{b}$, with disturbed input data $\delta\mathbf{A}$ and $\delta\mathbf{b}$, gives an upper bound for the influence of the condition number on linear solvers: If the condition number of $\mathbf{A}$ is proportional to 10^s and the relative errors $\|\delta\mathbf{A}\|/\|\mathbf{A}\|$ and $\|\delta\mathbf{b}\|/\|\mathbf{b}\|$ are proportional to 10^{-k} for some $k > s$, then the relative error in the computed solution vector of the system is proportional to 10^{s-k}, i. e., s digits are lost in the solution, or in other words, the data error of order k is amplified by s orders of magnitude.

For small system matrices $\mathbf{A}$ corresponding to moderately coarse levels of refinement, it is feasible to compute the spectral condition number by brute force, for instance using GNU Octave.[3] The following values are obtained for refinement levels $L = 2$ to $L = 6$: 4.92, 13.77, 52.22, 207.66 and 830. Finite element theory states that the condition number of the system matrix stemming from the discretisation of the Poisson problem is proportional to h^{-2}, where h denotes the mesh width [10]. The resulting factor of four per refinement level is clearly visible in the computed values, and we can extrapolate the condition numbers for the refinement levels $L = 7$ to $L = 10$ as follows: 3 320, 13 280, 53 120 and 212 480.

Putting everything together, we can estimate that the reduction from double to single precision, which introduces data errors of the order 10^{-7}, leads to a relative error in the computed solution of the order 10^{-2} for the highest level of refinement. We confirm this estimate with a numerical experiment, using the polynomial function $v(x,y) = 16x(1-x)y(1-y)$ to construct a Poisson problem. We assemble the system matrix and the right hand side in double precision, cast them once to single precision, cast them back to double precision and execute the solver in double precision; in other words, we introduce truncation errors into the input data and leave everything else unchanged. Table 4.1 confirms, only for the larger problem sizes, that the observed relative L_2 errors show differences in the second significant digit, as estimated. We note that the computed results are identical for all three solvers, all differences are on the order of floating point noise.

	double precision		truncated right hand side		truncated right hand side and matrix	
Level	L_2 error	Error reduction	L_2 error	Error reduction	L_2 error	Error reduction
7	6.9380072E-5	4.00	6.9397341E-5	4.00	6.9419428E-5	4.00
8	1.7344901E-5	4.00	1.7362171E-5	4.00	1.7384278E-5	3.99
9	4.3362353E-6	4.00	4.3535279E-6	3.99	4.3757082E-6	3.97
10	1.0841285E-6	4.00	1.1015042E-6	3.95	1.1239630E-6	3.89

Table 4.1: Influence of the condition number when introducing data errors of the order of single precision.

4.5.2. Influence of the Approximation Quality

For our second experiment, we employ the function $v(x,y) = \sin(k\pi x)\sin(k\pi y)$ with an integer parameter k to construct Poisson problems. The idea is that with increasing value of k, the function

[3] http://www.octave.org

is harder and harder to approximate due to its high-frequency oscillating behaviour. Correspondingly, we expect the discretisation error to dominate longer with increasing value of k. For large k, these problems are prototypical for complex problems in real-world scenarios. This experiment is performed using the SuperLU direct solver in both single and double precision.

Level	L_2 error	Red.	L_2 error	Red.	L_2 error	Red.	L_2 error	Red.
	$k=1$, double		$k=1$, single		$k=5$, double		$k=5$, single	
3	1.5175620E-2	3.98	1.5175401E-2	3.98	3.7078432E-1	2.85	3.7078434E-1	2.85
4	3.7994838E-3	3.99	3.8002450E-3	3.99	9.4017531E-2	3.94	9.4017554E-2	3.94
5	9.5022801E-4	4.00	9.5126203E-4	3.99	2.3686629E-2	3.97	2.3686710E-2	3.97
6	2.3757945E-4	4.00	2.4901750E-4	3.82	5.9350336E-3	3.99	5.9353746E-3	3.99
7	5.9396269E-5	4.00	1.2445919E-4	2.00	1.4846263E-3	4.00	1.4864333E-3	3.99
8	1.4849160E-5	4.00	4.2356988E-4	0.29	3.7121131E-4	4.00	3.8315635E-4	3.88
9	3.7123159E-6	4.00	2.1157079E-3	0.20	9.2806257E-5	4.00	1.6205892E-4	2.36
10	9.2816107E-7	4.00	8.8573806E-3	0.24	2.3201782E-5	4.00	3.4972053E-4	0.46
	$k=10$, double		$k=10$, single		$k=20$, double		$k=20$, single	
3	1.0563694E+0	1.11	1.0563694E+0	1.11	1.1687879E+0	0.86	1.1687879E+0	0.86
4	3.7078432E-1	2.85	3.7078434E-1	2.85	1.0563694E+0	1.11	1.0563694E+0	1.11
5	9.4017531E-2	3.94	9.4017551E-2	3.94	3.7078432E-1	2.85	3.7078434E-1	2.85
6	2.3686629E-2	3.97	2.3686768E-2	3.97	9.4017531E-2	3.94	9.4017575E-2	3.94
7	5.9350336E-3	3.99	5.9355176E-3	3.99	2.3686629E-2	3.97	2.3686792E-2	3.97
8	1.4846263E-3	4.00	1.4873870E-3	3.99	5.9350336E-3	3.99	5.9357198E-3	3.99
9	3.7121131E-4	4.00	3.8738947E-4	3.84	1.4846263E-3	4.00	1.4882195E-3	3.99
10	9.2806257E-5	4.00	1.7088372E-4	2.27	3.7121131E-4	4.00	3.8860304E-4	3.83

Table 4.2: L_2 error reduction rates for increasingly oscillating test functions, double and single precision sparse LU solver.

Table 4.2 summarises the computed L_2 errors and error reduction rates when solving in double and single precision, respectively. Looking at the double precision results first, the expected behaviour can be observed: The larger k is chosen, the finer the discretisation has to be to obtain a good approximation of the oscillating analytical functions in the finite element space. If the refinement level is too coarse, the computed solution is far from the exact one and the error reduction rates do not convey any meaningful information: Due to the estimate $\|u-u_h\|_{L_2} \leq ch^2\|u\|_{H_2}$ (u and u_h denoting the exact and computed solutions) and $\|u\|_{H_2} = \mathcal{O}(k^2)$ in this particular case, the L_2 error may even increase again if h^2 is not the dominant factor. Nonetheless, at some point all four configurations exhibit error reduction rates by a factor of four per refinement level. The fact that the oscillating functions ($k \geq 10$) are hard to approximate is further emphasised by the order of magnitude of the final L_2 errors: With a fixed amount of $1\,025^2$ degrees of freedom, the approximation of the 'hardest' function ($k = 20$) is more than three orders of magnitude worse than that of the low-frequency test problem ($k = 1$). The repetitive pattern that can be observed in the L_2 errors is a consequence of the particular test function used: A doubling of k and thus of the 'frequency' corresponds to a halving of the mesh width h. Finally, we note that exactly the same results are obtained with the double precision multigrid and conjugate gradient solvers, using a relative stopping criterion of 10^{-4}.

In single precision, the results are fundamentally different. Looking at the small refinement levels up to $L = 5$ and 6 first, the L_2 errors are reasonably close to their double precision counterparts. For $k = 1$, further refinement of the mesh rapidly deteriorates the computed approximation, and the L_2 error even increases again: The solution on refinement level 10 is even worse than the one on level 4, even though more than one million unknowns instead of only 289 are used: Clearly, the accumulated floating point errors (amplified by the increasing condition number) dominate all other sources of error. The same behaviour can be observed for the increasingly oscillating test functions, but to a lesser extent so that for $k = 20$, the final difference in the L_2 errors of the double and single precision solution are reduced to the unavoidable error due to the condition number: In this case, the discretisation error dominates, and all *accumulated* floating point roundoff and cancellation errors do not significantly influence the result. The loss in accuracy—of the order of differences in approximately the second significant digit of the L_2 errors as estimated in Section 4.5.1—is clearly visible in the single precision result.

Two observations indicate why single precision is increasingly inappropriate the 'easier' the test function can be approximated: First, very 'smooth' functions[4] imply that the discrete values at neighbouring grid points do not differ significantly. This means that the exponent of the floating point representation plays only a minor role, and roundoff errors occur because of the low resolution of the mantissas. In Section 5.6.3 on page 174 we demonstrate that even the combined 46 bit mantissa of the double-single emulated precision format described in Section 4.2 is insufficient. Second, increasing the refinement level even further so that the final approximation quality of the high-frequency test functions is of the same order as for the smooth one would rapidly break down in single precision as well, because of the proportional increase of the condition number. In practice, target quantities prescribed by the actual application (for instance, drag and lift values in fluid dynamics) have to be considered, and in general, a quantitatively accurate simulation result seems very unlikely in single precision. This is particularly important for real-world scenarios, where accuracy requirements are higher than, e. g., an error of 1 % which is often encountered in the engineering community.

Repeating this experiment again with a deformed domain which is not axis-aligned, and introducing mesh anisotropies towards one corner during the refinement (cf. Figure 5.24 on page 158), yields qualitatively identical results, and thus, the unitsquare domain is not a special case. Two different effects come into play: Introducing anisotropies and irregularities in the mesh increases the condition number of the resulting linear system; and a good approximation is harder to achieve because small refinement levels capture the prescribed analytical test function to a lesser degree.

It would be interesting to repeat this experiment using finite elements with approximation order higher than one, because such discretisations allow to achieve more accurate results with the same amount of degrees of freedom. At the time of writing, FEAST does not support higher-order elements and therefore, this is not possible without major re-implementations.

4.5.3. Influence on the Convergence Behaviour of Iterative Solvers

After establishing that single precision is insufficient for accurately solving linear systems stemming from the discretisation of Poisson problems with finite elements, we now focus on the practical aspects and consequences arising from the use of iterative solvers. All subsequent experiments are based on the polynomial test problem introduced in Section 4.5.1. Qualitatively, this function resembles the sinusoidal function for $k = 1$ in Section 4.5.2, which cannot be solved in single precision up to the finest refinement level: Table 4.3 depicts results obtained with the SuperLU solver in single and double precision.

In the following, two different iterative solvers are employed, conjugate gradients as a representative of Krylov subspace schemes, and multigrid. We empirically determine in double precision that the linear solver needs to reduce the initial residuals by at least seven digits in order to exhibit L_2 error reductions by a factor of four up to the largest problem size (refinement level $L = 10$), so one could expect single precision to exhibit even more problems than in the previous tests, because this threshold is close to the limitations of the s23e8 format. As the results using the sparse direct solver show, this seemingly demanding stopping criterion is not the dominant reason

[4]Smooth is set in quotation marks here because all test functions under consideration are in C^∞ and therefore arbitrarily smooth in the mathematical sense.

Level	Double precision L_2 error	Double precision Error reduction	Single precision L_2 error	Single precision Error reduction
3	1.7802585E-2	4.03	1.7802457E-2	4.03
4	4.4429149E-3	4.01	4.4435526E-3	4.01
5	1.1102359E-3	4.00	1.1111655E-3	4.00
6	2.7752803E-4	4.00	2.8704684E-4	3.87
7	6.9380072E-5	4.00	1.2881795E-4	2.23
8	1.7344901E-5	4.00	4.2133101E-4	0.31
9	4.3362353E-6	4.00	2.1034461E-3	0.20
10	1.0841285E-6	4.00	8.8208778E-3	0.24

Table 4.3: Polynomial test problem in double and single precision (relative L_2 errors, direct solver).

for the behaviour described in the following. The maximum admissible number of iterations for the conjugate gradient solver is set to N, the number of unknowns in the system, and to 32 for the multigrid solver.

Conjugate Gradients

In the first experiment, we use a conjugate gradient solver with Jacobi preconditioner. The results are summarised in Table 4.4: In double precision, the solver converges very well, and the iteration number increases proportionally to the mesh width h (which in turn is antiproportional to the square root of the condition number). The L_2 error reduces by a factor of four per refinement level. This is not the case when computing entirely in single precision instead: From refinement level 7 onwards, the single precision solver starts to need increasingly more iterations until it claims convergence: The solver does not exhibit a doubling of iterations in each refinement step any more. We say that the solver claims convergence because the computed results are simply wrong: Again from $L = 7$ onwards, the L_2 error reduces increasingly slower, until for the largest problem size, it even increases again: The accuracy of the solution on level 10 is between that on level 8 and 9, although four or even 16 times as many unknowns are used. Remarkably, the iterative solver yields correct results for larger levels of refinement than the direct solver. We suspect stability differences between the solvers, but emphasise that only one relevant observation can be made: As soon as the solution starts to be wrong due to floating point errors, these errors build up rapidly, and the L_2 errors obtained on even finer meshes are no longer comparable between solvers.

Level	Double precision Iterations	Double precision L_2 error	Double precision Error reduction	Single precision Iterations	Single precision L_2 error	Single precision Error reduction
3	9	1.78025845E-2	4.03	8	1.78026922E-2	4.03
4	18	4.44291488E-3	4.01	18	4.44304533E-3	4.01
5	34	1.11023592E-3	4.00	34	1.11055951E-3	4.00
6	68	2.77528031E-4	4.00	80	2.78169002E-4	3.99
7	136	6.93800722E-5	4.00	171	7.07162815E-5	3.93
8	273	1.73449005E-5	4.00	358	1.94255938E-5	3.64
9	551	4.33623132E-6	4.00	884	1.01070553E-5	1.92
10	1113	1.08411274E-6	4.00	2250	1.49700372E-5	0.68

Table 4.4: Polynomial test problem in double and single precision (relative L_2 errors, conjugate gradient solver).

Looking at the computed results in more detail reveals that the solver indeed converges to a wrong solution: This is easily determined (in addition to the deteriorating L_2 errors) by computing the exact defect after the solver has claimed convergence: For the largest problem size in this test, the solver reports a final norm of the defect of order 10^{-8}, and the (independently computed) norm of the defect (in single precision, using the computed result) is four orders of magnitude higher. An obvious 'scapegoat' for this behaviour and the observed disproportionate increase in iteration numbers lies in the construction of the conjugate gradient algorithm itself: In each iteration, the defect (used for convergence control) is computed incrementally via the recurrences determining the (locally) optimal search directions, cf. Figure 2.2 on page 25. One could believe that due to the reduced precision, the solver needs more iterations until convergence because the computed search directions are no longer $\mathbf{A}$-orthogonal, whereas in fact the computed results are simply wrong. To pinpoint this suspect, we modify the implementation to compute the defect directly and not incrementally in each iteration. As a result, the solver now diverges except for $L = 2$ and even computes `NaN` for $L \geq 5$. This behaviour is surprising, since many textbooks recommend computing the defect directly every other iteration to prevent floating point errors from building up to a level where they dominate.[5] A detailed analysis reveals that the convergence control is mostly responsible for this behaviour. Figure 4.5 depicts the progress of the norms of residuals after each step of the conjugate gradient solver for the first 250 solver iterations and refinement levels $L = 3, \dots, 6$ (the solver is terminated after reaching N iterations).

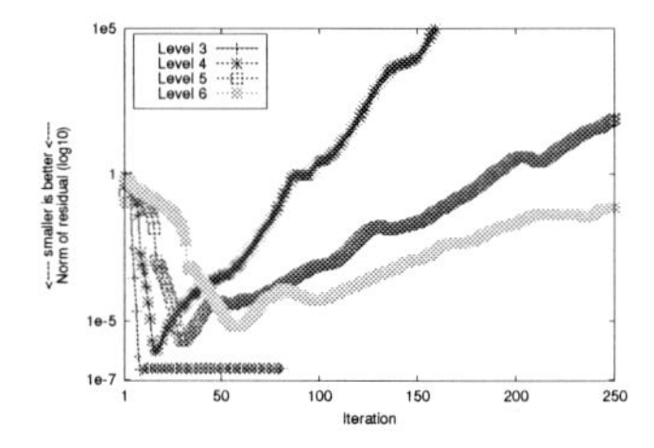

Figure 4.5: Progress of residuals in the conjugate gradient solver with direct defect calculation (single precision).

At some point before reaching the stopping criterion, floating point errors start to build up rapidly and lead to divergence. Looking at the convergence development in more detail, an important observation can be made: *In single precision, the minimal achievable norm of residuals increases with the problem size.*[6] This is inherent to the amplification of floating point errors by the increasing condition number of the systems, and consequently means that any advanced convergence control mechanism is bound to fail in a similar fashion: It does not make sense to terminate a solver once intermediate residual norms drop below, e. g., 10^{-7} or some corresponding (absolute) value which seems to be close to the limitations of single precision. For the modified solver, such a stopping criterion would never hold, and for the unmodified one, we demonstrate below that this is equivalent to not iterating the solvers until convergence. Analogously, it makes no sense to terminate a solver once the norms of the residuals do not reduce further, as the algorithm does not necessarily converge monotonically in floating point. Such convergence control mechanisms essentially boil down to gaining only two to four digits depending on the problem size, and consequently, the targeted reduction in the L_2 error norm cannot be achieved: The solution computed

[5] Computing the defect directly (in single precision) after every, say, 50th iteration does actually prevent divergence, but the solver still converges to a wrong solution.

[6] This observation also explains why the highly oscillating test problem did not have problems with single precision for the levels of refinement under consideration: The linear solver needed a comparatively weak stopping criterion of only four digits, and the impact of single precision did not show clearly yet due to the dominance of the discretisation error.

by the (prematurely terminated) iterative scheme is simply not accurate enough, because the (true) residuals cannot be reduced enough (cf. Figure 4.5). We nonetheless confirm this observation by empirically determining the iteration number that leads to the (absolutely) smallest residual and configuring the algorithm to perform exactly this amount of iterations, effectively disabling convergence control: The exact numbers are not shown in detail, we restrict ourselves to exemplary pointing out that the expected error reduction is no longer achieved from refinement level 6 onwards, identical to the results shown in Table 4.4.

	Double precision			Single precision		
Level	Iterations	L_2 error	Error reduction	Iterations	L_2 error	Error reduction
3	5	1.78025923E-2	4.03	5	1.78027112E-2	4.03
4	12	4.44291510E-3	4.01	12	4.44311540E-3	4.01
5	26	1.11023594E-3	4.00	26	1.11034233E-3	4.00
6	52	2.77528351E-4	4.00	52	3.14905825E-4	3.53
7	103	6.93808325E-5	4.00	93	7.95717888E-5	3.96
8	202	1.73450374E-5	4.00	213	6.01743367E-5	1.32
9	391	4.33717982E-6	4.00			
10	767	1.08396607E-6	4.00			

Table 4.5: Polynomial test problem in double and single precision (relative L_2 errors, BiCGStab solver.

To rule out eventual side effects due to the loss of orthogonality, we repeat the test corresponding to Table 4.4, but this time employ a BiCGStab solver. The results are shown in Table 4.5. This solver is in general more prone to suffer from floating point noise, and thus, we do not observe the expected halving of iterations in double precision. In single precision and for the two largest refinement levels, it even diverges, and for smaller levels of refinement, the iteration numbers are the same as in double precision.

Multigrid

We now repeat the previous experiments employing a multigrid solver. Table 4.6 lists the iteration numbers and relative L_2 errors obtained when executing the solver either in single or in double precision data and arithmetic.

	Double precision			Single precision		
Level	Iterations	L_2 error	Error reduction	Iterations	L_2 error	Error reduction
3	5	1.7802587E-2	4.03	32	1.7802623E-2	4.03
4	5	4.4429156E-3	4.01	32	4.4430374E-3	4.01
5	5	1.1102361E-3	4.00	32	1.1110643E-3	4.00
6	5	2.7752815E-4	4.00	32	2.8134801E-4	3.95
7	5	6.9379565E-5	4.00	32	6.2351369E-5	4.51
8	5	1.7344000E-5	4.00	32	4.5337956E-5	1.38
9	5	4.3351293E-6	4.00	32	2.6890305E-5	1.69
10	5	1.0829283E-6	4.00	32	5.7035827E-5	0.47

Table 4.6: Multigrid solver in double and single precision (relative L_2 errors). The maximum admissible number of iterations is fixed to 32.

Two important observations can be made: In double precision, the solver converges very well, independent of the refinement level, five iterations are performed. In single precision however, the solver fails to converge and always reaches the prescribed maximum number of iterations. Second, the results computed in double precision are accurate in terms of the L_2 norm, which is reduced by a factor of four per refinement step. This is not true in single precision: For refinement levels

3–6, the computed results are accurate even though convergence control is not possible any more. For refinement level 7, the error drops more than expected. We suspect a favourable cancellation of error components, as the convergence control is not able to capture this effect. For the three largest problem sizes, the error decreases much slower and increases for $L = 10$, so that the solution computed on refinement level 10 is only as accurate as the one computed on refinement level 7, even though 64 times as many unknowns are used.

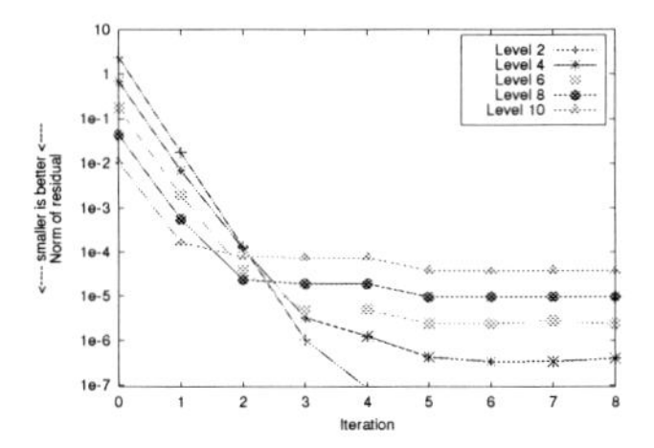

Figure 4.6: Progress of residuals in single precision after each multigrid cycle, only for every second level to improve clarity.

Figure 4.6 examines the convergence behaviour of the single precision solvers in more detail by plotting the norm of the residual after each multigrid cycle. We immediately see why the convergence control fails: The solvers stall, and more iterations do not reduce the norms of the residuals any further. In some sense, these results are better than those observed using the modified conjugate gradient algorithm (cf. previous section): The solver stalls instead of diverging and ultimately computing `NaN`. The plot does not include residuals after the eighth iteration to improve clarity, but we confirm that the tendency of stagnating continues for the remaining iterations. We conclude that neither fixing the solver to perform only five iterations nor iterating further would improve the achievable accuracy, in accordance with the conclusions drawn from the convergence behaviour of the (modified) conjugate gradient solver. The plots also demonstrate that gaining, e. g., only six digits does not alleviate the problem on coarser refinement levels: The solver stalls before the weakened convergence criterion holds. This is further emphasised by the convergence behaviour of the multigrid solver on the sinusoidal, highly oscillating test problems examined in Section 4.5.2. For these problems, reducing the initial residuals by four instead of seven digits is sufficient to observe the error reduction rates up to the finest refinement levels, but except for the smallest problem size, all solvers reach the maximum number of iterations: Convergence control fails even with this comparatively weak stopping criterion.

The norms of the residuals in Figure 4.6 also indicate why the L_2 errors increase again for higher levels of refinement: Comparing for example the plots for refinement level 4 and 10, we see that the initial residual of the level 10 solution is two orders of magnitude better than the one on level 4, whereas the final residual is two orders of magnitude worse. This is the same behaviour as observed for the conjugate gradient scheme, and explains again why the $k = 20$ case in Section 4.5.2 converges well.

As multigrid is our favourite solver for the types of problems in the scope of this thesis because of its independence of the mesh width h, we analyse its components separately, i. e., smoothing (in this case, Jacobi smoothing implemented using divisions), grid transfers (implemented as fixed-coefficient interpolation, the divisions use power-of-two denominators and are therefore exact) and the coarse grid solver (in this case, the preconditioned conjugate gradient solver without the modification, analysed previously): We first increase the amount of transfers and coarse grid corrections by executing the solver not in a V cycle configuration, but employ F and W cycles instead. The results (not shown) are qualitatively and quantitatively very close to the V cycle, and no conclusion can be drawn. Storing the inverse of the diagonal (computed in double precision and casted to single precision) in a separate array prior to solving and thus avoiding the division in the Jacobi smoother neither changes the observed convergence behaviour nor the accuracy of the solution. Next, we reduce the mesh hierarchy for the solver, by starting the coarse grid correction not on refinement level 1 but on finer grids. To focus on the coarse grid solver alone, we prescribe a maximum number of five iterations (performing more iterations does not change the results qualitatively and quantitatively). The coarse grid solver is configured to gain seven digits, and we empirically determine its optimal number of iterations as explained above. To gain further insight, only the coarse grid solver is executed in double precision in a second experiment, for which the modifications to the coarse grid solver are not necessary.

	coarse grid single		coarse grid double	
Number of multigrid levels	rel. L_2 error	rel. error	rel. L_2 error	rel. error
10	5.8619020E-5	53.1	7.26021692E-5	66.0
9	6.5100979E-5	59.1	6.35145511E-5	57.6
8	1.1376739E-4	104.1	6.25543553E-5	56.7
7	9.7346791E-5	88.9	4.96629621E-5	44.8
6	8.3395229E-5	76.0	5.66366511E-5	51.2
5	6.2642598E-5	56.9	6.08495555E-5	55.1
4	5.9498448E-5	53.9	5.37317057E-5	48.6
3	5.0064723E-5	45.2	6.77214048E-5	61.5
2	3.7507415E-4	345.4	6.18298783E-5	56.0

Table 4.7: Influence of the multigrid hierarchy and the coarse grid solver ($1\,024^2$ elements, relative L_2 errors and relative error compared to the double precision L_2 errors, single precision, conjugate gradient coarse grid solver either in single or in double precision).

Table 4.7 lists the results, only for the largest problem size (i. e., all rows in the table correspond to the same mesh with $1\,024^2$ elements). The columns labeled 'rel. error' contain the relative error of the single precision L_2 error against the one obtained in double precision. When executing the coarse grid solver in single precision, the results are inconclusive and do not allow to pinpoint one particular component of the multigrid algorithm. The depth of the multigrid hierarchy does not change the results significantly, with the exception of the two-grid case (last row in the table). This artefact does however not appear so strongly when using the SuperLU solver instead of conjugate gradients (not shown). The results are qualitatively identical, because both coarse grid solvers converge to wrong solutions. Interestingly enough, executing the coarse grid solver in double precision (and hence, accurately) still gives wrong results, independently of the depth of the mesh hierarchy. We can conclude that only the problem on the finest level is responsible for the behaviour we observe. In a way, this is the expected behaviour, because this problem determines the ill-conditioning of the system matrix. In summary, we can say that neither of the components (smoothing, coarse grid solver, depth of the mesh hierarchy) in the multigrid solver is responsible for giving wrong results, only the arithmetic work performed on the finest level of refinement causes the behaviour we observe. This experiment also confirms why single precision gives wrong results independently of the choice of the solver, all problems with iterative solvers that we have observed and described in this section are of practical nature, and none of these problems is responsible for generating wrong results.

4.5.4. Summary and Conclusions

All numerical experiments presented in this section are based on the solution of linear systems stemming from the discretisation of the prototypical Poisson problem, *the* fundamental model problem in the scope of this thesis. In all tests, computing in double precision gives the correct result for the considered levels of refinement (which of course does not imply that double precision is always sufficient for problems like the ones we examine), and can thus be used as a reference solution.

The condition number of the system matrix of the linear system stemming from the discretisation with finite elements amplifies data errors in the input, as well as roundoff and truncation errors in the course of the solution process. For Poisson and similar (elliptic Poisson-like) problems, the condition number of the system matrix increases proportionally to the refinement level of the discretisation, which is convenient in the experimental analysis of its impact. For a simple unitsquare domain, we have demonstrated this effect in a numerical experiment based on extrapolated condition numbers and established a loss of accuracy corresponding to two digits for one million elements, the largest problem size we consider, only by truncating the mantissa of the input data and leaving the computational precision unchanged. This result alone is technically sufficient to answer the question if single precision is sufficiently accurate to solve Poisson problems using finite element discretisations: More realistic cases involving larger problem sizes and anisotropies introduced by either the differential operator or the underlying mesh, generally increase the condition number further.

A detailed analysis of increasingly oscillating analytical test functions revealed that for functions that are hard to approximate in the finite element space, the discretisation error dominates accumulated and amplified floating point errors for higher levels of refinement, and single precision can be successfully applied for moderately large problem sizes. For the highest-frequency test function we evaluated, a mesh resolution of 513^2 grid points is required to reach stable convergence in the finite element approximation, i. e., a reduction in the L_2 norm by a factor of four per refinement step. For this function, the discretisation error dominates on all coarser refinement levels. However, at this refinement level, the impact of the condition number (which depends solely on the matrix) becomes visible, and single precision is not accurate enough.

Neither SuperLU as a representative of (sparse) direct solvers, nor the Krylov subspace conjugate gradient scheme, nor multigrid are able to give the correct answers as soon as floating point errors dominate in the course of the solution process. Except for the highly oscillating sinusoidal test function (Table 4.2, $k = 20$), no configuration gives a reasonably exact result in single precision up to the finest levels of refinement we consider, i. e., for refinement levels at which the unavoidable impact of the condition number starts to influence the computed results. The quantitative crossover point (in terms of problem size) from which on floating point errors dominate depends on the actual choice of the solver and its configuration parameters, but this observation does not allow any meaningful conclusion: As soon as floating point errors or the condition number dominate, further refining the mesh only amplifies errors and hence obscures any effect that might or might not allow hypotheses on solver behaviour.

SuperLU silently computes a solution that is wrong. We have examined the two iterative schemes in more detail and identified the failing convergence control to cause most of the problems in practice: The conjugate gradient scheme converges to a wrong solution or diverges, depending on the way it is implemented. One could say that multigrid behaves 'better' than the conjugate gradient algorithm, because it simply stalls. This, however, might not necessarily be true in all cases, in particular for very ill-conditioned problems, so one should not rely on this observation. We discussed a number of residual-based stopping criteria for these solvers, but—as the failing direct solver demonstrates and the argument in the previous paragraph underline—an improved (residual-based) convergence control mechanism would not alleviate the problems once single precision is insufficient and floating point errors build up rapidly.

One could argue that some of our test problems are ill-posed in single precision, because the grid resolution is too fine in the given floating point system and does not resolve additional frequency components any more: Colloquially speaking, the solvers then only compute on 'noisy' data. This is in so far correct as we always emphasise that only the crossover point allows any meaningful conclusion, but does not affect our argument that single precision fails for coarser levels of refinement than double precision. The theoretical error reduction rates we employ as a measurement have been proved assuming exact arithmetic, and for all tests we examine, double precision is 'exact enough', so to speak.

The experiments we presented underline why most practitioners employ double precision exclusively: Solver behaviour is simply much more robust in double precision. The additional advantage of this approach is that it requires no further analysis: The fact that single precision results may be arbitrarily wrong is irrelevant in this case.

In view of the (hardware-motivated) mixed precision iterative refinement schemes for iterative solvers that constitute an important contribution of this thesis, our experimental analysis yields a promising result: For all problems we examined, single precision is accurate enough to compute a comparatively 'good' approximate solution. Both the conjugate gradient and multigrid solvers seem to be sufficiently robust to compute an approximation via reducing the initial residuals by only one or two digits. This approximation can be used in the iterative refinement procedure (cf. Section 4.3) to successively advance the solution to high precision and thus, high accuracy. In Section 5.6 and Section 5.7, mixed precision iterative refinement schemes are experimentally analysed in detail.

Future Work

In practice, due to the failing convergence control that essentially prevents low-precision solvers of 'black-box type', computing error bounds of the results seems mandatory. Demmel et al. [60] suggest to use iterative refinement techniques (cf. Section 4.3) to achieve reliable error bounds at comparatively small computational cost. Another promising avenue for future research could investigate the use of stopping criteria that are not based on residual evaluations, but founded more in finite element approximation theory [7]. This, however, significantly increases the computational cost of convergence control, because residuals are readily available while derived quantities computed by (accurate) integration of some functional over the entire domain are not. In the following, we do not pursue these ideas further, as one main goal of this thesis is the evaluation of mixed precision methods, see Section 5.6 (tests on a single subdomain), Section 6.3.2 (large-scale Poisson problems), Section 6.4.2 (large-scale linearised elasticity) and Section 6.5.3 (stationary laminar flow) for numerical tests targeting the achievable accuracy. Furthermore, higher-order finite element discretisations will be interesting to evaluate, because they allow to achieve the same accuracy with a (much) coarser mesh.

5

Efficient Local Solvers on Graphics Hardware

In this chapter, we describe, analyse and evaluate our GPU implementation of efficient linear solvers specifically tailored to systems resulting from the discretisation of scalar PDEs on generalised tensor product grids. These meshes correspond to individual subdomains in FEAST, and furthermore, the solvers we present and evaluate in this chapter constitute the local solvers (Schwarz preconditioners) of the two-layer SCARC approach in FEAST. This chapter is thus dedicated to the single most important building block of the SCARC scheme in the scope of this thesis, and the thorough experiments we present lay the groundwork for the acceleration of full applications we discuss in Chapter 6. Moreover, we are interested in gaining insights into GPU performance characteristics for iterative linear solvers and sparse equation systems in contrast to conventional CPU implementations, the various tradeoffs in parallelism and the performance improvements of GPUs over CPUs. We use numerical experiments to evaluate the mixed and emulated precision techniques developed in the previous chapter in detail, both in terms of numerical and runtime performance: We demonstrate that these techniques can deliver the same accuracy as standard high precision implementations, while at the same time delivering significant performance improvements, on both CPUs and GPUs.

Section 5.1 outlines the structure of the entire chapter and contains a list of recommendations for the reader who wants to focus on particular sub-topics.

5.1. Introduction and Overview

As this chapter is rather long, Section 5.1.1 summarises its structure to make the presentation more accessible to the reader. It also provides a general outline and highlights the topics we cover in this chapter.

5.1.1. Chapter Outline

This chapter is roughly divided into two parts. We start with a high-level overview of our implementation in Section 5.1.2 and discuss various design decisions from a software engineering point of view. In Section 5.2 and 5.3 we describe the efficient implementation of numerical linear algebra operations and multigrid solver components, both in terms of 'legacy GPGPU' programming techniques based on graphics APIs and for the CUDA architecture, the most important environment for 'modern' GPU programming at the time of writing. Section 5.4 is then devoted to the efficient parallelisation of preconditioners and multigrid smoothers specifically tailored to systems resulting from generalised tensor product patches. It also presents our implementation of the cyclic reduction algorithm for the solution of tridiagonal systems of equations, that outperforms previous implementations significantly. Furthermore, we develop an alternative multicolouring-based parallel Gauß-Seidel smoother that can yield competitive convergence rates compared to the natural-order sequential standard approach on CPUs. Each of these sections concludes with detailed microbenchmarks of the individual operations on CPUs and GPUs, across a wide range of different hardware generations.

The second part of this chapter comprises a series of numerical studies, demonstrating the performance, efficiency and accuracy of native, emulated and mixed precision linear solvers, executing on the CPU, the GPU, or hybridly on both. In Section 5.5 we analyse the numerical and runtime efficiency of the parallelised smoothers and preconditioners developed in Section 5.4. In particular, we are interested in how much efficiency is lost by partially relaxing the coupling between the unknowns via mesh colouring, our strategy to recover parallelism for inherently sequential Gauß-Seidel type preconditioners. Section 5.6 demonstrates that even for very ill-conditioned problems, the mixed precision iterative refinement method computes identical results compared to solvers executing entirely in double precision. In Section 5.7 we examine if the mixed precision solvers only achieve the same accuracy at the cost of a loss in runtime efficiency. Most of the tests in these sections are focused on multigrid solvers, but we always discuss implications for solvers of Krylov subspace type. Finally, Section 5.8 examines the remaining free algorithmic parameter in multigrid, the choice of the cycle type, and, closely related, the configuration of the coarse grid solver.

Due to the variety of the topics covered in this chapter, we deliberately do not present a summary of the entire chapter. Instead, each section provides its own motivation, and its own self-contained summary. The casual reader who is mainly interested in our general findings is thus referred to these introductions and summaries. Section 5.9 lists, as a reference for the interested reader, GPU-related performance guidelines which we develop throughout this chapter. In the following, we briefly provide pointers to the contents of this chapter, but instead of a linear outline, we focus on the different topics that are covered.

Purely Numerical Results

Admittedly, the majority of the tests presented in this chapter focus on runtime efficiency improvements of GPUs over conventional CPUs, simply because this constitutes the main line of argument of this thesis. Nonetheless, we derive a number of interesting results from our experiments: In Section 5.5.2 we discuss the impact of the decoupling of dependencies induced by the parallelisation of inherently sequential preconditioners of Gauß-Seidel type via mesh colouring.

In Section 5.8.2 (double precision) and Section 5.8.8 (single precision) we identify a fundamental problem when using Krylov subspace type schemes as coarse grid solvers in multigrid cycles, a problem that only occurs in finite floating point number systems and not in theory. We present several techniques to alleviate this problem in practice.

Numerical results based on our mixed precision iterative refinement solvers specifically tailored to sparse linear systems are outlined in the next paragraph.

Mixed Precision Iterative Refinement

In the literature, mixed precision iterative refinement techniques are analysed theoretically mainly for dense matrices and solvers that employ some kind of a precomputed decomposition of the matrix, e. g., LU decompositions, to be able to solve the same system for varying right hand sides. In our setting of large, *sparse* linear systems of equations, this approach is not feasible, and we perform numerical experiments to gain insight into the convergence behaviour of these schemes when the low precision update is computed by iterative solvers of multigrid or Krylov subspace type.

Generally speaking, the mixed precision iterative refinement procedure is guaranteed to converge if the auxiliary problems to be solved are 'not too ill-conditioned' in low precision, cf. Section 4.3 on page 90. In Section 5.6 we demonstrate that mixed precision iterative refinement solvers using iterative solvers for the low precision update deliver exactly the same accuracy in the result as if executing the entire computation in high precision, even for extremely ill-conditioned problems, i. e., problems that can barely be assembled in double precision.

In terms of efficiency, the user-defined stopping criterion of the 'inner' iterative solvers is crucial. The 'optimal' choice of this parameter depends on the hardware under consideration, and the concrete performance characteristics. Nonetheless, we identify a setting that yields 'black-box' character for all configurations we evaluate. Detailed measurements can be found in Section 5.7 and Section 5.8.8.

GPU Implementation Techniques

Implementational aspects of multigrid and Krylov subspace solvers on GPUs are presented in Section 5.2, 5.3 and 5.4. The first two sections focus on fundamental operations from numerical linear algebra, operations that we use to construct full solvers. We choose four representative operations that cover four of the five most important parallel programming primitives, *map*, *reduce*, *gather* and *scatter*. Many of the implementational details are typically not covered in the literature, and hence, we describe them in the hope that the reader may find them useful as a starting point to working in the emerging field of GPU computing. Section 5.2 is concerned with 'legacy GPGPU' techniques using graphics APIs to program the device, and Section 5.3 presents CUDA implementations. We want to emphasise that the techniques we apply in our CUDA implementation are equally valid when using the OpenCL programming environment, as the underlying abstract model of the hardware is identical: In essence, a given data-parallel computation is partitioned into independent groups of threads (*thread block* or *cooperative thread array (CTA)* in CUDA speak, *work group* in OpenCL speak), threads in each group can communicate via lightweight barrier synchronisation primitives and local, shared scratchpad memory, and groups cannot communicate at the scope of the kernel launch but have to use global memory to transfer results from one kernel launch to the next.

All these operations exhibit a high degree of independent data-parallelism (potentially only after suitable reformulations of the mathematical algorithm), which makes them well-suited for throughput-oriented parallel processors, at least in theory. Nonetheless, it is a challenging task to devise an efficient mapping to the GPU architecture that achieves good performance.

Robust and 'strong' preconditioners and smoothers exhibit an inherently sequential coupling between the unknowns. We present several techniques to recover parallelism for these operations in Section 5.4. For Gauß-Seidel type preconditioners, we apply mesh colouring techniques that relax the coupling between the unknowns. Tridiagonal systems are solved with cyclic reduction, parallel cyclic reduction and recursive doubling techniques. Our cyclic reduction implementation significantly outperforms previous implementations reported in the literature. Here, we also introduce the remaining fifth important parallel programming primitive, the *scan (parallel prefix sum)* operation.

GPU Performance Characteristics and Speedups of GPUs over CPUs

GPU performance characteristics and the speedups we achieve over conventional cache-optimised CPU implementations are covered in all our *microbenchmarks* and *solver efficiency studies*, and thus in fact in each section of this chapter. We present three different sets of microbenchmarks, in Section 5.2.4 (legacy GPGPU implementations of numerical linear algebra operations), Section 5.3.3 (CUDA implementations thereof) and Section 5.4.7 (preconditioners specifically tailored to generalised tensor product grids). Similarly, efficiency studies of multigrid and Krylov subspace solvers are covered in Section 5.5.3 (preconditioners), Section 5.7.3 and 5.7.4 (mixed precision iterative refinement solvers on GPUs and hybridly on GPUs and CPUs), and in Sections 5.8.3–5.8.8 (multigrid cycle type and coarse grid solver).

We embrace many different aspects in our experiments, the most important ones are: The spatial locality of memory references is much more important on GPUs than on CPUs, simply because GPUs do not employ hierarchies of large caches to counteract the memory wall problem. GPUs are throughput-oriented while CPUs minimise latencies of a single instruction stream, which means that GPUs only excel for sufficiently large problem sizes; in other words, the device has to be reasonably saturated in order to achieve good efficiency. In the scope of multigrid solvers, this is particularly relevant: While this behaviour is also observed in the microbenchmarks, F and W cycle multigrid that perform a lot of work on small problem sizes suffer disproportionately. On the CPU, the work on fine levels of the mesh hierarchy that do not fit into cache anymore dominates the runtime, and all work on coarser levels is essentially 'for free'. This heuristic assumption (backed up by experimental data) is not valid on GPUs, see also Section 5.9 for a summary of GPU performance guidelines.

GPU Performance Improvements Across Hardware Generations, Energy Efficiency

A common claim by GPU manufacturers is that GPUs are 'getting faster' at a higher pace than CPUs. Similarly, GPUs are said to be more efficient than conventional CPUs in terms of performance per Watt and performance per Euro. Due to the lack of proper measuring equipment, we only briefly discuss these topics based on reasonable estimates in Section 5.2.4 (OpenGL 'legacy GPGPU') and Section 5.3.3 (CUDA 'modern GPGPU'). In these sections, we also examine the increase in performance over previous hardware generations.

Publications

The first publication by the author of this thesis on revisiting the well-known mixed precision iterative refinement procedure not in terms of increased accuracy but in terms of performance improvements [82] predates the independent work of Langou et al. [142] which received a lot of attention in the community. For the most part, this chapter constitutes a significantly updated and revised version of previously published work, and we list the relevant publications in the introduction of each section.

5.1.2. A Brief Overview of the Code Base

Prior to writing this thesis, the author decided, for a number of reasons, to reorganise his collection of test codes into one common, unified code base which incorporates all improvements made to the algorithms and implementations in the course of the pursuit of this thesis. This code base provides CPU and GPU implementations of numerical linear algebra building blocks, and iterative solvers for sparse linear systems. It is realised in the form of a static library called `libcoproc`, which can either be linked into FEAST or used for 'standalone' tests. In particular, all experimental codes used in earlier publications [79, 82, 85] are now available only from within this framework.

We use the nomenclature of *targets*, short for target hardware, or synonymously *backends*, for the individual implementations. Section 5.2.3 describes the implementation of the OpenGL backend, and Section 5.3.2 is dedicated to the implementation for the CUDA architecture. In single precision, all low-level operations have been realised in a sufficiently tuned, cache-optimised CPU variant, in OpenGL and in C for CUDA. In native double precision, only CPU and C for CUDA implementations are available, the OpenGL implementation instead uses emulated precision (see Section 4.2 on page 86). Emulated precision is not available for the other targets, because it is only a substitute workaround. We consider our CPU implementation sufficiently tuned, because it delivers the same performance as FEAST's SBBLAS library for all experiments presented in this chapter. The CPU backend relies on the GotoBLAS[1], which is to the author's best knowledge the fastest BLAS library available on commodity CPUs; and the OpenGL code uses the GLEW library[2] for extension management.

The implementation is fully flexible with respect to mixing execution targets and computational precision, and follows an *explicit-copy idiom*: Memory copies and format conversions always have to be triggered explicitly by the user of the library, there are neither hidden format conversions nor copies. The actual realisation is implementation-dependent and may involve an explicit call to a copy routine, or a setting in a parameter file for a solver.[3] In addition, all memory copies between different targets are implemented asynchronously if supported by the hardware and language, allowing to overlap independent computation with data transfers. Finally, a memory manager maintains a pool of available arrays (vectors): This realisation avoids expensive allocations during computations as some operations need temporary intermediate vectors, and GPU memory is in general a precious resource. Vectors that are no longer needed are returned to the pool, and memory is only actually allocated when there is no instance available for the given combination of target, precision and size that are passed to the memory manager to request a new vector.

For the CUDA and CPU backends, all vectors are zero-padded at both start and end, trading a small additional memory overhead for a significant simplification and performance improvement of matrix-vector operations (cf. Section 5.3.2). The concrete amount of padding depends on the architecture and is always at least half the bandwidth of the matrix, i. e., the offset of the left- or rightmost offdiagonal band. The OpenGL backend does not need padding, the same result is achieved using graphics-specific constructs.

At the time of writing, the `libcoproc` library provides (preconditioned) conjugate gradients, BiCGStab and multigrid solvers, cf. Section 2.2 on page 23. The OpenGL backend only provides the JACOBI preconditioner, while the other two support the full range of operations available in FEAST. Even though only CPU, OpenGL and C for CUDA are supported as backends, the implementation of our framework is open to future target architectures: Adding new backends only requires the reimplementation of the linear algebra building blocks and some infrastructure such as data copies between memory spaces. Our implementation in fact provides a full API to use various co-processors from within FEAST. All solvers are built from sequences of API calls encapsulating numerical linear algebra operations, and the solvers are in turn available through a high-level API to FEAST, which is discussed in Chapter 6.

Unfortunately, not all machines that have been used in previous publications for the microbenchmarks and local solver benchmarks presented in this chapter are available anymore. Some systems have been upgraded, and some have been decommissioned because of hardware failures. As the underlying implementation has also been fully rewritten from scratch to incorporate improvements made over the years, the results presented in this chapter are not always exactly comparable to previously published ones.[4] In the scope of this thesis, this is actually an advantage, as it enables a direct comparison of the results based on exactly the same code.

[1] `http://www.tacc.utexas.edu/resources/software/#blas`

[2] `http://glew.sourceforge.net/`

[3] A concrete example of a format conversion within one solver is using double precision for the coarse grid solver within a single precision multigrid scheme, which is achieved by a simple flag in a configuration file.

[4] Additionally, the unified code uses GLSL as shading language for the OpenGL backend instead of Cg [154], which has been used in earlier implementations. As explained in Section 3.1.3, this is only a technical difference, neither functionality nor performance differ for these two dialects.

5.2. Legacy GPGPU: OpenGL Implementation

This section covers the OpenGL 'legacy' implementation of numerical linear algebra operations on GPUs. Section 5.2.1 briefly lists the implementational specialisations we applied to the general legacy GPGPU programming model as outlined in Section 3.1.1 on page 47. In Section 5.2.2 we introduce four instructive exemplary numerical linear algebra building blocks, a scaled vector update, a norm calculation, a defect calculation and a prolongation. These operations represent different important parallel programming primitives, namely the *map*, *reduce*, *gather* and *scatter* techniques. We describe our implementation of these operations in Section 5.2.3. This approach allows us to present general techniques for different classes of operations, and simultaneously discuss technical and implementational aspects based on concrete examples. Section 5.2.4 assesses the performance of our implementation with a series of *microbenchmarks* of these kernels on different CPUs and GPUs spanning two hardware generations. These benchmarks are artificial in the sense that they only indicate general performance characteristics of graphics processors compared to conventional CPUs. Numerical studies of various local solver configurations are presented in Section 5.5, 5.7 and Section 5.8.

5.2.1. Implementational Specialisations

Our implementation uses the following (OpenGL-specific) specialisations of the legacy GPGPU programming model of interpreting the graphics pipeline as a data-parallel computational engine (cf. Section 3.1.1). Readers interested in implementing GPGPU programs with OpenGL are referred to the hands-on tutorials[5] written by the author of this thesis, which cover the exact use and parameters of OpenGL commands, and other technical details.

Vectors of length N on the CPU are mapped to $M \times M$ square, single-channel, single precision floating point textures, in a rowwise fashion ($M = \sqrt{N}$). This mapping always fits in our case, because all vectors we are dealing with in FEAST have the length $(2^L+1)^2$ for some refinement level L: The rowwise storage corresponds to the linewise numbering of the underlying logical tensor product mesh. For the precision format based on the unevaluated sum of two single precision values (cf. Section 4.2 on page 86), two separate textures are used for the high and low order components. The texture target is `GL_TEXTURE_RECTANGLE_ARB`, the type is `GL_LUMINANCE` and the internal format is `GL_FLOAT_R32_NV`. Texture filtering is set to `GL_NEAREST`, and the wrapping mode is `GL_CLAMP`.

Band matrices are stored as individual vectors, with appropriate zero-padding for off-diagonal bands. We also experimented with mapping each vector to an RGBA texture instead of a single-channel texture, but found no performance improvement on the hardware we used. Similarly, the band matrix layout suggested by Krüger and Westermann [140, 141] did not improve performance; on the contrary, the involved padding required nontrivial data transformations in the interface between FEAST and our `libcoproc` library.

5.2.2. Example Operations: Parallel Programming Primitives

We describe the implementation of numerical linear algebra operations using the following set of example kernels. They cover, in this order, the important parallel programming primitives *map*, *reduce*, *gather* and *scatter*.[6]

[5] see `http://www.mathematik.tu-dortmund.de/~goeddeke/gpgpu`

[6] The fifth type of operation, a *scan*, is only used in Section 5.4 and is thus explained there, as we did not implement stronger, more robust elementary smoothing and preconditioning operators in OpenGL due to the limitations of the underlying programming model.

- The `axpy` kernel computes, for two vectors $\mathbf{x},\mathbf{y} \in \mathbb{R}^N$ and a scalar $\alpha \in \mathbb{R}$, the scaled vector sum $\mathbf{y} = \alpha\mathbf{x} + \mathbf{y}$. The operation is trivially parallel, as each entry of the vectors is independent of all others. In the parallel computing literature, kernels of this type are commonly referred to as *map operations*: A given function, in this case $f(\mathbf{x},\mathbf{y}) = \alpha\mathbf{x} + \mathbf{y}$, is applied to every set of elements (matching indices) in the input and output arrays. `axpy` requires $2N$ flops (floating point operations), and $3N+1$ floating point values are read from and written to memory. This kernel is representative for vector update operations.

- The `nrm2` kernel computes the Euclidean norm of a given vector $\mathbf{x} \in \mathbb{R}^N$: $\alpha = \|\mathbf{x}\|_2 = \left(\sum_{i=0}^{N-1} x_i^2\right)^{1/2}$. This kernel is an example of a *reduction operation*, a computation of a single scalar value from an input vector. Other important applications of this technique are computations of dot products, maximum and minimum values etc. This operation performs $2N-1$ flops (plus the root operation) and $N+1$ memory transfers.

- The `defect` kernel computes the expression $\mathbf{d} = \mathbf{b} - \mathbf{A}\mathbf{x}$ for vectors $\mathbf{d}$, $\mathbf{b}$ and $\mathbf{x}$ and a matrix $\mathbf{A}$. We employ a matrix corresponding to one subdomain and a scalar problem in FEAST (see Section 2.3.1 on page 29), i. e., the matrix has nine diagonal bands at fixed positions and corresponds to a linewise numbering of the mesh points on a (logical) tensor product grid. This kernel is an example of a *gather operation* as it reads in data at different positions in the input vectors. It requires $18N-1$ flops and $12N$ memory transactions.

- The `prolongate` kernel implements a prolongation operation, i. e., the transfer of a correction term from a coarse to a fine grid within a multigrid solver. As explained in Section 2.2.3 on page 26, we confine ourselves to a constant coefficient prolongation on a logical tensor product grid, as this is the case we are using in our real applications. In addition, we always apply homogeneous Dirichlet boundary conditions, as this is the test case we use in many of our solver tests later in this chapter. This operation serves as an example to reformulate an inefficient *scatter operation* to an efficient *gather operation*: The underlying hardware does not support scatter until late 2006 due to the architecture being restricted to the CREW parallel machine model, cf. Section 3.3.1 on page 74. Denoting the size of the coarse array with N_{coarse} and the size of the fine array with N_{fine}, then this operation reads in N_{coarse} floating point values, writes out N_{fine}, and we count the required floating point operations as follows: For N_{coarse} array entries, no operation is necessary, these values are simply copied from the coarse into the fine array. The values in the fine array that correspond to element centres in the coarse grid are interpolated from four values, which requires three additions and one division. The number of elements in the coarse grid, $N_{\text{elemCoarse}}$ is related to N_{coarse} via $N_{\text{elemCoarse}} = (\sqrt{N_{\text{coarse}}} - 1)^2$. All remaining values correspond to edge midpoints of the coarse grid and are interpolated from two coarse values, which requires two flops. In total, we have $4N_{\text{elemCoarse}} + 2((2(\sqrt{N_{\text{coarse}}} - 1) + 1)^2 - N_{\text{elemCoarse}} - N_{\text{coarse}})$ flops.

We employ an *abstract performance model* which allows us to compare different implementations of different operations on all architectures under consideration: In essence, we only count the 'minimally mathematically required' floating point operations and memory accesses. All implementational and hardware-specific details are deliberately excluded, such as index arithmetic, or repeatedly accessing the same element in a data vector, for instance the coefficient vector in the `defect` kernel. One could think of this model as assuming an infinite amount of registers; repeatedly accessing a datum is for free once it has been moved from memory to a register. Only after the computation is completely finished, all result data is written back to memory, and we assume reading from memory costs as much as writing to memory. Obviously, this performance model does not allow any conclusions relative to theoretical peak performance, except for special cases like fully bandwidth-limited operations.

5.2.3. Implementation of Numerical Linear Algebra Operations

In the following, we describe our implementation of the four example numerical linear algebra operations outlined in Section 5.2.2.

AXPY: Scaled Vector Update

Figure 5.1 illustrates our OpenGL implementation of the `saxpy` (single precision `axpy`) kernel.

Input: Vectors $\mathbf{x}$ and $\mathbf{y}$, stored in $M \times M$ textures named `texX` and `texY`, scalar α
Output: Vector $\mathbf{y}$, stored in $M \times M$ texture `texY`

1. Bind 'passthrough' vertex program, cf. Listing 3.2 on page 56.
2. Bind fragment program, a trivial modification of the program given in Listing 3.1 on page 55. Bind `texX`, `texY` and α as input parameters to this fragment program.
3. Request a temporary $M \times M$ texture from the memory manager, and bind it as output.
4. Set viewport to exactly cover the quadrilateral $(0,0),\dots,(M,M)$.
5. Render the quadrilateral $(0,0)-(M,M)$.
6. Swap the temporary output texture with `texY` and return it to the memory manager for future reuse.

Figure 5.1: OpenGL implementation of a `saxpy` kernel.

The first steps are clear. The notation $(0,0)-(M,M)$ is used to abbreviate a quadrilateral with the four vertices $(0,0)$, $(M,0)$, (M,M) and $(0,M)$. As textures cannot be read from and written to in one rendering pass, an auxiliary output texture is used in step 3, i. e., rather than computing $\mathbf{y} = \alpha\mathbf{x} + \mathbf{y}$ directly, the operation $\mathbf{z} = \alpha\mathbf{x} + \mathbf{y}$ is performed in step 5. Our implementation allows to swap this temporary result buffer with the 'real' texture `texY` at no cost (a simple swapping of pointers), so that this hardware limitation is completely transparent to the user. This technique is commonly known as *ping-pong*, textures are alternately used as render targets and input.

NRM2: Euclidean Vector Norm

The calculation of the Euclidean norm of a given vector is an example of a *parallel reduction operation*. We illustrate the procedure with a sum-reduction, i. e., the calculation of the sum of all elements in an array. For simplicity, we first restrict ourselves to a square texture with dimensions that are powers of two. The algorithm, depicted in Figure 5.2, proceeds as follows: For an input texture of size $2^K \times 2^K$, the quadrilateral $(0,0),\dots,(2^{K-1},2^{K-1})$ is rendered. The corresponding vertex program doubles these coordinates and writes them to the first texture coordinate interpolant, so that after rasterisation the generated texture coordinates cover the full input texture. Each fragment then computes the sum of four values in the input texture and writes them to an auxiliary output buffer, which is then bound as input: Proceeding recursively for subsequently smaller textures yields the final result. To improve performance, the process can be stopped before reaching the trivial input size 2×2, and the last pass directly computes the sum of the remaining values. Furthermore, the halving per dimension in each step is often replaced by letting each fragment compute the maximum of larger regions to reduce computational overhead.

Our implementation of the `nrm2` kernel essentially follows the same approach. As we are only dealing with texture dimensions that are 'off-by-one' from power-of-two dimensions, we first reduce the input to power-of-two dimensions and simultaneously compute the squares of each input value. The generic reduction algorithm implemented for instance by Harris [109] or in BrookGPU (cf. Section 3.1.5) turned out to impose too much overhead for this particular case. The resulting texture is then used as input to the above sum-reduction. Two intermediate auxiliary buffers are used in a ping-pong fashion to avoid overwriting the input array. The reduction is

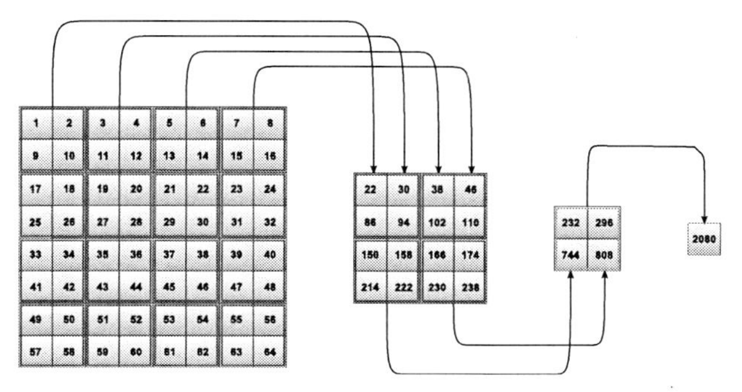

Figure 5.2: Conceptual illustration of a sum-reduce operation. The texture coordinate range for each fragment per pass is highlighted.

stopped as soon as the number of elements to be reduced is less than 1 024, or if the input size is smaller than 1 089 (corresponding to refinement level $L = 5$) and the final result is computed on the CPU. This threshold is determined by calibration runs and is a good compromise on all test platforms we consider.

Krüger and Westermann [140, 141] suggest to keep the result of the reduction in a 1×1 texture to avoid the comparatively high latency of halting the pipeline and reading a single scalar value back to the host. In our implementation, we abstain from using this technique for two reasons: First, experiments on the hardware under consideration indicate that the performance improvement for dot product calculations is almost negligible. Second, for norm calculations, we typically need the result on the host in a subsequent step, e. g., to perform convergence checks of iterative solvers.

Defect Calculation

The implementation of a matrix-vector multiplication has two challenging aspects: First, neighbouring elements in the coefficient vector need to be gathered. Due to our rowwise mapping of one-dimensional arrays to two-dimensional textures, the underlying tensor product structure of the mesh is recovered, and to compute the value at index (i, j), the neighbouring values $(i \pm 1, j \pm 1)$ are required. This is easily performed for indices corresponding to 'interior' texels: A vertex program precomputes the corresponding nine sets of texture coordinates, and the fragment program samples the matrix textures (one texture per band) and the coefficient vector correspondingly and essentially performs one multiply-add operation per matrix band. Consequently, the texture cache is utilised in an optimal way, as it performs best for data access patterns that exhibit spatial locality in two dimensions. However, this approach does not give correct results at texture boundaries: The left neighbour of index i is located at index $i - 1$ in one-dimensional addressing. In two-dimensional addressing, the index $(0.5, j)$ must be mapped to $(M - 0.5, j - 1)$, and analogously for the right boundary. In our implementation, we use three different vertex programs to generate appropriate texture coordinates for the same fragment program. The computation is triggered by rendering, with the corresponding vertex program activated, first a single line that covers the leftmost column of texels in the output, then the quadrilateral $(1, M), \ldots, (M - 1, M)$ to compute all 'inner' values, and finally a single line covering the rightmost column. The approach is an example of *static branch resolution*, a powerful performance optimisation technique, as the naïve approach involves the evaluation of a conditional for every fragment. Listing 5.1 depicts GLSL code for the fragment program.

```
 1 #extension GL_ARB_texture_rectangle : enable
 2 uniform sampler2DRect textureX, textureB;
 3 uniform sampler2DRect textureLL, textureLC, textureLU;
 4 uniform sampler2DRect textureCL, textureCC, textureCU;
 5 uniform sampler2DRect textureUL, textureUC, textureUU;
 6 void main()
 7 {
 8   // y = Ax
 9   float m = texture2DRect(textureLL, gl_TexCoord[0].xy).x;
10   float x = texture2DRect(textureX,  gl_TexCoord[3].xy).x;
11   float y = m*x;
12   m = texture2DRect(textureLC, gl_TexCoord[0].xy).x;
13   x = texture2DRect(textureX,  gl_TexCoord[2].xy).x;
14   y = y + m*x;
15   m = texture2DRect(textureLU, gl_TexCoord[0].xy).x;
16   x = texture2DRect(textureX,  gl_TexCoord[3].zw).x;
17   y = y + m*x;
18   m = texture2DRect(textureCL, gl_TexCoord[0].xy).x;
19   x = texture2DRect(textureX,  gl_TexCoord[1].xy).x;
20   y = y + m*x;
21   m = texture2DRect(textureCC, gl_TexCoord[0].xy).x;
22   x = texture2DRect(textureX,  gl_TexCoord[0].xy).x;
23   y = y + m*x;
24   m = texture2DRect(textureCU, gl_TexCoord[0].xy).x;
25   x = texture2DRect(textureX,  gl_TexCoord[1].zw).x;
26   y = y + m*x;
27   m = texture2DRect(textureUL, gl_TexCoord[0].xy).x;
28   x = texture2DRect(textureX,  gl_TexCoord[4].xy).x;
29   y = y + m*x;
30   m = texture2DRect(textureUC, gl_TexCoord[0].xy).x;
31   x = texture2DRect(textureX,  gl_TexCoord[2].zw).x;
32   y = y + m*x;
33   m = texture2DRect(textureUU, gl_TexCoord[0].xy).x;
34   x = texture2DRect(textureX,  gl_TexCoord[4].zw).x;
35   y = y + m*x;
36   // y = b-Ax
37   m = texture2DRect(textureB, gl_TexCoord[0].xy).x;
38   gl_FragColor.x = m-y;
39 }
```

Listing 5.1: Fragment program for defect calculation.

The second important aspect of our implementation is the handling of out-of-bounds accesses. The matrix bands are appropriately padded with zeros, but the coefficient vector is not, and for certain index ranges, is accessed out-of-bounds. We configure the texturing subsystem to *clamp* these out-of-bounds accesses, i. e., a texture lookup always returns the corresponding value at the texture boundary. This value is well-defined, and multiplying it with a zero which is read from a matrix band always yields the correct result. This optimisation technique is called *moving branches up the pipeline*, as again, the fragment program remains completely free of conditionals.

In emulated double precision, the defect calculation must be performed in two passes, as on the employed hardware, the maximum number of different textures is limited to 16 per rendering pass.

Prolongation

Our last example operation is the prolongation of a given coarse grid vector to a fine grid vector. A viable implementation on CPUs, which is realised in FEAST, is to loop with strided access over the elements of the coarse grid vector and scattering them, with appropriate weights, into the fine array. OpenGL and the underlying older hardware generations do not support scattering[7] operations, so we have to *reformulate the scatter as a gather operation*. To achieve this, we

[7]To be technically correct, scattering can be realised by rendering many single points, but this is very inefficient.

configure the graphics pipeline with a viewport corresponding to the fine vector, and render a matching screen-sized quadrilateral. We have three different cases to consider:

1. A node in the fine grid lies in the centre of an element and must be interpolated from four coarse grid nodes.
2. A node in the fine grid lies on a coarse edge and must be interpolated from two coarse grid nodes.
3. A node in the fine grid lies on top of a coarse grid node and takes its value.

In general, such a distinction is nontrivial to implement efficiently in OpenGL. Extensive branches to decide which case applies for the current fragment are prohibitively expensive, at least on our target hardware for legacy GPU computing with its high branch granularity (cf. Section 3.1.7). A common solution in parallel computing would be to implement a multipass algorithm using some kind of colouring technique to decouple the texels. In our implementation however, we take advantage of the texture coordinate interpolation of OpenGL, and the fact that texture coordinates are not stored as integers but as floating point values: Instead of going fully to the next texel in the coarse texture by adding or subtracting 1.0 to the texture coordinates in the vertical and horizontal direction, we only add or subtract 0.25 and rely on the OpenGL texture coordinate interpolation rules to access the correct values. This results in the same nodes being read two or four times in the second and third case, but as these values are in the texture cache anyway after the first access, it does not incur a performance penalty. Consequently, the corresponding fragment program is free of conditionals, see Listing 5.2. The setting of homogeneous Dirichlet boundary values where necessary is realised by calling up to eight additional trivial kernels; we omit the corresponding code to improve clarity, but emphasise that the performance measurements presented in the next section include these operations.

5.2.4. Performance Microbenchmarks

In this section, we evaluate the performance of the four example kernels on three different host systems. Unfortunately, the machines that the author has used in previous publications are not available any more. For the majority of the tests in this section, we use a system that marks a typical high-end workstation as of June 2006. It comprises an AMD Athlon64 X2 4200+ CPU (Windsor, 512 kB level-2 cache, 2.2 GHz), and fast DDR2-800 memory in a dual-channel configuration. We only use one core in our tests. The GPU is an NVIDIA GeForce 7900 GT (G71 chip), with 256 MB of GDDR3 memory connected via a 256 bit bus for a theoretical peak memory bandwidth of 42.2 GB/s. The GPU features seven vertex processors and 20 fragment processors, and is connected to the host system's AM2 chipset via PCIe x16. This GPU implements the multiply-add operation in a fused way (the intermediate result of the multiplication is not rounded prior to the addition), which allows for significant performance improvements of the emulated precision arithmetic (cf. Section 4.2). None of these benchmarks except of the `nrm2` kernel include bus transfers.

As the parameter space of the experiments in this section is rather high-dimensional[8], a detailed performance analysis is only presented for this test system. The specifications of the remaining two test platforms are given below as soon as they are needed.

The system runs 32 bit OpenSuSE 11.1 and the NVIDIA driver 180.51. All codes are compiled with the Intel C compiler, version 10.1.021, using tuned compiler settings tailored to the specific microarchitecture.

[8]In fact, it comprises the cross product of four different operations with six different input sizes, on three different test platforms in single and double precision on the CPU and the GPU.

```
1 #extension GL_ARB_texture_rectangle : enable
2 uniform sampler2DRect textureC; // coarse texture
3 void main()
4 {
5   //   0.25 snaps back to original position
6   //    |
7   //    |     0.25 snaps precisely to neighbouring nodes because we address
8   //    |     |     the middle between two "real" texels in the coarse mesh
9   //    |     |
10  //   <X>  <->  <X>  <->  X        coarse array
11  //    |\        /|\       /|
12  //    | \      / | \     / |
13  //    |  \    /  |  \   /  |
14  //    X    X    X    X    X       fine array
15
16  // convert from fine to coarse coordinates and compute lookup stencil
17  vec2 pos = 0.5*floor(gl_TexCoord[0].st)+0.5;
18  vec4 ccoord = vec4(pos, pos);             // (coarse xyxy)
19  vec2 len = vec2(0.25, 0.25);
20  vec4 off = vec4(len, -len);               // (0.25 0.25 -0.25 -0.25)
21  vec4 dcoord;                              // diagonal
22  vec4 acoord;                              // antidiagonal
23  dcoord.xz = ccoord.xz + off.xz;           // (x+0.25,x-0.25)
24  dcoord.yw = ccoord.yw + off.yw;           // (y+0.25,y-0.25)
25  acoord.xz = ccoord.xz - off.xz;           // (x-0.25,x+0.25)
26  acoord.yw = ccoord.yw + off.yw;           // (y+0.25,y-0.25)
27  // sample the coarse array at the computed locations
28  float res = texture2DRect(textureC, dcoord.zw).x;
29  res = res + texture2DRect(textureC, acoord.xy).x;
30  res = res + texture2DRect(textureC, acoord.zw).x;
31  res = res + texture2DRect(textureC, dcoord.xy).x;
32  gl_FragColor.x = 0.25*res;
33 }
```

Listing 5.2: Fragment program for the prolongation kernel.

Detailed Analysis of the High-End Test System

Figure 5.3 depicts the absolute performance in MFLOP/s and GB/s we measure for the four example operations. Looking at the CPU results first, we see the expected behaviour: For large input sizes, the `axpy` kernel exhibits a throughput of 4.5–4.7 GB/s in single and double precision, and consequently, single precision is twice as fast (in MFLOP/s) as double precision. This confirms that performance of this operation is exclusively limited by the available memory bandwidth for the larger problem sizes. Absolute performance (in MFLOP/s) of the `defect` kernel is similar to the `axpy` kernel. For the less interesting smaller problem sizes, the data fits entirely into cache and consequently, performance is close to peak. The point at which the cache has no advantageous effect on performance anymore is particularly pronounced in the GB/s diagram: A problem size of 257^2 ($L = 8$) just barely exceeds cache capacity in single precision, and requires twice the available cache memory in double precision. Consequently, the measured performance in GB/s does not change when further increasing the problem size.

The `nrm2` operation performs best for all problem sizes, reaching a remarkable performance of 2.5 GFLOP/s in single and 1.7 GFLOP/s in double precision, respectively. This is due to the extremely well-tuned implementation of the underlying GotoBLAS library which exploits the new 'horizontal' instructions of SSE3 if available[9], and the fact that this operation exhibits an arithmetic intensity of 2 instead of 2/3 as the `axpy` operation (see Section 5.2.2). The `prolongate` kernel performs relatively poor.

The most important observation of the GPU results is that the general performance behaviour is reversed: On the CPU, smaller problem sizes that fit entirely into cache perform best, while larger problem sizes suffer from the relatively low memory bandwidth. On the GPU, this is ex-

[9] GotoBLAS is always compiled for the particular microarchitecture.

actly the other way round: Small input sizes yield very poor performance, because first, there is simply not enough work available to saturate the hardware-managed threads. Second, there is a substantial amount of overhead associated with configuring the GPU to perform an actual computation, overhead mostly due to the fact that a graphics API is used to access the device.

We look at the single precision results first: For the `axpy` and `defect` kernels, we measure 5.5 GFLOP/s for the largest problem size, and 3.4 GFLOP/s for the `nrm2` and `prolongate` kernel. This corresponds to more than 32 GB/s for the `axpy` operation, more than 75 % of the card's theoretical maximum bandwidth. It is worth mentioning that the GPU performance (for large problem sizes) exceeds the in-cache performance of the CPU, so in a way the 256 MB device memory can be seen as a 'gigantic' cache memory, because blocking for locality is equally important. Furthermore, performance still improves quite significantly for the three largest problem sizes, indicating that the hardware is not entirely saturated yet. Similar to the CPU measurements, the `nrm2` operation is a special case: While the other kernels reach competitive performance, i. e., the device is sufficiently saturated already for medium-sized input data, the `nrm2` kernel is only fast for the two largest problems. This is due to the *reduction* character of the operation, which requires multiple rendering calls on subsequently smaller input domains: The time complexity of a parallel reduction is $\mathcal{O}(\log N)$, so the work of $\mathcal{O}(N \log N)$ is not cost-efficient. The device is not saturated even for the largest problem size.

Our OpenGL implementation emulates double precision via a double-single format (see Section 4.2 on page 86). The GeForce 7 series GPUs all support a fused multiply-add, making the emulation significantly less expensive. To enable a more honest comparison with the CPU results, we do not count all native single precision operations required for the emulation, but count a double-single addition as one flop, as in double precision. The bandwidth requirements of the emulated format are identical to native double precision. This explains the—at first sight—surprisingly low MFLOP/s rates of the emulated format, which otherwise could be expected closer to being compute bound with an arithmetic intensity of two (addition) and three (multiplication). On average, the emulated precision kernels are 5–7 times slower than their single precision counterparts on the device.

Figure 5.4 summarises the results by depicting the achieved speedup of the OpenGL implementation versus the baseline CPU implementation. For the largest problem sizes and single precision, the GPU executes the test kernels `axpy` and `defect` seven times faster. The `nrm2` kernel just barely outperforms its CPU counterpart, while the `prolongate` kernel is 35 times faster. As explained above, the former is a consequence of a particularly well-tuned CPU code and a GPU code that involves the overhead of performing a parallel reduction in multiple rendering passes; and the latter is in fact a comparison of a sophisticated GPU implementation with a CPU implementation that uses strided memory accesses. The break-even point at which the GPU starts to outperform the CPU is at a problem size of $N = 127^2$ ($L = 7$) for `axpy` and a surprisingly small $N = 65^2$ ($L = 6$) for `defect` and `prolongate`. The `nrm2` kernel is only faster for the largest problem size in single precision, and slower otherwise. Only for the `defect` kernel, the speedup is constant for sufficiently large problem sizes, for all other kernels, it still increases significantly.

Despite its relatively low performance compared to single precision, the emulated precision kernels reach a stable speedup by a factor of 2.5 compared to their native double precision CPU counterparts. Here, the `nrm2` operation is not able to achieve faster performance at all.

GPU Performance Increase Across Hardware Generations

To evaluate the performance improvements of GPUs across hardware generations, we use two other systems. The first one comprises an AthlonXP 2400+ processor (Thoroughbred, 256 kB level-2 cache, 2.0 GHz), DDR-333 memory in a single-channel configuration, and a 'vanilla' NVIDIA GeForce 6800 GPU (NV40 chip, 5 vertex- and 12 fragment processors) with 256 MB

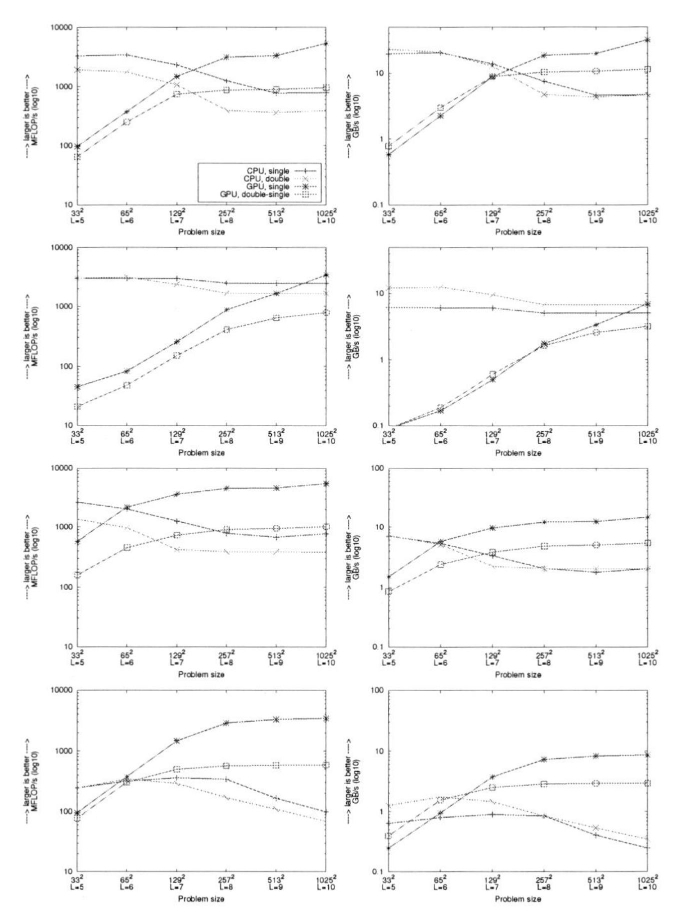

Figure 5.3: Microbenchmark results on the high-end test platform, absolute performance in MFLOP/s (left) and GB/s (right). From top to bottom: `axpy`, `nrm2`, `defect` and `prolongate`. All plots share the same legend, and the y-axis uses a logarithmic scale.

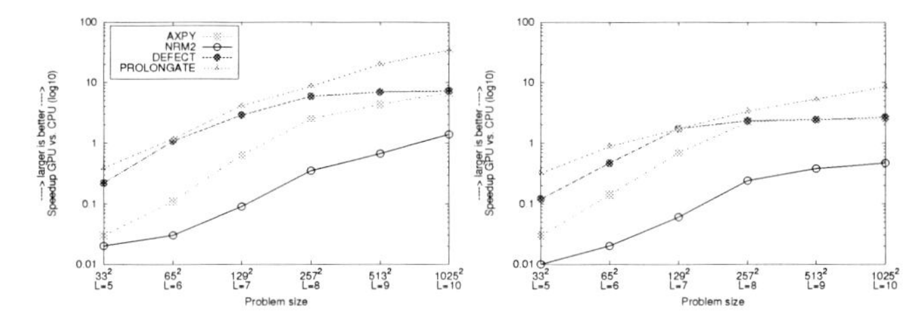

Figure 5.4: Microbenchmark results on the high-end test platform, speedup of the GPU implementation vs. the CPU implementation in single (left) and double (or emulated double-single) precision (right). The same legend is used in both plots, and the y-axis uses a logarithmic scale.

DDR memory (256 bit interface, 22.4 GB/s bandwidth). The GPU is connected via the AGP 8x bus. The second system is based on an Intel Pentium 4 processor (Prescott, 1 MB level-2 cache, 3.0 GHz), uses DDR-400 memory in a dual-channel configuration, and the GPU is an NVIDIA GeForce 7600 GT (G73 chip, 5 vertex- and 12 fragment processors) with 256 MB GDDR2 memory (128 bit interface, 22.4 GB/s bandwidth). It is also connected via the AGP 8x slot.

The AthlonXP processor is one generation behind the GeForce 6800 GPU, and the configuration marks a typical desktop PC in 2003/2004 which has been upgraded with a high-end GPU 1.5 years later. This is the oldest GPU available to us, and as explained in Chapter 3.1.3 on page 51, it comes from the second generation of GPUs supporting single precision floating point. We refer to this system colloquially as 'outdated' and include it to enable a historical comparison. The Pentium 4 machine is a typical representative of a 'mid-range' workstation put into operation in mid 2006. The NV40 GPU does not support fused multiply-add, while the G73 chip does. Both systems run the exact same operating system and compiler version as the high-end workstation in the previous tests.

We do not present results for emulated and double precision in the following. The corresponding experiments and performance measurements have however been performed, and we confirm that the previously established line of argument is equally valid.

Figure 5.5 depicts the performance (in MFLOP/s) of the four test kernels on the three test systems. The most important result is that all GPUs exhibit the same general performance characteristics. Comparing the measurements in detail, a couple of interesting observations can be made.

The mid-range GPU is always approximately as fast as the GeForce 6800 for the `axpy` and `nrm2` kernels, and faster for the two other operations. This means that the narrower bus to off-chip memory is compensated by the faster memory; in fact, the peak bandwidth of the two GPUs is identical. The GB/s rates of the `axpy` kernel underline this (not shown due to space constraints). Both the outdated as well as the mid-range systems suffer equally bad from the reduction overhead of the `nrm2` operation. The `defect` kernel executes almost 50 % faster on the GeForce 7600 GT, even though it has the same theoretical peak bandwidth. This is surprising at first, because this kernel is also extremely bandwidth-bound with an arithmetic intensity less than one. However, the larger amount of resources available to each fragment processor and, more importantly, the higher amount of concurrent threads managed by the hardware allow to hide stalls due to DRAM latency more efficiently in the newer design. This is a very important observation: Single-threaded, bandwidth-limited CPU codes are limited in performance by the available bandwidth to off-chip memory as soon as the computation is not performed in cache. For GPUs as examples of fine-grained massively multithreaded designs, the ability of an architecture to hide stalls due to DRAM

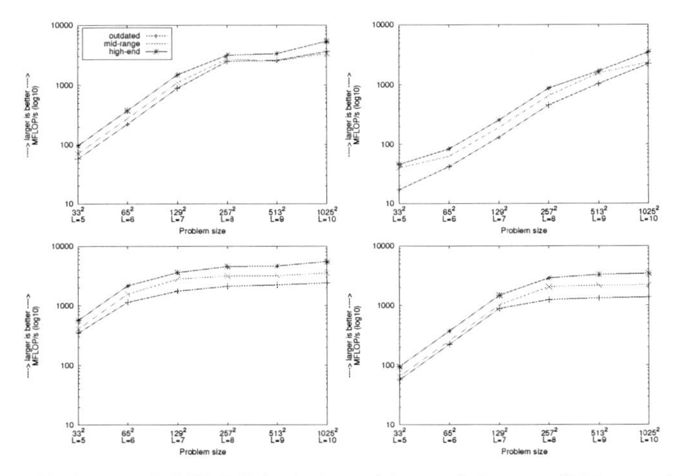

Figure 5.5: Performance in MFLOP/s in single precision on all three test GPUs, `axpy` (top left), `nrm2` (top right), `defect` (bottom left) and `prolongate` (bottom right). All plots share the same legend, and the y-axis uses a logarithmic scale.

latencies is much more important than raw bandwidth, see also Section 3.3.2 on page 75. The same argument holds true for the `prolongate` operation.

The high-end GPU almost doubles the theoretical peak bandwidth and significantly increases the amount of fragment processors, and consequently performs best. We currently cannot explain why the factor of two is not visible in the performance of the (fully bandwidth-bound) `axpy` kernel, which executes (for the largest problem size) at 21.7, 20.1 and 32.2 GB/s, corresponding to 97, 90 and 76 % of the theoretical peak bandwidth. The curves in Figure 5.5 indicate that all GPUs are not completely saturated, though. Nonetheless, for the largest problem size, the 7900 GT GPU outperforms the vanilla 6800 by factors of 1.5, 1.6, 2.3 and 2.5 for the four test kernels, and the mid-range model of the same series by a remarkably constant factor of 1.6.

We again confirm that the same performance improvements are observed in emulated double precision (results not presented here).

Speedup Over Hardware Generations

A common claim of GPU manufacturers is that GPUs are 'getting faster, faster' than CPUs. While this is certainly true for the raw, theoretical peak compute and memory performance, we aim at examining the achievable speedup by the GPUs versus the CPUs for the four test kernels within the limitations of our three test systems. By emphasising problem sizes typically encountered, we actually evaluate the (for us) more practically relevant situation for multilevel and multigrid type algorithms. We again omit double precision results due to space constraints and confirm that detailed measurements show the same characteristic behaviour as in single precision.

Figure 5.6 depicts the speedup factors for single precision computations. For the `axpy` and `nrm2` kernels the outdated system achieves the highest speedups, which is expected because we compare a CPU with a GPU from a newer hardware generation. This observation justifies one of our proposed strategies: The addition of a mid-range GPU or even a GPU that comes from a previous hardware generation into a low-end system significantly improves performance, cf.

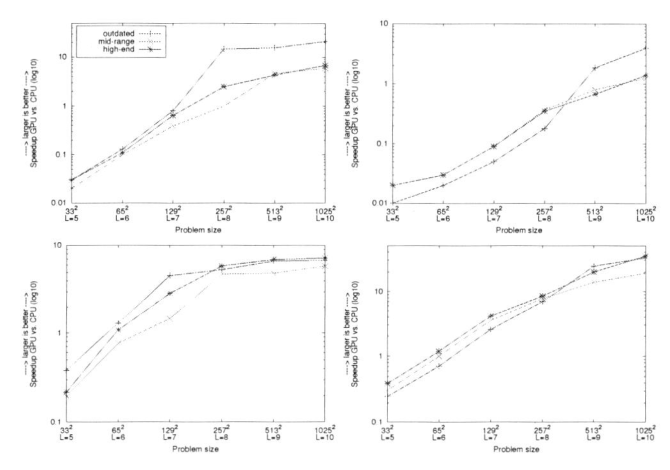

Figure 5.6: Speedup of the GPU over the CPU in single precision for the four example operations on all three test platforms. All plots share the same legend, and the y-axis uses a logarithmic scale, but different ranges to enable a better comparison.

Section 6.3.4 on page 234. For the other two test kernels, the speedup on the outdated system is very close to the high-end system, in line with the observations outlined in the previous paragraph: The achievable fraction of the peak GPU bandwidth decreases, the faster main memory in the two systems translates perfectly into speedup of the newer CPU (exact numbers not presented here), and consequently, the obtained speedup levels are observed. The slope of the plots correlates with the cache size of the CPUs, steepest gradients are observed as soon as a problem size does not fit into cache anymore, corresponding to problem sizes for which the GPU only starts to become reasonably saturated. This is particularly emphasised by the mid-range configuration, the Prescott CPU has twice the amount of level-2 cache memory than the high-end system.

It is interesting to note that the break-even point, i. e., the smallest problem size at which the GPU outperforms the respective CPU, is at least one refinement level higher for the mid-range system than for the other test platforms. This is again explained by the doubled (quadrupled) amount of level-2 cache memory on the Prescott system.

Finally, factoring out the 'unfair' comparison of the outdated system, the speedup is always maximised by the high-end configuration, which within the limitations of our test configurations confirms that GPU performance indeed increases faster than that of CPUs.

Performance per Watt and Performance per Euro

We conclude our microbenchmark evaluation with a short analysis of derived metrics, as raw, absolute performance improvements do not tell the whole story. Performance per Watt has become very important both ecologically and economically ('green computing'), and in particular high-end graphics processors are known to be very power-hungry. Performance per Euro can be used to balance budgets, and is especially relevant in the scenario where an existing workstation or server is upgraded with a fast GPU. The arguments are prototypical in the sense that we do not fully use all resources, the CPU is idle while the GPU computes and vice versa. We only focus on the CPU and the GPU and ignore all remaining components in the system. While this approach is potentially simplistic, it is justified because the GPU cannot be used without an existing system.

The AthlonXP processor in the 'outdated' system has a TDP[10] of 60 W, and the GeForce 6800 GPU consumes 100 W under full load. The baseline variant using only the CPU thus achieves $353/60 \approx 6$ MFLOP/s per Watt, the addition of the GPU increases this value to $2404/(60+100) \approx 15$ MFLOP/s per Watt, a factor of 2.5. The 'mid-range' and 'high-end' workstations' CPUs have a TDP of 89 W, the GPUs consume 40 and 80 W under full load, respectively. Without GPUs, the systems achieve 6.8 and 8.6 MFLOP/s per Watt, with GPUs, these values increase to 27.4 and 32.6 MFLOP/s per Watt, i. e., by factors of 4.0 and 3.8. The absolute increase in performance is 6.8×, 5.9× and 7.2× for the three test systems, which means that the additional power requirements are well justified; in fact, from a peak power consumption perspective, the addition of GPUs is justified if the increase in performance is higher than the higher power requirements.

At the time of acquisition, the three workstations cost, without the GPUs, approximately 1 000, 500 and 1 000 Euro respectively, and the retail price of the GPUs was 250, 100 and 200 Euro. Without GPUs, the measured MFLOP/s rates translate to 0.35, 1.21 and 0.77 MFLOP/s per Euro. The addition of GPUs increases these values to 1.9, 5.9 and 4.6 MFLOP/s per Euro, i. e., by factors of 5.5, 4.9 and 6.0 for the 'outdated', 'mid-range' and 'high-end' systems, respectively.

In summary, these derived metrics indicate that increasing the acquisition costs of the workstations by 20–25 % (and the power requirements by 45–266 %) yields a speedup between 600 and 720 %.

5.2.5. Summary

We discuss important conclusions about GPU performance characteristics that can be drawn from these first microbenchmarks of individual operations in Section 5.3.4, after presenting our CUDA implementation and the corresponding microbenchmarks.

[10] thermal design power, the typical power consumption when running real applications

5.3. Modern GPGPU: CUDA Implementation

In this section, we describe our implementation of numerical linear algebra operations using C for CUDA and the corresponding runtime API. Section 5.3.1 briefly summarises the underlying parallel programming model and its realisation in hardware, and then presents important features and optimisation techniques that are relevant in the scope of this thesis. Section 5.3.2 describes CUDA kernels implementing the four instructive exemplary operations (see Section 5.2.2), in the same way as the OpenGL implementation presented in the previous section. Finally, Section 5.3.3 evaluates the performance of our implementation on a number of GPUs, spanning all three CUDA hardware generations released to date and different price-performance regimes. We conclude with a performance comparison of our CUDA and OpenGL implementation on the same hardware, and summarise the most important conclusions that we can draw from our findings in the microbenchmarks of the legacy (Section 5.2.4) and CUDA implementations in Section 5.3.4.

5.3.1. CUDA Programming Techniques

An approach to interpret the graphics pipeline (as exposed in the OpenGL API) as a computational engine has been presented in Section 3.1.4 on page 53. In contrast, CUDA is explicitly designed as a compute API and consequently exposes the hardware more directly. Thus, it is not necessary to devise an analogous mapping of graphics concepts to general-purpose computations, and we only briefly summarise the analogies in the CUDA programming model and its implementation in hardware and refer to Section 3.2.2 on page 64 and the references therein for the details we omit in the following. In the remainder of this section, basic and advanced CUDA features relevant for our implementation are introduced, features that have not been discussed previously.

CUDA Programming Model Summary

In Section 3.2.2 we have described the CUDA hardware-software model in an unbiased way. CUDA-related publications, the official CUDA documentation [171] and in particular press releases vary dramatically in how the programming model is mapped to the actual hardware. Enthusiastic reports classify the stream/thread processors as actual 'cores', using the terminology of multicore CPU designs. The author of this thesis disagrees with this terminology. In the following, our presentation builds upon the classification of Fatahalian and Houston [68] as outlined in Section 3.3.2:

CUDA *kernels* are executed by partitioning the computation into a *grid* of *thread blocks*. The 'atomic' execution unit in our view on CUDA is thus the thread block, which corresponds to a virtualised multiprocessor (SM, in Section 3.2.2 we used the official NVIDIA terminology of 'streaming multiprocessors'). The same multiprocessor can execute several blocks, the order in which blocks are assigned to multiprocessors is undefined, but blocks do not migrate from one SM to the next once their execution has begun. Threads within one block may coordinate via a 16 kB, low-latency on-chip *shared memory*, and hardware barrier synchronisation instructions. Synchronisation between blocks is only possible at the kernel scope, i. e., there is always an implicit synchronisation between kernel calls on dependent data (the output of one kernel is used as input to the next). It is not possible (except via slow atomic memory operations which are currently not available for floating point data) to synchronise blocks within the execution of one grid.

All threads within one block are executed in a SIMT/SIMD fashion, and the SIMD granularity is the *warp*. The threads within one block are assigned subsequently to warps, i. e., the first 32 (warp size on current hardware) threads form the first warp etc. The ALUs ('streaming processor cores' or 'CUDA cores' in NVIDIA nomenclature) within a multiprocessor execute an entire warp in lockstep, using a shared instruction pointer. Branch divergence within a warp should therefore be avoided, the performance penalty can be large, because the multiprocessor computes both sides of the conditional and masks out threads, respectively. In the worst case (32 conditionals to select each thread separately), the instruction throughput is 1/32-th of the maximum throughput.

Each multiprocessor (multithreaded hardware unit) is capable of keeping 768 threads (1 024 on newer GPUs) per multiprocessor in flight simultaneously, by switching to the next available warp when the currently executed one stalls due to memory requests. This approach maximises the utilisation of the ALUs, minimises the effective latency of off-chip memory requests, and thus maximises the overall throughput.

In summary, CUDA employs a *blocked single program multiple data (SPMD)* programming model.

Launch Configuration Parameter Space Pruning

Ideally, several thread blocks reside on one multiprocessor simultaneously. Its resources, in particular the shared memory and the register file, can easily become a limiting factor because they are partitioned among the resident thread blocks. Additionally, a high *multiprocessor occupancy* is essential to allow the hardware to adequately hide off-chip memory latency. This is achieved by prescribing a partitioning of the computation (an *execution configuration* into a grid of blocks) so that the number of independent warps is maximised, to increase the probability that at any given stall, a warp exists that can be computed immediately. For low-resource kernels, this task is trivial, but for kernels that put significant pressure onto the register file and in particular onto shared memory, it means a partitioning into small enough thread blocks, i. e., trading the degree of intra-block communication via shared memory and synchronisation for a larger amount of blocks in the *grid*. In all our kernels, the shared memory which is split disjunctly among resident blocks is the relevant factor, the amount of available registers is never a bottleneck.

Ryoo et al. [189, 190] discuss the impact of various techniques to maximise performance in this context based on a dense matrix multiply kernel. We adapt their nomenclature of 'parameter space pruning', and whenever applicable, our CUDA kernels are parameterised with the launch configuration: A special variable per operation determines the number of threads per block, and the dimension of the resulting grid is automatically calculated depending on the problem size. The optimal setting for each kernel, which obviously depends on the problem size, is determined by a small benchmarking code that measures performance for all block dimensions in multiples of the warp size and automatically stores this configuration in a small database, for each GPU model evaluated so far. This approach is analogous to FEAST's SBBLAS and SBBLASBENCH functionality (cf. Section 2.3.1), and therefore the name SBCUBLASBENCH has been adopted.

Coalescing of Memory Transactions

In general, the threads of a warp can access arbitrary addresses in off-chip (global) memory, with both load and store operations. As the instruction pointer of the warp is shared, a concrete situation in the following—assuming no warp divergence for simplicity—is for instance a load instruction from an array with offsets (indices) depending on the thread number. Memory requests are serviced for 16 threads (a 'half-warp') at a time. In the most general case each thread accesses an arbitrary location in the array, which results in so-called memory divergence: The memory controller has to issue one memory transaction per thread. If the locations being accessed are sufficiently close together, the memory requests can be *coalesced* by the hardware into as few as a single transaction per half-warp for greater memory efficiency. One can think of this mechanism as short-vector loads and stores, even though the analogy to vectorised load and store operations for instance in SSE is only partly valid. In the most recent GPU generation based on the GT200 architecture, the following (simplified) rules hold: When accessing single or double precision floating point values, global memory can be conceptually organised into a sequence of 128-byte segments. The number of memory transactions performed for a half-warp is equal to the number of segments touched by the addresses used by that half-warp. Fully coalesced memory requests and thus maximum performance are achieved if all addresses within a half-warp touch precisely one segment. This is trivially achieved for linear addressing, but also for 1:1 permutations or if not all threads participate. On the other hand, in the worst case the effective bandwidth utilisation is only 1/16-th, when each thread of a half-warp touches a separate segment. Older hardware (the first and second generation of CUDA capable GPUs) has much stricter coalescing rules. Another issue stems from the fact that global memory is partitioned, and accessed through several memory controllers operating in parallel. Data is assigned to memory partitions in chunks of 256 bytes, in a round-robin fashion. For best performance, simultaneous global memory accesses should be distributed evenly among the partitions. If this is not the case, one speaks of *partition camping*. For technical details, we refer to the CUDA documentation [171].

Arranging memory accesses and the addresses touched by each half-warp to maximise coalescing is in general the single most important performance tuning step for memory-bound kernels and thus for our entire implementation. Bell and Garland [22] demonstrate the impact of non-coalesced memory requests with a very simple yet illustrative example, a strided `axpy` kernel: $\mathbf{y}_{i\cdot\text{stride}} = \alpha\mathbf{x}_{i\cdot\text{stride}} + \mathbf{y}_{i\cdot\text{stride}}$. Any stride greater than one results in an increase of memory transactions issued, and hence, a reduction of the achieved memory bandwidth up to a order of magnitude. Figure 5.7, based on a quick re-implementation of their test case and measured on two of our test platforms (cf. Section 5.3.3), illustrates this effect.[11]

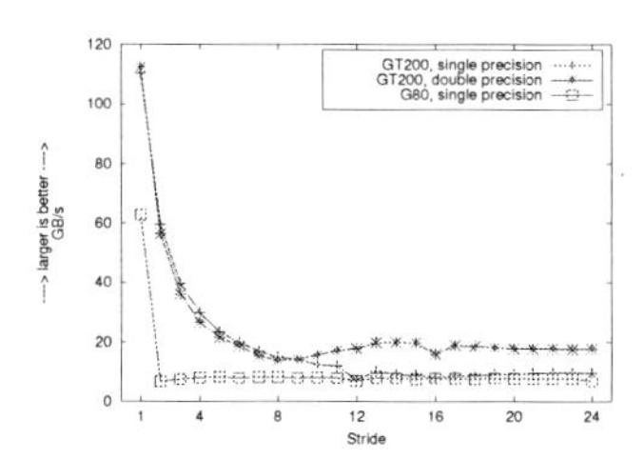

Figure 5.7: Impact of memory coalescing for a strided `axpy` kernel, in single and double precision.

An important technique to improve coalescing for both load and store operations is to *stage loads and stores through shared memory*. This effectively means that different threads within the block are used to load data into shared memory (and to write out data back to global memory) than for the actual computation. Synchronisation via barriers is required inbetween these three stages of the execution of the kernel. We use this technique in many of our kernels, Section 5.3.2 presents a concrete example for the `defect` operation.

Texture Cache

As explained in Section 3.2.2, the texture cache is optimised for streaming hits exhibiting two-dimensional spatial locality. All grid transfer (prolongation and restriction) and matrix-vector multiplication kernels exhibit a high degree of spatial locality when neighbouring values are gathered, due to the tensor product structure of the underlying mesh. CUDA allows to *route memory*

[11] See Section 5.3.2 (discussion of the performance of the `axpy` kernel) why double precision achieves higher throughput than single precision.

requests through the texture cache, reusing cached data. This technique can only be used for load operations from memory, as the hardware does not provide cache coherence and no sophisticated cache replacement policies. CUDA essentially supports two different approaches to using texture memory: In the simple case, an array in global memory is bound to a texture object, routing accesses to it through the texture cache. This approach only requires a few additional lines of code and is therefore practically for free, the data just travels differently from off-chip memory to the processing cores. One can argue that this approach to using the texture cache amounts to an increase of the effective memory bandwidth, but as it increases latency, it is not beneficial when memory accesses can be fully coalesced into one memory transaction per half-warp already. The full functionality of (OpenGL) textures, in particular interpolation of neighbouring values, mipmapped accesses and address warping at texture boundaries (cf. our OpenGL implementation of a defect calculation in Section 5.2.3), is only available if the data to be accessed is either already available in, or manually copied to a special CUDA array type. In our implementation, only the former variant is feasible: As textures are read-only on a per-kernel basis, the implied memory copies and data re-arrangements into the special CUDA array type proved to be too costly. We finally note that double precision textures are not natively available. It is however possible to load a double precision value as a pair of signed integers, which is a supported format for texture memory, and recombine the two integers into a double precision value using the intrinsic `__hiloint2double()` device function.

Asynchroneity: Stream API

The *stream API*, a subset of the CUDA runtime API, provides support for overlapping data transfers from host to device memory (and vice versa) with the execution of kernels on the device. Simple test codes demonstrate the potential of this technique to hide the PCIe bottleneck entirely. From a practical perspective, streaming from or to host memory is only supported if the host memory is allocated using a special CUDA variant of the `malloc` command which allocates so-called *page-locked* memory, enabling DMA transfers performed by the device driver. Pageable memory cannot be copied asynchronously since the operating system may move it or swap it out to disk before the GPU is finished using it. In the case of FEAST, it is nontrivial to employ this mechanism, because all host memory is allocated outside the `libcoproc` library. Allocating all memory in FEAST using the CUDA runtime is not a feasible option, because the allocation of too much page-locked memory causes system instabilities.

Bank Conflicts in Shared Memory

As many threads access shared memory simultaneously, it is realised in hardware as a multi-banked memory. On the GPUs we employ throughout this thesis, the number of banks is 16, the size of a half-warp. Successive 32-bit words are assigned to successive banks. Each bank can serve one memory request per cycle, and shared memory is then as fast as registers. A valid conceptual model is to view bank conflicts as the shared memory analogue of partition camping in global, off-chip memory. If all threads in a half-warp access the same value, no performance penalty occurs because of a built-in broadcast mechanism. Simultaneous accesses to the same bank result in *bank conflicts*, and memory requests are then serialised. By construction, accesses of independent half-warps are free of bank conflicts. Optimal performance is achieved trivially when using sequential (or random 1:1 permutated) addressing within each half-warp for 32-bit words, because each thread addresses a consecutive bank and all requests are serviced concurrently. In double precision, there are consequently always at least two-way conflicts, even for linear addressing. For strided accesses, performance drops significantly, and in the worst case of stride-16 addressing, all shared memory accesses are serialised, resulting in a 16-fold slowdown. In our advanced kernels that extract most of their performance from shared memory usage (see Section 5.4 on page 142 for our implementation of strong smoothers), we rearrange the data layout when loading data from global to shared memory to allow for sequential addressing.

5.3.2. Implementation of Numerical Linear Algebra Operations

In the following, we describe our implementation of the four example numerical linear algebra operations summarised in Section 5.2.2. We restrict ourselves to the single precision variants of the operations, as double precision requires only trivial modifications.

AXPY: Scaled Vector Update

Listing 5.3 depicts the host and device code (the kernel) for the single precision `axpy` operation. We do not use CUBLAS, because we do not need support for strided memory accesses; this yields a small performance improvement because it needs less index arithmetic.

```
1 // host code
2 dim3 grid;
3 dim3 block;
4
5 block.x = getNumThreads(OP_SAXPY, PREC_SINGLE, N);
6 grid.x = (int)ceil(N/(double)(block.x));
7 saxpy_kernel<<<grid,block,0,0>>>(y,x,alpha,N);
8
9 // device code
10 __global__ void saxpy_kernel (float *y, float *x, const float alpha, const int N)
11 {
12   int idx = blockDim.x * blockIdx.x + threadIdx.x;
13   if (idx < N)
14     y[idx] = y[idx] + alpha * x[idx];
15 }
```

Listing 5.3: Host and device code for the single precision `axpy` kernel.

Each hardware thread computes one element in the output vector, based on one-dimensional addressing for both the grid and the thread block. On the host, the number of threads per block is read from the database of benchmarking results, and the number of blocks is calculated accordingly (lines 5–6). Then, the kernel is launched, array parameters are assumed to be resident in device memory already, the scalar parameters are residing on the host. Due to the fully general memory model, there is no need to use an auxiliary vector as in the OpenGL implementation, the update can be computed in-place. In the kernel, we first compute the global index from the built-in variables storing the number of threads per block (`blockDim`), the number of the current block (`blockIdx`) and the local index of the current thread within the block (`threadIdx`). As the total number of threads exceeds that of the elements in the vectors (the last block is partially 'empty' except for vector lengths that are multiples of the block size), the conditional in line 13 is obligatory to prevent out-of-bounds reads and in particular writes. Because of the sequential addressing, all reads and writes are trivially fully coalesced.

NRM2: Euclidean Vector Norm

The basic idea of implementing reduction operations in CUDA is the same as in the OpenGL implementation: Portions of the array(s) are recursively reduced, using a tree-like approach to—in case of the `nrm2` kernel—sum up partial results. In the blocked SPMD model CUDA provides, this applies to both the level of threads within each block, and to the level of blocks: The threads within each block compute partial results using a tree-like pattern, and the blocks corresponding to independent partial results are then further reduced analogously. There are two main challenges: First, communication is only possible using hardware barrier functions within each block, there is no global communication except at the granularity of the kernel launch. The computation must therefore be partitioned into several kernel launches, each one corresponding to a successively coarser level of the reduction tree. Second, on these coarser levels, not all threads within each block or not all multiprocessors are busy as the partial reductions and the computation as a whole proceed. Furthermore, during the summations within one block, the data access patterns on coarser levels of the tree are challenging to arrange in a way to achieve full coalescing due to the interleaved addressing imposed by the (binary) tree structure. One solution to this problem is to perform the per-block reductions in shared memory, which unfortunately gives rise to an analogous problem: Uncoalesced accesses to global memory have a similar impact as accesses to shared memory that introduce bank conflicts. Shared memory is implemented as a multi-banked memory in hardware, and if several addresses touch the same bank simultaneously, the accesses are serialised. Access patterns that improve coalescing can easily lead to bank conflicts, and additional rearrangements of the actual work that individual threads perform are necessary.

Furthermore, detailed experiments by NVIDIA engineers[12] reveal that on coarser levels of the reduction tree, the instruction overhead dominates. Performance can be improved further by unrolling, i. e., by letting each thread sum more values: Unrolling the last warp or even the entire block has the additional advantage that a sufficient amount of work remains on the coarser levels, reducing the depth of the reduction tree. Once the global reduction size is small enough, the intermediate results are transferred to the CPU, which computes the final result.

As the `nrm2` and `dot` kernels are readily available in CUBLAS, NVIDIA's optimised implementation of the BLAS subroutine collection, we call these functions in our implementation instead of using self-implemented variants. From a practical point of view, no additional dependencies are added because CUBLAS is an integral component of the CUDA toolkit, so we only have to link against CUBLAS and include its header file.

Defect Calculation

Listing 5.4 depicts the CUDA kernel for the single precision `defect` operation.

On the host, the same one-dimensional indexing and partitioning is used as for the `axpy` kernel, cf. lines 1–6 in Listing 5.3. The result vector is computed rowwise, using one thread per row. The current row index is computed in line 13. The matrix bands are stored as individual vectors, like in our CPU and OpenGL implementations, see Figure 2.7 on page 30 for notation. Each thread *gathers* nine distinct values from the matrix bands, and the reads are fully coalesced (lines 45–49) because of the sequential addressing. The same holds true when loading the right hand side vector, and storing the result. The coefficient vector however is not accessed contiguously, as demonstrated by the following equivalent but naïve code snippet:

```
y[idx] =  b[idx] - (ll[idx]*x[idx-M-1] + lc[idx]*x[idx-M] + lu[idx]*x[idx-M+1] +
                    cl[idx]*x[idx-1]   + cc[idx]*x[idx]   + cu[idx]*x[idx+1]
                    ul[idx]*x[idx+M-1] + uc[idx]*x[idx+M] + uu[idx]*x[idx+M+1]);
```

Here, N is the number of matrix rows, and $M = \sqrt{N}$, cf. Section 2.3.1. We use the shared memory of each multiprocessor to coalesce the reads into larger memory transactions, and route these memory requests through the texture cache to take advantage of the spatial and temporal locality of the accesses: Each thread reads one element from the index region corresponding to the lower three bands, one for the centre three bands, and one for the upper three bands, which is fully coalesced. The first and last threads additionally load the missing border cases, which

[12] See for instance the parallel reduction case study in NVIDIA's Supercomputing 2007 CUDA tutorial, available at `http://gpgpu.org/static/sc2007/SC07_CUDA_5_Optimization_Harris.pdf` and also included in the CUDA SDK version 2.3 'reduction' example.

```
1 texture<float, 1, cudaReadModeElementType> tex;
2
3 __global__ void sdefect_kernel(float *y, float *b,
4                                float *ll, float *lc, float *lu,
5                                float *cl, float *cc, float *cu,
6                                float *ul, float *uc, float *uu,
7                                const int N, const int M, const int offset)
8 {
9   // declare shared memory (allocated at kernel launch time)
10  extern __shared__ float sdefect_cache[];
11
12  // global and local indices
13  int idx = blockDim.x * blockIdx.x + threadIdx.x;
14  int lindex = threadIdx.x;
15
16  // "partition" shared memory into three chunks for easier index arithmetic
17  // C refers to the three centre diagonals etc.
18  float* Ccache = sdefect_cache;
19  float* Lcache = sdefect_cache + blockDim.x+2;
20  float* Ucache = sdefect_cache + 2*(blockDim.x+2);
21
22  // prefetch chunks from coefficient vector:
23  // each thread loads three elements via the texture cache, the first
24  // and last one additionally load the border cases
25  Ccache[lindex+1] = tex1Dfetch(tex, idx + offset);
26  Lcache[lindex+1] = tex1Dfetch(tex, idx-M + offset);
27  Ucache[lindex+1] = tex1Dfetch(tex, idx+M + offset);
28  if (lindex == 0)
29  {
30    Ccache[0] = tex1Dfetch(tex, blockDim.x*blockIdx.x-1 + offset);
31    Lcache[0] = tex1Dfetch(tex, blockDim.x*blockIdx.x-M-1 + offset);
32    Ucache[0] = tex1Dfetch(tex, blockDim.x*blockIdx.x+M-1 + offset);
33  }
34  if (lindex == blockDim.x-1)
35  {
36    Ccache[blockDim.x+1] = tex1Dfetch(tex, blockDim.x*(blockIdx.x+1) + offset);
37    Lcache[blockDim.x+1] = tex1Dfetch(tex, blockDim.x*(blockIdx.x+1)-M + offset);
38    Ucache[blockDim.x+1] = tex1Dfetch(tex, blockDim.x*(blockIdx.x+1)+M + offset);
39  }
40  // synchronise and compute
41  __syncthreads();
42  if (idx < N)
43  {
44    y[idx] = b[idx] -
45            ( ll[idx]*Lcache[lindex]   + lc[idx]*Lcache[lindex+1] +
46              lu[idx]*Lcache[lindex+2] + cl[idx]*Ccache[lindex]   +
47              cc[idx]*Ccache[lindex+1] + cu[idx]*Ccache[lindex+2] +
48              ul[idx]*Ucache[lindex]   + uc[idx]*Ucache[lindex+1] +
49              uu[idx]*Ucache[lindex+2]);
50  }
51 }
```

Listing 5.4: Device code for the `defect` kernel.

is not coalesced but only introduces a small performance penalty that is fully alleviated by the texture cache. The amount of shared memory used by each thread block is set to three times the number of threads per block plus two floating point values, as part of the launch configuration on the host. The host also binds the coefficient vector to the texture object `tex`. As the threads load different subsets of the coefficient vector[13] than they compute on, a synchronisation barrier is necessary (line 41). The actual computation is then performed using data from both global and shared memory. Besides achieving almost full coalescing, a major benefit of this approach is that the shared memory serves as an additional cache, enabling limited data reuse of the coefficient vector in each thread block. Figuratively speaking, we cache three regions in the coefficient vector, corresponding to a batch of subsequent matrix rows.

More generally, our approach is inspired by viewing the shared memory as a software-managed cache in addition to the texture cache. Small experiments based on the `axpy` kernel already demonstrate that staging reads through shared memory that are already fully coalesced does not yield performance improvements: In fact, it is consistently slower due to the increased instruction overhead and the higher latency of texture reads compared to fully coalesced memory transactions. The values from the matrix bands are read once and then discarded, so we use all available shared/cache memory to store portions of the coefficient vector. On the CPU, preventing the matrix bands from occupying valuable cache lines is only possible by explicit SSE programming, in the author's experience, compilers are often not able to realise that some data need not be cached because it is not reused in the current working set.

A word of caution is necessary: Our kernel accesses the coefficient vector out-of-bounds, with a maximum offset of $M \pm 1$. It is not guaranteed that multiplying these arbitrary values with zero (from the corresponding matrix band) evaluates to zero. As explained in Section 5.1.2, correctness is ensured by manually padding all vectors on both sides with zeros: While $M+1$ zeros are technically sufficient, it violates data alignment rules and thus significantly reduced memory throughput. We therefore pad by rounding the offset $M+1$ up to the nearest multiple of the block size. This approach is a good compromise between (slightly) increased storage size in global memory and a high-performance implementation: The only viable alternative approach to prevent out-of-bounds accesses is to evaluate a conditional for each band in each matrix row, which is much too expensive.

Prolongation

The prolongation operation, including the enforcement of homogeneous Dirichlet boundary conditions where necessary, is implemented using the texture cache, but entirely different from the OpenGL version.

Listing 5.5 presents the corresponding CUDA kernel code. Each thread computes one element in the output array corresponding to the fine grid, so similar to the OpenGL version, we reformulate an inefficient scatter operation into a more efficient gather operation. First, the row and column indices in both the coarse and fine array are computed based on the index of the current thread. The correct case (copy, interpolate across an edge, interpolate at element centre) is then determined by classifying each thread based on odd or even row and column indices. The result is computed by gathering the necessary values from the 'coarse array', and finally, homogeneous Dirichlet values are enforced at the boundary if necessary. The implementation relies, similar to the `defect` kernel, on an appropriate zero-padding of the arrays.

[13] The additive `offset` parameter to the texture read function `tex1Dfetch` is a consequence of how the texture cache is exposed in CUDA, see the programming guide for details [171].

```
1 texture<float ,1 ,cudaReadModeElementType> tex;
2 __global__ void sprolong_kernel(float *fine, const int Nfine, const int Ncoarse,
3                     const int Mfine, const int Mcoarse, const int offset
4                      const int node1, ..., const int edge1, ...)
5 {
6   int ifine = blockDim.x*blockIdx.x+threadIdx.x;  // output index
7   if (ifine < Nfine) {
8     int rfine = ifine/Mfine;                    // row index in the fine array
9     int cfine = ifine - rfine*Mfine;            // column index in the fine array
10    int rcoarse = (int)floorf(0.5f*rfine);      // row index in the coarse array
11    int ccoarse = (int)floorf(0.5f*cfine);      // column index in the coarse array
12    int icoarse = rcoarse*Mcoarse + ccoarse; // base index in the coarse array
13
14    // compute odd/even information (==0 = even)
15    int rodd = rfine & 1;
16    int codd = cfine & 1;
17
18    // manually factor out common expression
19    float val = tex1Dfetch(tex, icoarse+offset);
20    if (rodd==0) {
21      if (codd==0) // case 1: node is present in coarse and fine array
22        fine[ifine] = val;
23      else         // case 2: node on coarse edge, horizontal
24        fine[ifine] = 0.5f*(val + tex1Dfetch(tex, icoarse+1+offset));
25    } else {
26      // manually factor out common expression
27      float val2 = val + tex1Dfetch(tex, icoarse+Mcoarse+offset);
28      if (codd==0) // case 3: node on coarse edge, vertical
29        fine[ifine] = 0.5f*val2;
30      else         // case 4: inner node
31        fine[ifine] = 0.25f*(val2 + tex1Dfetch(tex, icoarse+1+offset)
32                              + tex1Dfetch(tex, icoarse+Mcoarse+1+offset));
33    }
34    // apply homogeneous Dirichlet BCs if necessary
35    // quick coarse branch to minimise warp divergence
36    if (rfine==0 || rfine==Mfine-1 || cfine==0 || cfine==Mfine-1) {
37      if (ifine==0 && node1==DIRICHLET)           // bottom left node
38        fine[ifine] = 0.0f;
39      else if (ifine==Mfine-1 && node2==DIRICHLET) // bottom right node
40        fine[ifine] = 0.0f;
41      else if ...
42    }
43  }
44 }
```

Listing 5.5: Device code for the `prolongation` kernel.

5.3.3. Performance Microbenchmarks

In this section, we evaluate the performance of the four example kernels on four different test systems. The majority of the tests in this section is carried out on a system that marks a typical high-end workstation as of June 2008, from now on referred to as the 'high-end system'. It comprises an Intel Core2Duo E6750 CPU (Conroe, 4 MB level-2 cache, 2.66 GHz), and fast DDR2-800 memory in a dual-channel configuration for a theoretical peak bandwidth of 10.6 GB/s. We only use one core in our tests, as FEAST uses MPI on each core of a multicore compute node, see the tests in Chapter 6. The GPU is an NVIDIA GeForce GTX 280 (GT200 chip), with 1 024 MB of GDDR3 memory connected via a 512 bit bus for a theoretical peak memory bandwidth of 141.7 GB/s. The GPU features 30 multiprocessors, supports double precision, and is connected to the host system via PCIe x16 Gen2. Similarly to the OpenGL microbenchmarks, a detailed performance analysis is only presented for this test system, and the specifications of the remaining test systems are given as soon as they are needed.

The test system runs 32 bit OpenSuSE 11.1 and the NVIDIA driver 190.18. All codes are compiled with the Intel C compiler (version 11.1.046), and the CUDA toolkit version 2.3.

Influence of the Partitioning into Blocks and Threads

The partitioning of a kernel into a grid of thread blocks, in particular the number of threads per block which in our case always determines the total number of scheduled blocks, has potentially significant impact on performance, see Section 5.3.1 for details: A sufficiently high amount of concurrently active threads is necessary to hide the latency of off-chip memory accesses, but there are not necessarily enough resources (register file, shared memory) available per microprocessor. We demonstrate this effect by measuring the runtime of the `defect` kernel (refinement level $L = 10$) in double precision, increasing the number of threads per block from 32 to 512 in multiples of 32, the size of the warp. Figure 5.8 presents the results of this experiment. The influence of the number of threads per block is clearly visible, the difference between the slowest and fastest configuration is approximately 25 %.

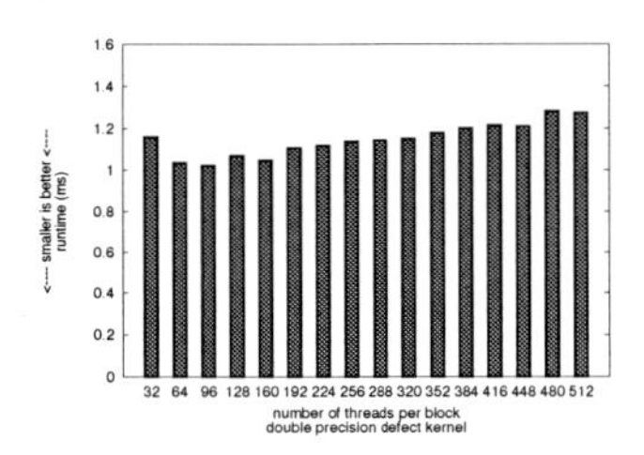

Figure 5.8: Influence of the number of threads per block.

Further experiments reveal that for certain operations, the influence is much more pronounced when varying input problem sizes. An optimal partitioning for a large problem instance can sometimes lead to a substantial slowdown for small problems, if the number of scheduled blocks is smaller than two or three times the amount of multiprocessors on the device: With low occupancy, the hardware is no longer able to adequately hide off-chip memory latency.

In the remainder of this thesis, we always use an 'optimal' launch configuration on each test platform, empirically determined by the SBCUBLASBENCH tool.

Detailed Analysis of the High-End Test System

Figure 5.9 depicts the absolute performance in MFLOP/s and GB/s we measure for the four example operations. The CPU exhibits the same qualitative behaviour as the CPU in the test system used to evaluate the OpenGL implementation in Section 5.2.4, cf. Figures 5.3 on page 120 and 5.5 on page 122. We only note that, mostly due to the much larger level-2 cache memory, the kernels are executed on average twice as fast if the cache can be exploited, i. e., for small problem sizes. In particular, the refinement level that fits entirely into cache is now higher. For larger problem sizes, and for operations without any data reuse such as the `axpy` kernel, performance barely improves, because both systems use the same, fast DDR2-800 memory.

On the GTX 280 GPU however, performance is overall very fast as soon as the problem size is sufficiently large. The `axpy` kernel is able to extract 85 % and 87 % of the theoretical peak bandwidth for the largest input vectors in single and double precision, respectively. Consequently, the MFLOP/s rate in double precision is half that of single precision as soon as the chip is entirely saturated. It is worth mentioning that for smaller problem sizes, double precision is more efficient than single precision (cf. for instance 101.4 and 113.6 GB/s for refinement level $L = 9$ or 65.4 and 85.9 GB/s for $L = 8$). A similar observation of more efficient 64 bit loads has been reported by Billeter et al. [25]. In contrast to the OpenGL implementation, the overhead due to the parallel reduction in the `nrm2` kernel is almost entirely alleviated by the carefully tuned implementation

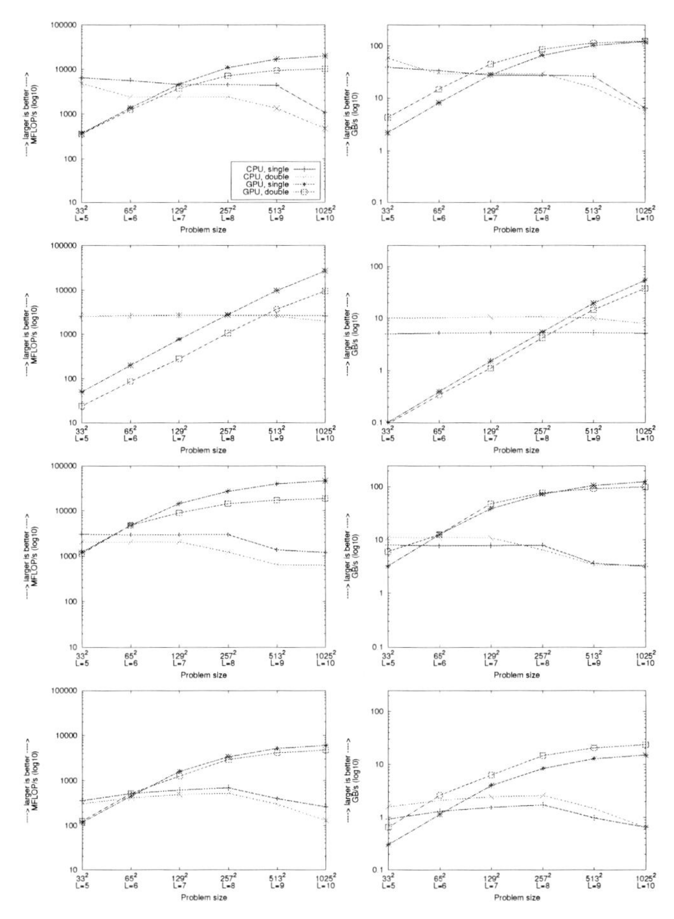

Figure 5.9: Microbenchmark results on the high-end test platform, absolute performance in MFLOP/s (left) and GB/s (right). From top to bottom: `axpy`, `nrm2`, `defect` and `prolongate`. All plots share the same legend, and the y-axis uses a logarithmic scale.

using shared memory to compute intermediate results, and we observe more than 27 GFLOP/s in single precision. The runtime in double precision is more than twice as long, with a maximum rate of 9.5 GFLOP/s. As the CUBLAS code is not publicly available, we cannot explain why, but suspect that the computation of the intermediate results is faster in single precision simply because eight ALUs are available per microprocessor in contrast to only one double precision unit, and that the block size can be chosen larger so that the reduction can be performed with fewer kernel launches. This argument is emphasised by the measured memory throughput, which is 40 % higher in single than in double precision: If the computation were entirely bandwidth bound (like the `axpy` kernel), the throughput would be the same. The carefully tuned implementation of the `defect` kernel results in excellent performance, reaching as much as 46 GFLOP/s in single and 18.5 GFLOP/s in double precision. Without using the texture cache, we achieve 36 and 16 GFLOP/s, respectively. This result underlines impressively that the texture cache is very beneficial for operations for which memory accesses cannot be fully coalesced but exhibit some kind of spatial locality. The `prolongation` kernel performs worst. The implementation seems to be limited by instruction overhead and warp divergence due to the 'checkerboard' pattern, as the similar MFLOP/s rates in single and double precision indicate, compare 6 to 4.75 GFLOP/s. We tried many different implementation variants, such as splitting the computation into four phases resulting in linear thread addressing in each of the four cases, using a two-dimensional launch configuration to reduce the amount of index arithmetics etc., but none performed better.[14] The version of the code we present simply benefits best from data reuse through the texture cache. Finally, the absolute runtime of all four test kernels does not change when the problem size is increased from 1 089 to 4 225 elements ($L = 5, 6$), it is always approximately 8 µs. This means that smaller problem sizes do not make sense on this GPU, since the runtime is dominated by the overhead of launching the kernel. This measurement is in line with the launch overhead reported by other researchers [237].

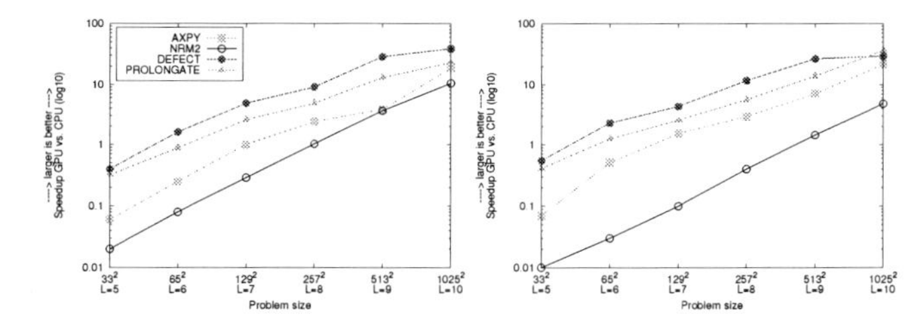

Figure 5.10: Microbenchmark results on the high-end test platform, speedup of the GPU implementation over the CPU implementation in single (left) and double precision (right). The same legend is used in both plots, and the *y*-axis uses a logarithmic scale.

Figure 5.10 summarises the results by depicting the achieved speedup of the CUDA implementation versus the baseline CPU implementation. For the largest problem sizes and single precision, the GPU executes the test kernels between 10.5 (`nrm2`) and 39 (`defect`) times faster. Despite its comparatively low speed, the `prolongate` kernel is still up to 23 times faster. In double precision, the factors are similar, reaching for instance 29 for the `defect` kernel. The break-even point at which the GPU starts to outperform the CPU is at a problem size of 65^2 ($L = 4$) for the `defect` kernel, 129^2 ($L = 7$) for `nrm2` and `prolongate` and 257^2 ($L = 8$) for `axpy`.

[14] The odd-even split-and-merge techniques developed for the multicoloured Gauß-Seidel preconditioner (see Section 5.4.3 on page 143) will be evaluated in future work.

CUDA vs. OpenGL Comparison

In this experiment, we compare the performance of the legacy OpenGL implementation with the CUDA implementation on the same hardware, i. e., the 'high-end' test system based on the NVIDIA GeForce GTX 280. Figure 5.11 summarises the obtained MFLOP/s rates.

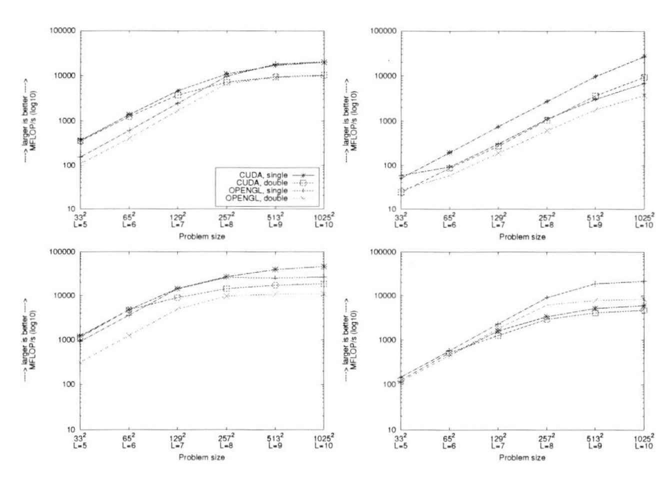

Figure 5.11: OpenGL vs. CUDA implementation on the high-end test platform, absolute performance in MFLOP/s: `axpy` (top left), `nrm2` (top right), `defect` (bottom left), `prolongate` (bottom right). All plots share the same legend, and the y-axis uses a logarithmic scale.

The following observations are valid in both single and double precision: The `axpy` kernel is significantly faster for small problem sizes, which confirms that CUDA exposes the hardware more directly; the smaller overhead of launching computations is clearly visible. For the larger problem sizes, both implementations achieve the same performance, because this operation is only limited by the available memory bandwidth. The CUDA implementation of the `nrm2` and `defect` kernels is between 1.5 and four times faster than the equivalent OpenGL implementation, even for small test problems. This highlights the significant benefit of the small shared memory per multiprocessor, which is not exposed in OpenGL. The speedup obtained with the CUDA implementation is particularly large for the `nrm2` kernel in single precision. Finally, the `prolongate` kernel is (for the larger problem sizes) roughly three and 1.75 slower than the OpenGL implementation in single and (emulated) double precision, respectively. This is a consequence of exploiting graphics constructs in OpenGL compared with a suboptimal CUDA variant as explained in the discussion of the above benchmark results. Technically, CUDA exposes the full functionality of OpenGL texture objects as explained in Section 5.3.1, and consequently, the technique used in the OpenGL implementation could be ported to CUDA. However, we did not implement such a variant yet.

In summary, with the exception of the `prolongate` operation, a modern CUDA implementation of numerical linear algebra building blocks is very beneficial: It is much less cumbersome and error-prone, and at the same time at least as fast as an equivalent OpenGL variant. In case of small problem sizes or if the shared memory can be exploited optimally, CUDA is significantly faster.

GPU Performance Increase Across Hardware Generations

To evaluate the performance improvements of GPUs across hardware generations, we use three other systems. The first one comprises an Intel Core2Duo E4500 processor (Allendale, 2 MB level-2 cache, 2.2 GHz), DDR2-667 memory, and an NVIDIA GeForce 8600 GT GPU (G84 chip, 4 multiprocessors) with 256 MB GDDR3 memory (128 bit interface, 22.4 GB/s bandwidth). This machine marks a typical 'low-end' workstation as of July 2007. The second system comprises an AMD Opteron 2214 processor (Santa Rosa, 2×1 MB level-2 cache, 2.2 GHz), DDR2-667 memory, and an NVIDIA GeForce 8800 GTX GPU (G80 chip, 16 multiprocessors) with 768 MB GDDR3 memory (384 bit interface, 86.4 GB/s bandwidth). This machine is part of our small GPU cluster (cf. Section 6.3.1), and we refer to it as the 'server' system. It constitutes the state of the art as of mid 2007. The third system is based on an Intel Core2Duo E6750 CPU (Conroe, 4 MB level-2 cache, 2.66 GHz), fast DDR2-800 memory, and an NVIDIA GeForce 9600 GT GPU (G94 chip, 4 multiprocessors) with 512 MB GDDR3 memory (256 bit interface, 57.6 GB/s bandwidth). This machine marks a typical 'mid-range' workstation as of mid 2008.

All GPUs only support single precision, and are included into the host systems via PCIe x16.

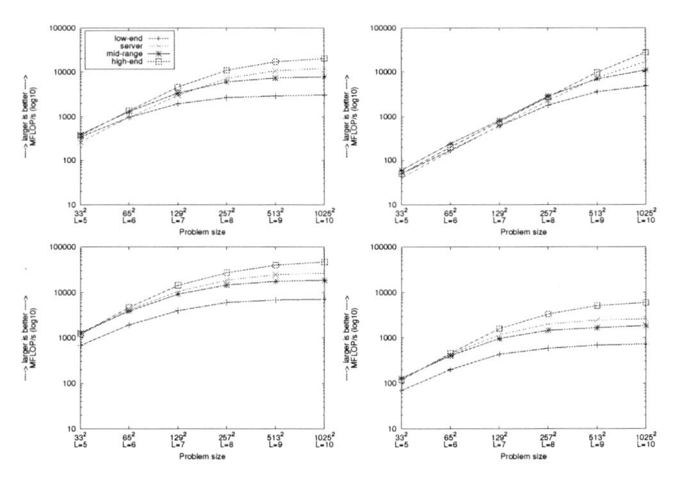

Figure 5.12: Performance in MFLOP/s in single precision on all four GPUs under consideration, `axpy` (top left), `nrm2` (top right), `defect` (bottom left) and `prolongate` (bottom right). All plots share the same legend, and the y-axis uses a logarithmic scale.

Figure 5.12 depicts the measured MFLOP/s rates (single precision) of the four test kernels obtained on the four test systems. A couple of important observations can be made:

Due to their superior memory bandwidth, the two high-end GPUs (GeForce 8800 GTX and GTX 280) exhibit the best performance for all four tests. These two models are the initial releases of the first and third generation of CUDA-capable boards, and despite spanning the development over three hardware generations, the GTX 280 is on average only 1.6 to 2.3 times faster. The `axpy` kernel which is fully limited by the available bandwidth to off-chip memory achieves almost exactly the expected speedup when comparing the improvement of the theoretical memory bandwidth. The other three kernels benefit to a (slightly) higher extent from microarchitecture improvements (the shader clock is almost identical on both GPUs).

The number of multiprocessors does not have a significant impact on performance except for the `prolongate` kernel which is limited in performance by instruction overhead. The 8800 GTX and the 8600 GT come from the same hardware generation, and for the largest problem size, the better GPU is faster by a factor of 3.5. The 9600 GT is the direct successor of the 8600 GT, and on the new hardware generation, all kernels execute approximately 2.5 times faster.

Speedup Over Hardware Generations

Figure 5.13 depicts the speedup factors we obtain in single precision. The speedup is maximised on the two high-end systems for all four test kernels. With the exception of the `defect` kernel, the other two systems exhibit very similar speedup factors, which is the expected behaviour because the mid-range workstation comprises not only a better GPU, but also a faster CPU with twice the amount of level-2 cache and in particular faster memory.

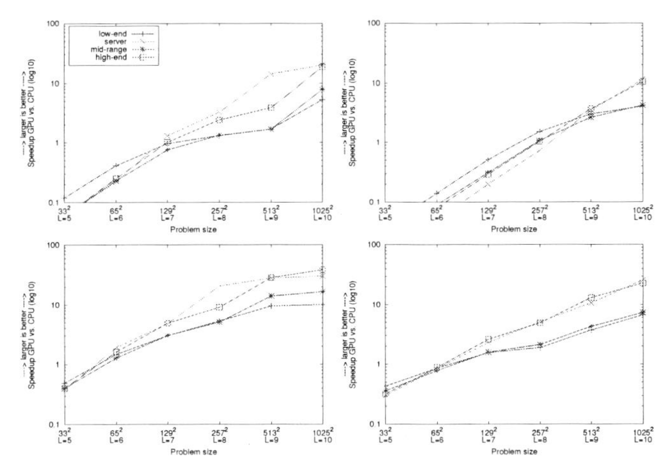

Figure 5.13: Speedup of the GPU over the CPU in single precision for the four example operations on all three test platforms: `axpy` (top left), `nrm2` (top right), `defect` (bottom left), `prolongate` (bottom right). All plots share the same legend, and the y-axis uses a logarithmic scale.

The break-even point, i. e., the smallest problem size at which the GPU outperforms the respective CPU, does not change for all four test systems, but of course depends on the specific operation. On all four systems, remarkable speedups are achieved, even a very cheap entry-level GPU can deliver reasonable performance improvements, close to an order of magnitude except for the `nrm2` kernel.

Performance per Watt and Performance per Euro

Similarly to the OpenGL benchmarks in Section 5.2.4, we briefly analyse the derived metrics performance per Watt and performance per Euro. All calculations are based on the MFLOP/s rate of the single precision `defect` kernel and the largest problem size, and the TDP of the CPU as a baseline reference value.

The Core2Duo processor in the 'low-end' system has a TDP of 65 W; the GeForce 8600 GT GPU consumes 40 W under full load. The baseline variant using only the CPU thus achieves $710/65 \approx 11$ MFLOP/s per Watt, the addition of the GPU increases this value to $7151/(65+$

40) $\approx$ 68.1 MFLOP/s per Watt, a factor of 6.2. The CPU in the 'server' test system a TDP of 95 W, and the GPU consumes 185 W. Without GPUs, the measured MFLOP/s rate translates to 9.1 MFLOP/s per Watt, with the GPU, this value increases to 94.2 MFLOP/s per Watt, i. e., by a factor of 10.4. The 'mid-range' and 'high-end' workstations' CPUs have a TDP of 65 W, the GPUs consume 100 and 236 W under full load, respectively. Without GPUs, the systems achieve 17.2 and 18.4 MFLOP/s per Watt, with GPUs, these values increase to 112.6 and 154.2 MFLOP/s per Watt, i. e., by factors of 6.5 and 8.4. The absolute increase in performance is 10.1×, 30.6×, 16.7× and 38.9× for the four test systems.

At the time of acquisition, the four systems cost, without the GPUs, approximately 500, 3 000, 600 and 800 Euro, and the retail price of the GPUs was 100, 500, 130 and 570 Euro. Without GPUs, the measured MFLOP/s rates translate to 1.4, 0.3, 1.9 and 1.5 MFLOP/s per Euro. The addition of GPUs increases these values to 11.9, 7.5, 25.5 and 33.9 MFLOP/s per Euro, i. e., by factors of 8.5, 25, 13 and 22.6 for the 'low-end', 'server', 'mid-range' and 'high-end' systems, respectively.

In summary, especially when comparing these values with the ones obtained on the older systems used in the OpenGL benchmarks in Section 5.2.4, the derived metrics indicate that performance per Watt and performance per Euro increases steadily over hardware generations. CPUs are becoming more energy-efficient without sacrificing performance. GPU power requirements remain the same for mid-range models and increase for high-end variants, but this increase translates to significant performance improvements. The same holds true for the acquisition costs. Both effects are in essence a consequence of the shrinking semiconductor manufacturing process. Phrased differently, these derived metrics indicate that increasing the acquisition costs of the workstations and servers by 15–75 % (and the power requirements by 60–460 %) yields a speedup between 1 000 and 3 900 %. In all systems, the addition of GPUs (unless a low-end model such as the 8600 GT is already included in the system anyway) is very beneficial in all three metrics.

5.3.4. Summary

In Section 5.2.3 and Section 5.3.2 we have described our implementation of numerical linear algebra building blocks. We have selected four example operations, corresponding to the important parallel programming primitives *map*, *reduce*, *gather* and *scatter*. This approach allowed us to present different techniques for general problem classes and discuss technical aspects based on concrete examples at the same time. Our implementation has been evaluated by means of performance microbenchmarks of these operations in terms of the MFLOP/s and GB/s rates, and the achieved speedup over an equivalent (singlecore) CPU implementation, see Section 5.2.4 and 5.3.3 for detailed results.

The most important observation of these microbenchmark measurements is that the general performance characteristic is reversed: On the CPU, smaller problem sizes that fit entirely into cache perform best, while larger problem sizes suffer from the relatively low memory bandwidth. On the GPU, this is exactly the other way round: Small input sizes yield very poor performance, because there is simply not enough work available to saturate the device; GPUs are designed to maximise throughput over latency. If the problem size is too small, using the GPU instead of the CPU can lead to a slowdown by more than an order of magnitude, and this does not include any bus transfers yet. On the other hand, we always observe more than an order of magnitude speedup if the problem size is sufficiently large. The best speedup we obtain is an impressive factor of 40 for the single precision defect computation.

This is a very remarkable observation: Single-threaded CPU codes with limited data reuse scale almost exactly with the available bandwidth to off-chip memory as soon as the computation can no longer be performed in cache. For GPUs as examples of fine-grained massively multithreaded designs, the ability of an architecture to hide stalls due to DRAM latencies is much more important than raw bandwidth. Furthermore, due to the lack of large caches, arranging memory access patters carefully for spatial and temporal locality is even more important on GPUs than on CPUs.

CUDA allows for a much better control over how a given computation is actually performed on the device. By partitioning the available resources per microprocessor (register file, shared memory) specifically for each individual operation, performance can be influenced significantly. This is however not a trivial task, as certain algorithms require careful balancing. For the defect calculation which uses the on-chip shared memory as a software-managed cache, we could launch one block of threads per multiprocessor and allocate the entire shared memory for it. However, this would mean that the entire multiprocessor stalled at the synchronisation barrier, resulting in a significant degradation of performance. Balancing the amount of communication between threads, the resource usage and the goal to ensure a high number of independent warps (the SIMD granularity in CUDA) to maximise throughput can be quite an 'art', and carefully tuning the implementation is necessary. In our experience, getting correct results is easy, getting them fast is challenging.

Our comparison of CPUs and GPUs spanning five hardware generations confirms two common claims: First, GPUs are becoming significantly faster with each hardware generation, at a higher pace than CPUs. Second, even entry-level GPUs can give an order of magnitude speedup over modern CPUs.

GPUs also excel in terms of performance per Watt and performance per Euro, we established this using reasonable estimates on energy consumption. The former metric is becoming more and more important, as practitioners and vendors in the HPC field are increasingly aware of the energy consumption of large-scale cluster installations, which can easily require several megawatts. This is known as *green computing*. For instance, any given supercomputer that ranks among the top 20 in the Top500 lists of supercomputers at the time of writing requires approximately one million Euro per year just for energy. Due to the lack of measurement equipment, we do not pursue this interesting topic further in the remainder of this thesis.

In summary, our microbenchmark results are very promising. In Sections 5.5, 5.7 and 5.8 we analyse how these findings carry over to Krylov subspace and in particular multigrid solvers.

5.4. Parallelisation and Implementation of Preconditioners

In this section, we present parallelisation techniques for preconditioners specifically tailored to the banded matrix structure stemming from the linewise numbering of generalised tensor product meshes. Section 5.4.1 introduces the notation and the exact matrix structure we are concerned with, which is exploited in the implementation. Here, we also explain the difference between preconditioning and smoothing (in multigrid solvers), and discuss the limitations of our current implementation for the CUDA architecture. Sections 5.4.2–5.4.6 present a number of preconditioners, their parallelisation and various implementational challenges. We conclude with performance microbenchmarks in Section 5.4.7 and a summary in Section 5.4.8.

The efficient implementation we present in Section 5.4.4 has been co-developed with Robert Strzodka, and has been accepted for publication in IEEE Transactions on Parallel and Distributed Systems [81]. The idea of recovering stronger coupling in the multicoloured parallelisation of Gauß-Seidel type smoothers has also been co-developed with Robert Strzodka, during the work for a textbook chapter [80]. All other contributions of this section have not been published previously.

5.4.1. Introduction and Preliminaries

As throughout the entire chapter, we consider the case of one subdomain in FEAST, i. e., the underlying mesh has the logical tensor product structure as outlined in Section 2.3.1 on page 29. The unknowns, which for bilinear finite elements coincide with the grid points, are numbered linewise starting in the bottom left corner. The mesh comprises M rows of M unknowns each, $N = M \times M$.

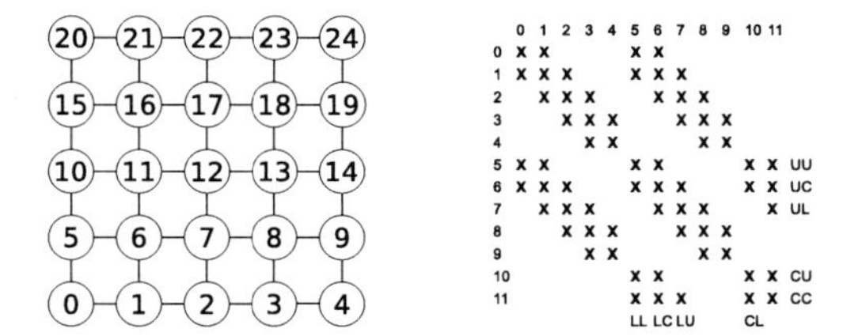

Figure 5.14: Numbering of the unknowns (left) and illustration of the sparsity structure of the system matrix corresponding to one subdomain in FEAST (right, only the first half of the unknowns shown).

Figure 5.14 depicts the numbering of a 5×5 mesh and the sparsity structure of the matrix. Figuratively speaking, we observe that the first subdiagonal realises the coupling of each entry with its left neighbour (its predecessor in the numbering), the first superdiagonal the coupling to the right, and analogously, the lower three bands and the upper three bands realise the coupling to the bottom and top, respectively. It is important to note that all off-diagonals contain zeros at those entries corresponding to the beginning and end of each row in the mesh. This simple observation is crucial for the design of efficient implementations of the preconditioners. We adopt the following formal decomposition of the system matrix into its nine bands as introduced in Section 2.3.1, labeled from bottom left to top right (see also Figure 5.14 on the right):

$$\mathbf{A} = (\mathbf{LL} + \mathbf{LC} + \mathbf{LU}) + (\mathbf{CL} + \mathbf{CC} + \mathbf{CU}) + (\mathbf{UL} + \mathbf{UC} + \mathbf{UU})$$

Smoothers and Preconditioners

The underlying goal of preconditioning is to reduce the condition number of the system matrix. A good preconditioner $\mathbf{C}$ is a (spectrally) good approximation to $\mathbf{A}^{-1}$, so that $\text{cond}(\mathbf{C}^{-1}\mathbf{A})$ is much smaller than $\text{cond}(\mathbf{A})$, because $\text{cond}(\mathbf{A}^{-1}\mathbf{A}) = \text{cond}(\mathbf{I}) = 1$. In the algorithms presented in Section 2.2.1 and Section 2.2.2 and more generally in the textbook by Barrett et al. [13], it can be seen that the application of a *preconditioner* can be written in the form $\mathbf{x} = \mathbf{C}^{-1}\mathbf{d}$, where $\mathbf{C} \in \mathbb{R}^{N\times N}$—in our case—is a submatrix of the banded system matrix $\mathbf{A}$ which is 'easier to invert' than $\mathbf{A}$. In multigrid, *smoothing* is realised by few steps of a (damped), preconditioned defect correction iteration, $\mathbf{x}^{(k+1)} = \mathbf{x}^{(k)} + \omega\mathbf{C}^{-1}(\mathbf{b} - \mathbf{A}\mathbf{x}^{(k)})$, see Section 2.2.3. This means that we can use the terms preconditioner and smoother synonymously in the remainder of this section: One preconditioning step is the formal *application* of $\mathbf{C}^{-1}$ to a given vector $\mathbf{d}$. Note that $\mathbf{C}^{-1}$ is never assembled explicitly, as in general it does not exhibit the sparsity pattern (band structure) of the original system matrix $\mathbf{A}$.

Limitations of the Current Implementation

It is certainly possible to implement some of the advanced preconditioners we present in the following in terms of 'legacy GPGPU', i. e., by using graphics constructs. This has been done for instance by Kass et al. [124]. However, the `libcoproc` library does not provide such implementations in its OpenGL backend. On the CPU, the most efficient sequential algorithms have been implemented, their performance is on par with the highly tuned implementations in the SBBLAS library. On the GPU, different schemes are necessary to recover parallelism from these inherently sequential algorithms, and we have chosen the CUDA backend for the implementation because of the much higher flexibility of the underlying programming model. In any case, both double and single precision variants are provided. Due to limitations of current hardware (in particular on the size of the small on-chip shared memory), not all preconditioners can be applied for arbitrary problem sizes in the current implementation.

5.4.2. Jacobi Preconditioner

The JACOBI preconditioner is defined as the diagonal of the system matrix, i. e.,

$$\mathbf{C}^{\text{JAC}} := \mathbf{CC}.$$

Its application is a pointwise operation and reduces to scaling each entry of the vector $\mathbf{d}$ with the reciprocal of the corresponding entry of the main diagonal of the matrix. An efficient implementation in CUDA is straightforward and has the same structure as the code shown in Listing 5.3 on page 129. We map one thread to each entry in the result vector, and due to the linear addressing, all memory accesses are automatically fully coalesced into one memory transaction per half-warp. Consequently, the bandwidth to off-chip memory—which is the decisive performance factor for this operation—is fully saturated for sufficiently large problem sizes.

When used within a multigrid solver, it is even possible to merge the scaling and damping steps of the defect correction with the defect calculation into one kernel for one damped, preconditioned Richardson step to further improve performance, using Listing 5.4 on page 131 as a starting point.

5.4.3. Preconditioners of Gauß-Seidel-Type

The GSROW preconditioner has the following form (see Section 5.4.6 for a motivation of the suffix 'row'):

$$\mathbf{C}^{\text{GSROW}} := (\mathbf{LL} + \mathbf{LC} + \mathbf{LU} + \mathbf{CL} + \mathbf{CC})$$

$\mathbf{C}^{\mathrm{GSROW}}$ is a lower triangular matrix (cf. Figure 5.14), and consequently, its application is inherently sequential. Starting with the first row of the matrix which only affects one unknown, the CPU implementation performs a standard forward substitution sweep. All values $\mathbf{x}_i$ depend on the previously updated values $\mathbf{x}_{i-1}, \mathbf{x}_{i-M+1}, \mathbf{x}_{i-M}$ and $\mathbf{x}_{i+M-1}$ corresponding to the coupling by the matrix bands $\mathbf{CL}, \mathbf{LU}, \mathbf{LC}$ and $\mathbf{LL}$.

To recover parallelism in this operation, we have two general possibilities: *Wavefront techniques* parallelise the operation in an exact manner, and *multicolouring* decouples some of the dependencies. The term wavefront refers to maximally independent sets of values that can be computed in parallel. Figure 5.15 (left) depicts such wavefronts of independent work, in the order red, green, blue, yellow, magenta, cyan, orange and so on. It becomes clear that this parallelisation technique does not lead to a sufficient amount of independent work for the applicability on GPUs: We would need $2M+3$ sweeps through the domain, and the maximum amount of work per sweep is $\lceil M/2 \rceil$. We therefore relax the preconditioning step and introduce a multicolouring scheme. The common approach in the literature is to use a *red-black* Gauß-Seidel scheme, alternatingly updating all odd-indexed and then all even-indexed values in parallel, see for instance the textbook by Barrett et al. [13]. Closer inspection reveals that two colours are only sufficient in case of a five-point stencil. In our setting we have a nine-point stencil, and we need four colours to decouple all dependencies.

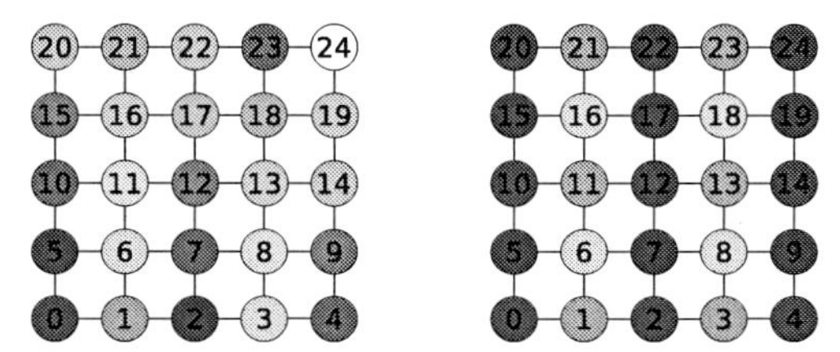

Figure 5.15: Illustration of the wavefront (left) and multicolouring (right) schemes to recover parallelism in the GSRow preconditioner. See online version for colours.

We split the nodes into four disjoint index sets by alternating colours between rows in the mesh, and additionally alternating colours between adjacent grid points per row. The application of the GSRow preconditioner thus decomposes into four sweeps over the domain, each sweep is trivially parallel. We illustrate the procedure exemplarily for the nodes 12, 13, 17 and 18, see Figure 5.15 on the right. During the treatment of node 12 (coloured red), no neighbours have been updated so far, and it is updated with a standard Jacobi step. Node 13 (green) can incorporate the previously updated node 12. Node 17 (blue) takes into account the 'green' nodes 11 and 13 and the 'red' node 12, and finally, node 18 (yellow) can include all four neighbours. These updates are realised by shifting new contributions to the right hand side as soon as they become available. Figure 5.16 summarises the multicoloured algorithm to apply the GSRow preconditioner.

Comparison with Natural Ordering Gauß-Seidel

The Gauß-Seidel preconditioner is not independent of the order of the unknowns. In fact, we obtain an entire family of preconditioners, one for each ordering. A common interpretation is that the multicoloured version corresponds to a different enumeration of the unknowns, see also the following paragraph. However, the 'optimal' order is highly dependent on the concrete problem at hand, and the natural ordering is not necessarily the best one. It is obvious that the multicoloured variant is not equivalent to the *natural ordering* which is exploited in the sequential forward substitution on the CPU. Only the fourth sweep resolves the same dependencies, and thus, this first

1. For all red nodes in parallel:
   $\mathbf{x}_i = \mathbf{d}_i / \mathbf{CC}_i$
2. For all green nodes in parallel:
   $\mathbf{x}_i = (\mathbf{d}_i - \mathbf{CL}_i \mathbf{x}_{i-1}) / \mathbf{CC}_i$
3. For all blue nodes in parallel:
   $\mathbf{x}_i = (\mathbf{d}_i - \mathbf{LL}_i \mathbf{x}_{i-M-1} - \mathbf{LC}_i \mathbf{x}_{i-M} - \mathbf{LU}_i \mathbf{x}_{i-M+1}) / \mathbf{CC}_i$
4. For all yellow nodes in parallel:
   $\mathbf{x}_i = (\mathbf{d}_i - \mathbf{CL}_i \mathbf{x}_{i-1} - \mathbf{LL}_i \mathbf{x}_{i-M-1} - \mathbf{LC}_i \mathbf{x}_{i-M} - \mathbf{LU}_i \mathbf{x}_{i-M+1}) / \mathbf{CC}_i$

Figure 5.16: Multicoloured GSRow preconditioner.

decoupled variant is not equivalent to a renumbering of the unknowns. In fact, it corresponds to the blockwise application of the Jacobi preconditioner, interwoven with defect calculations.

One idea to alleviate the comparatively high amount of unknowns that are treated with lesser coupling is to change the underlying colouring in each application of the preconditioner. There are many possibilities to implement this idea, for instance by shifting colours circularly, or by switching from treating even rows with the first two colours to treating odd rows with the first two colours. We evaluate these ideas in our numerical and efficiency studies in Section 5.5.2 and in Section 5.5.3.

Increasing the Coupling in the Multicoloured Algorithm

Once the implicit neighbourhood dependencies are decoupled by a multicolour numbering, the standard Gauß-Seidel update as interpreted from the point of view of the mesh (i. e., the inclusion of all already available values to the left and bottom) is a suboptimal choice with unnecessarily restricted explicit coupling between the colours. Instead, a *fully coupling multicoloured* algorithm can be used in which the four sweeps over the domain incorporate *all* already available values. This idea has been co-developed by Robert Strzodka and the author of this thesis [80].

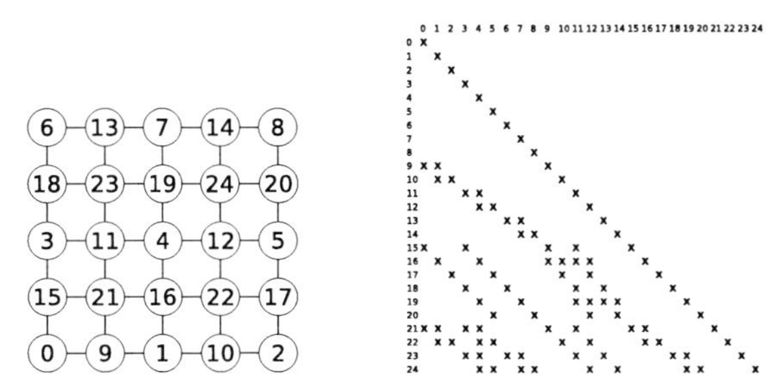

Figure 5.17: Numbering of the unknowns (left) and illustration of the sparsity structure of the preconditioner $\mathbf{C}^{\mathrm{GSFULLROW}}$.

As demonstrated in Figure 5.17, the approach now corresponds to a renumbering of the unknowns, and thus, all theoretical properties of the Gauß-Seidel preconditioner, in particular the *smoothing property* when used in multigrid solvers, directly carry over. All 'red' nodes (node 12 in the example) are treated without modification. The 'green' nodes are updated with both their left and right neighbours (node 13 in the example incorporates the 'red' nodes 12 and 14). The remaining two sweeps not only include the previous line, but also the next one, both of which have been treated with the first two colours. These updates are realised by shifting new contributions to the right hand side at the beginning of each sweep. Figure 5.18 summarises the multicoloured algorithm to apply the resulting parallel GSFULLROW preconditioner.

1. For all red nodes in parallel:
   $\mathbf{x}_i = \mathbf{d}_i / \mathbf{CC}_i$
2. For all green nodes in parallel:
   $\mathbf{x}_i = (\mathbf{d}_i - \mathbf{CL}_i\mathbf{x}_{i-1} - \mathbf{CU}_i\mathbf{x}_{i+1})/\mathbf{CC}_i$
3. For all blue nodes in parallel:
   $\mathbf{x}_i = (\mathbf{d}_i - \mathbf{LL}_i\mathbf{x}_{i-M-1} - \mathbf{LC}_i\mathbf{x}_{i-M} - \mathbf{LU}_i\mathbf{x}_{i-M+1}$
   $- \mathbf{UL}_i\mathbf{x}_{i+M-1} - \mathbf{UC}_i\mathbf{x}_{i+M} - \mathbf{UU}_i\mathbf{x}_{i+M+1})/\mathbf{CC}_i$
4. For all yellow nodes in parallel:
   $\mathbf{x}_i = (\mathbf{d}_i - \mathbf{CL}_i\mathbf{x}_{i-1} - \mathbf{CU}_i\mathbf{x}_{i+1}$
   $- \mathbf{LL}_i\mathbf{x}_{i-M-1} - \mathbf{LC}_i\mathbf{x}_{i-M} - \mathbf{LU}_i\mathbf{x}_{i-M+1}$
   $- \mathbf{UL}_i\mathbf{x}_{i+M-1} - \mathbf{UC}_i\mathbf{x}_{i+M} - \mathbf{UU}_i\mathbf{x}_{i+M+1})/\mathbf{CC}_i$

Figure 5.18: Multicoloured GSFULLROW preconditioner.

Implementational Details

The efficient implementation of the multicoloured Gauß-Seidel type preconditioners is surprisingly challenging due to the inherent stride-two memory access patterns implied by the 'checkerboard' numbering. Our solution is to fuse the computations per mesh line into one CUDA kernel that treats both associated colours. We use blocks of 128 (or less) threads for each line in the mesh, and let each thread compute two values. The last thread additionally treats the last value as the size of each mesh line, M, is a power of two plus one. All loads from global memory are staged through shared memory, and with linear addressing, these loads are fully coalesced. Unfortunately, this mapping now introduces bank conflicts in shared memory when reading the values to compute the final result for each colour. This can be avoided by separating the values during the loads, so that all even and odd indexed values are stored contiguously. With an appropriate padding between the two arrays, these and all subsequent shared memory accesses are completely free of bank conflicts. To guarantee correctness, each block loads data with a halo of one to each side. The resulting kernels (sketched in Listing 5.6) are thus completely free of bank conflicts and off-chip memory is accessed in a fully coalesced fashion. We again rely on zero-padding of the arrays to avoid special treatment of out-of-bounds accesses in the coefficient vector for indices corresponding to the bottom row of the mesh (and the top row in case of GSFULLROW), and at the left and right boundaries.

5.4.4. Tridiagonal Preconditioner

The TRIDIROW preconditioner has the following form:

$$\mathbf{C}^{\text{TRIDIROW}} := (\mathbf{CL} + \mathbf{CC} + \mathbf{CU})$$

$\mathbf{C}^{\text{TRIDIROW}}$ couples each degree of freedom with its immediate left and right neighbour. We make one simple yet important observation (cf. Figure 5.14): Each row in the mesh is completely independent of the other ones because at the left and right boundary, there are no neighbours. This can also be seen in the matrix, which has zero entries at the corresponding positions in the first sub- and superdiagonal. In other words, TRIDIROW is a *linewise preconditioner* in contrast to JACOBI which acts entirely pointwise. We exploit this observation as follows: The tridiagonal preconditioner does not treat one large tridiagonal system with N equations, but rather M smaller, independent systems, each of which is tridiagonal and has M unknowns.

```
1 // grid and block dimensions, exemplarily for M=1025
2 dim3 grid;  grid.x  = 513; grid.y = 4;
3 dim3 block; block.x = 128;
4 // padding and shared memory requirements in bytes
5 int padOffset  = ceil((block.x+1) / 8.0) * 8;
6 int lineOffset = padOffset + block.x;
7 int shmem_in_bytes = lineOffset * 3 * sizeof(float);
8
9
10
11 // kernel code, for the first two colours (the loads of the bottom
12 // and top row are the same for both colours and are thus trivially
13 // coalesced. It suffices that each thread reads two values into
14 // shared memory per row rather than three, reducing bandwidth.
15 extern __shared__ float shmem[];
16 float *deven  = shmem + 0*lineOffset;
17 float *dodd   = shmem + 0*lineOffset + padOffset;
18 float *ceven  = shmem + 0*lineOffset;
19 float *codd   = shmem + 0*lineOffset + padOffset;
20 float *CCeven = shmem + 0*lineOffset;
21 float *CCodd  = shmem + 0*lineOffset + padOffset;
22
23 // indices to coalesce reads and to do the odd/even splitting
24 int blockStart = 2*blockIdx.x*m + m + blockIdx.y*2*blockDim.x;  // global
25 int tid        = threadIdx.x; // just a shortcut
26 int tidOdd     = (tid%2);     // these do the magic of splitting
27 int tidHalf    = tid/2;       // into odd and even shared memory locations
28 int ind        = tidHalf + tidOdd*padOffset; // final index in shared memory,
29                               // alternates between the even and odd pointers
30
31 // load d fully coalesced and perform the splitting
32 deven[ind]                = d[blockStart + tid];
33 deven[blockDim.x/2 + ind] = d[blockStart + blockDim.x + tid];
34 if (tid==blockDim.x-1)
35   deven[blockDim.x]  = d[blockStart + 2*blockDim.x];
36 __syncthreads();
37
38 // compute first color, without bank conflicts
39 ceven[tid] = deven[tid] / CCeven[tid];
40 if (tid==blockDim.x-1)
41   ceven[blockDim.x] = deven[blockDim.x] / CCeven[blockDim.x];
42 __syncthreads();
43
44 // compute second color into codd, using ceven as input [...]
45 codd[tid] = (dodd[tid] - CL[tid]*ceven[tid] - CU[tid]*ceven[tid+1]) / CCeven[tid];
46 if (tid==blockDim.x-1)
47   codd[blockDim.x] = (dodd[blockDim.x] - CL[blockDim.x]*ceven[blockDim.x] -
48                        CU[blockDim.x]*ceven[blockDim.x+1]) / CCeven[blockDim.x];
49 __syncthreads();
50
51 // undo the splitting and store back to global memory
52 x[blockStart + tid]              = ceven[ind];
53 x[blockStart + blockDim.x + tid] = ceven[blockDim.x/2 + ind];
54 // only last halo value per line is written back
55 if (tid==blockDim.x-1 && blockIdx.y*(blockDim.x*2)+2*tid+2 == m-1)
56   x[blockStart+ 2*blockDim.x] = ceven[blockDim.x];
```

Listing 5.6: Avoiding uncoalesced global and bank-conflict shared memory accesses in Gauß-Seidel preconditioners.

The CPU implementation is a straightforward application of the Thomas algorithm [221], which in turn is Gaussian elimination in the tridiagonal matrix case. It comprises two phases, first a forward sweep to eliminate the lower subdiagonal by distributing it to the upper one and the right hand side. The values are also scaled by the main diagonal, effectively eliminating it as well. We only show the necessary computations for the first tridiagonal system to improve clarity; the solution of all M systems is performed sequentially:

$$\begin{aligned} \widehat{\mathbf{CU}}_0 &= \mathbf{CU}_0/\mathbf{CC}_0 \\ \widehat{\mathbf{d}}_0 &= \mathbf{d}_0/\mathbf{CC}_0 \\ \widehat{\mathbf{CU}}_i &= \frac{\mathbf{CU}_i}{\mathbf{CC}_i - \widehat{\mathbf{CU}}_{i-1}\mathbf{CL}_i} && i = 1,\dots,M-1 \\ \widehat{\mathbf{d}}_i &= \frac{\mathbf{d}_i - \widehat{\mathbf{d}}_{i-1}\mathbf{CL}_i}{\mathbf{CC}_i - \widehat{\mathbf{CU}}_{i-1}\mathbf{CL}_i} && i = 1,\dots,M-1 \end{aligned}$$

The resulting bidiagonal system with the superdiagonal $\widehat{\mathbf{CU}}$, a main diagonal containing only ones, and the right hand side $\widehat{\mathbf{d}}$ is then solved in the second phase by a backward substitution sweep:

$$\begin{aligned} \mathbf{x}_{M-1} &= \widehat{\mathbf{d}}_{M-1} \\ \mathbf{x}_i &= \widehat{\mathbf{d}}_i - \widehat{\mathbf{CU}}_i \mathbf{x}_{i+1} && i = M-2,\dots,0 \end{aligned}$$

We refer to the diploma thesis by Altieri [1] for tuning strategies, such as precomputing an LU decomposition and minimising the number of divisions. His work served as the basis for the implementations available in the SBBLAS library.

The entire solution process is inherently sequential. In contrast to the GSRow preconditioner presented in the last section, it is impossible to recover parallelism by a multicolour decoupling, because each individual value depends not only on previously computed neighbours in the 'downstream' direction, but also on 'upstream' values: The tridiagonal solve constitutes a three-term recurrence, the dependencies are implicit, and thus, the application of the preconditioner requires the inversion of a tridiagonal system.

Our first step to parallelise this preconditioner is to exploit the independence between the individual M-unknown systems in each row trivially, by mapping each M-unknown system to a thread block and thus to (virtualised) multiprocessors in CUDA. The remaining problem for an efficient implementation of the TriDiRow preconditioner on GPUs is thus the question on how to solve one such comparatively small tridiagonal problem in parallel using the threads within one thread block.

Previous Work

Parallel methods for solving tridiagonal systems have been developed since the 1960s, the most important approaches are *cyclic reduction* [46, 117], *parallel cyclic reduction* [118] and *recursive doubling* [204]. These algorithms have mostly been targeted at vector supercomputers, and are thus good candidates for the implementation on GPUs, because GPUs have wide-SIMD characteristics (cf. Section 3.3.2).

Kass et al. [124] present the first GPU implemention of the alternating direction implicit (cf. Section 5.4.6) stencil-based tridiagonal solver using shading languages. Sengupta et al. [197] use on-chip shared memory with CUDA to obtain faster results. Both papers use classical cyclic reduction. Shortly before our publication [81], Zhang et al. [247] published a paper in which they discuss multiple parallelisation strategies for the tridiagonal solve. They carefully analyse various algorithmic and implementational bottlenecks. In the following, we briefly present standard and parallel cyclic reduction as well as recursive doubling, and summarise the main results from their paper. This is important to assess our implementation, which we present afterwards. Zhang et al. also map each system to one thread block, and coordinate the threads within each block via shared memory: They copy the three matrix bands and the right hand side to shared memory, and allocate an additional array for the solution vector.

Cyclic Reduction (CR)

Cyclic reduction has been proposed by Hockney [117]. The algorithm proceeds by recursively halving the number of equations and the number of unknowns (*forward elimination*), until a system of two unknowns is reached. This system is then solved, and in the *backward substitution* phase, the remaining half of the unknowns (per level of the reduction tree) is determined using the previously solved values.

The procedure starts by copying the three matrix bands and the right hand side into temporary vectors, denoted by $\widehat{\cdot}$. In each step of the forward reduction, all odd-indexed equations are updated in parallel; equation i of the current system is a linear combination of the equations $i-1$, i and $i+1$:

$$\begin{aligned}
k_1 &= \frac{\widehat{\mathbf{CL}}_i}{\widehat{\mathbf{CC}}_{i-1}} \qquad k_2 = \frac{\widehat{\mathbf{CU}}_i}{\widehat{\mathbf{CC}}_{i+1}} \\
\widehat{\mathbf{CL}}_i &= -k_1 \widehat{\mathbf{CL}}_{i-1} \\
\widehat{\mathbf{CC}}_i &= \widehat{\mathbf{CC}}_i - k_1 \widehat{\mathbf{CU}}_{i-1} - k_2 \widehat{\mathbf{CL}}_{i+1} \\
\widehat{\mathbf{CU}}_i &= -k_2 \widehat{\mathbf{CU}}_{i+1} \\
\widehat{\mathbf{d}}_i &= -\widehat{\mathbf{d}}_i - k_1 \widehat{\mathbf{d}}_{i-1} - k_2 \widehat{\mathbf{d}}_{i+1}
\end{aligned}$$

After the solution on the coarsest level of this reduction, all even-indexed values are solved in parallel by substituting the already solved $\mathbf{x}_{i-1}$ and $\mathbf{x}_{i+1}$ into equation i:

$$\mathbf{x}_i = \frac{\widehat{\mathbf{d}}_i - \widehat{\mathbf{CL}}_i \mathbf{x}_{i-1} - \widehat{\mathbf{CU}}_i \mathbf{x}_{i+1}}{\widehat{\mathbf{CC}}_i}$$

Figure 5.19 illustrates the data flow and communication pattern to solve one tridiagonal system with $M = 17$ unknowns. Cyclic reduction performs $23M$ arithmetic operations overall, almost three times as many as the Thomas algorithm. However, it only requires $2\log_2 M - 1$ steps in parallel. Cyclic reduction can be implemented entirely in-place, which is important due to the size constraints of fast on-chip shared memory.

Zhang et al. [247] analyse their implementation of cyclic reduction in detail, and conclude that the algorithm suffers significantly from high-degree bank conflicts in shared memory in later stages of the elimination and early stages of the substitution sweeps: The shared memory access stride is doubled in each step of the reduction. Furthermore, there is fewer and fewer work to be done on coarse levels of the reduction tree, and more and more threads are inactive: Their implementation uses contiguously numbered threads which significantly improves performance as it minimises warp divergence. Nonetheless, for fewer than 32 equations in the reduced system, only one warp remains active, and this leads to low vector hardware utilisation: The algorithm proceeds to perform four steps in the elimination phase, the direct solution of the final two-unknown system, and four backward substitution steps with only one warp. While the latter problem cannot be avoided in cyclic reduction, we present an implementation that has much more favourable memory access patterns, at the cost of not being completely in-place.

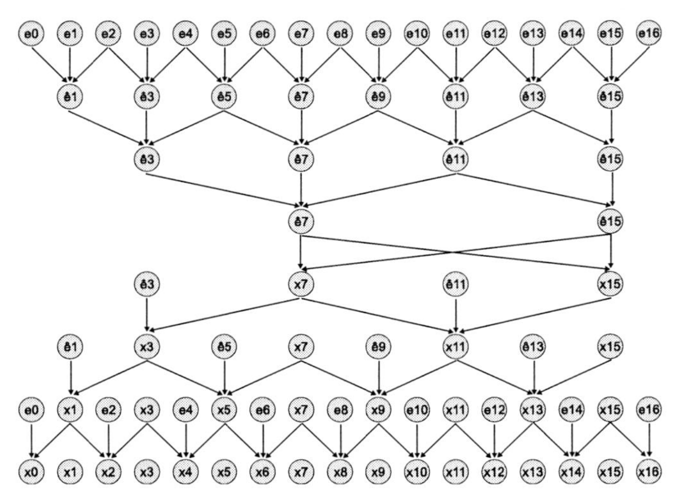

Figure 5.19: Data flow and data placement in shared memory cyclic reduction, $M = 17$. Superscript hat stands for updated equations, and data in the same column occupies the same storage.

Parallel Cyclic Reduction (PCR)

Parallel cyclic reduction is a variant that has been developed by Hockney and Jesshope [118]. In contrast to classical cyclic reduction, PCR only has the forward reduction phase. Instead of halving the number of equations in each step, each system is reduced to two systems of half the size. The reduction mechanism and formulae are the same as in CR. PCR requires fewer algorithmic steps, but each step performs asymptotically more work: PCR takes $12M\log_2 M$ operations and $\log_2 M$ steps to compute the solution of a tridiagonal system with M unknowns. Zhang et al. [247] report that PCR takes roughly half the time of CR despite performing more operations, for two reasons: First, it comprises only one phase, and second, it is free of bank conflicts in shared memory and there are always as many active threads as unknowns in the original system.

Recursive Doubling (RD)

Recursive doubling has been proposed by Stone [204]. Eğecioğlu et al. [64] reformulate the algorithm in terms of the *scan primitive* (parallel prefix sum), exploiting the fact that a trilinear system constitutes a three-term recurrence relation. We refer to Blelloch [26] for a general discussion on solving recurrences using parallel prefix sums. The recursive doubling algorithm is well suited for the implementation on GPUs, because the scan primitive can be implemented efficiently, see Section 3.1.5, Section 3.2.4 and the references therein.

Defining $\alpha_i = -\mathbf{CC}_i/\mathbf{CU}_i$, $\beta_i = -\mathbf{CL}_i/\mathbf{CU}_i$ and $\gamma_i = \mathbf{d}_i/\mathbf{CU}_i$, the recurrence of the tridiagonal system,

$$\mathbf{x}_{i+1} = \alpha_i \mathbf{x}_i + \beta_i \mathbf{x}_{i-1} + \gamma_i$$

can be written in a convenient matrix form:

$$\begin{pmatrix} \mathbf{x}_{i+1} \\ \mathbf{x}_i \\ 1 \end{pmatrix} = \begin{pmatrix} \alpha_i & \beta_i & \gamma_i \\ 1 & 0 & 0 \\ 0 & 0 & 1 \end{pmatrix} \begin{pmatrix} \mathbf{x}_i \\ \mathbf{x}_{i-1} \\ 1 \end{pmatrix} =: \mathbf{B}_i \mathbf{X}_i$$

We define $\mathbf{CL}_0 = \mathbf{CU}_{M-1} = 1$ and $\mathbf{x}_{-1} = \mathbf{x}_M = 0$ so that the recurrence expressions hold for $0 \le i \le M-1$. $\mathbf{B}_0$ is then duplicated, and the sequence of matrices $\mathbf{B}_i$ is generated by performing a parallel scan in the forward direction, using matrix-matrix multiplication as the binary, associative operator. We obtain the following formal sequence of chained matrix multiplications:

$$\begin{aligned}
\mathbf{X}_1 &= \mathbf{B}_0\mathbf{X}_0 \\
\mathbf{X}_2 &= \mathbf{B}_1\mathbf{X}_1 = \mathbf{B}_1\mathbf{B}_0\mathbf{X}_0 \\
&\vdots \\
\mathbf{X}_M &= \mathbf{B}_{M-1}\mathbf{B}_{M-2}\cdots\mathbf{B}_1\mathbf{B}_0\mathbf{X}_0
\end{aligned}$$

The last equation reads more explicitly,

$$\begin{pmatrix} \mathbf{x}_M \\ \mathbf{x}_{M-1} \\ 1 \end{pmatrix} = \begin{pmatrix} c_{00} & c_{01} & c_{02} \\ c_{10} & c_{11} & c_{12} \\ 0 & 0 & 1 \end{pmatrix} \begin{pmatrix} \mathbf{x}_0 \\ \mathbf{x}_{-1} \\ 1 \end{pmatrix},$$

where the c_{kl} depend on the α_i, β_i and γ_i for all equations i. Since $\mathbf{x}_{-1} = \mathbf{x}_M = 0$, the first entry of the solution vector can be computed immediately by $\mathbf{x}_0 = c_{02}/c_{00}$. With this 'seed value', we can compute all entries of the solution vector in parallel using the matrix multiplication chains, i. e., the output of the scan.

Zhang et al. [247] use a step-efficient scan implementation, and report that their implementation of RD requires $20M\log_2 M$ arithmetic operations and $\log_2 M + 2$ steps. Timings reveal that RD takes slightly longer than PCR, the implementation is also free of bank conflicts in shared memory.

Hybrid Variants

Based on their detailed performance analysis of the three parallel tridiagonal solvers, Zhang and his coworkers propose hybrid variants. The idea is that CR is the best algorithm from a computational complexity point of view, it is $\mathcal{O}(M)$ while PCR and RD are $\mathcal{O}(M\log M)$. On the other hand, CR suffers from a lack of parallelism at coarser levels of the elimination and substitution phases, while PCR and RD always have more parallelism. Their hybrid methods thus switch from CR to PCR and RD to solve intermediate, reduced systems when there is not enough parallelism to keep the device busy, allowing to finish the inefficient intermediate steps in less time. Best performance is achieved for a hybrid combination of CR and PCR. They also perform accuracy experiments using diagonally dominant but random test matrices and conclude that RD delivers significantly worse accuracy, while CR, PCR and the hybrid CR+PCR scheme are all on par with a sequential solver using the Thomas algorithm.

This concludes our summary of parallel tridiagonal solvers and the work by Zhang et al. [247]. In the following, we describe our implementation of cyclic reduction, which achieves the same performance as their best variant, at the cost of slightly increased shared memory usage and a more complicated addressing to minimise bank conflicts.

An Alternative, Efficient Implementation of Cyclic Reduction

Let superscript k denote the steps of the reduction and substitution. We first note that it is only necessary to store the three matrix bands and the right hand side in shared memory, the solution is computed incrementally by storing values in the corresponding entries of the right hand side vector as soon as they are available. After the backward substitution has finished, the result is written back to the vector $\mathbf{x}$ in global memory. If all updates in cyclic reduction occur in-place, i. e., $\mathbf{CL}^k$,$\mathbf{CC}^k$, $\mathbf{CU}^k$ and $\mathbf{d}^k$ all occupy the same shared memory storage for all k, then the reading and writing occurs with a 2^k stride in step k, see Figure 5.19. This causes bank conflicts in the on-chip memory and greatly reduces the internal bandwidth, as accesses to the same bank are serialised (cf. Section 5.3.2).

Our solution to this problem is to group the indices always in two contiguous arrays corresponding to odd and even indices. When the initial data is loaded into shared memory, the even- and odd-indexed data are already separated. With an appropriate padding between the arrays this read operation from global to shared memory is fully coalesced and free of bank conflicts. The output of each elimination step again writes into separate even and odd arrays. The even-indexed values of level $k+1$ overwrite the location of the odd-indexed ones of level k in a contiguous fashion, whereas the odd-indexed values of level $k+1$ are written into a new array. With appropriate padding there are absolutely no bank conflicts in the involved read and write operations. As a tradeoff to reduce the amount of memory needed, we can remove the padding and introduce two-way conflicts for the writes only.

Unfortunately, any out-of-place CR scheme makes the backward substitution more difficult, because now the even-indexed $\mathbf{d}_i^k$ cannot be updated in-place. Instead, they are written with stride-two into the odd array on level k. The second stride-two write into this array copies the already known odd-indexed $\mathbf{d}_i^k$ values. Only on the last level 0 all even-indexed $\mathbf{d}_i^0$ can be updated in place, before the result is written back into global memory. Because of the special treatment of the finest level the additional storage requirements of our efficient implementation are $M \cdot (1/4 + 1/8 + \ldots) = M/2$. Figure 5.20 summarises the data placement of our implementation, the data flow remains unchanged.

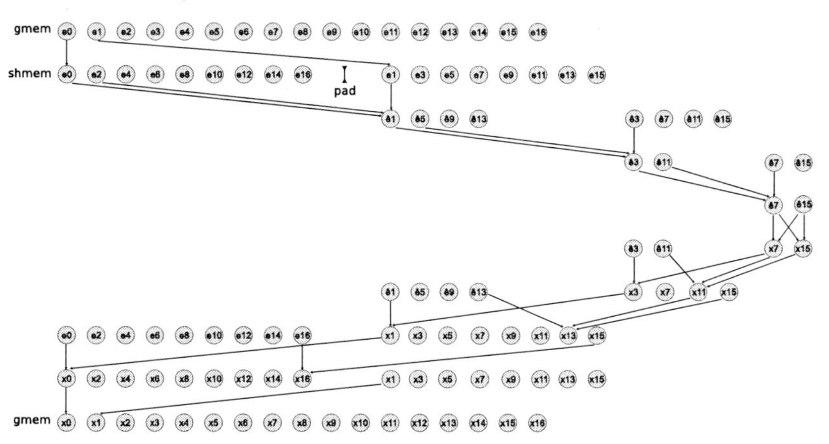

Figure 5.20: Data placement in our implementation of cyclic reduction, $M = 17$. Superscript hat stands for updated equations (we store updated solution values in the right hand side vector), and data in the same columns occupy the same storage.

With 16 kB on-chip memory per multiprocessor on current CUDA-capable GPUs we can thus solve a tridiagonal system with $M = 513$ unknowns (refinement level $L = 9$), as the storage requirement for all four vectors in single precision is $4 \cdot \frac{3}{2} \cdot 513 \cdot 4\,\text{B} = 12\,312\,\text{B}$ plus a few bytes of padding and local variables. Accordingly, in double precision $M = 257$ ($L = 8$) is the current maximum. This underlines that the mixed-precision approach analysed in this thesis not only leads to faster transfers and computations, but, very importantly, enables fast on-chip solvers for larger problem sizes. Clearly, we could implement an algorithm for large M when all data does not fit into on-chip memory and we perform multiple transfers to global memory, but this variant would be significantly slower. Zhang et al. [247] report at least a $3\times$ performance improvement when all updates are performed in shared memory instead of global memory. Our current implementation does not support larger problem sizes yet, as we have focused on the improvements inside shared memory.

In September 2009, NVIDIA announced the technical capabilities of their next-generation architecture named 'Fermi' (chip name GT300) [172]. According to the disclosed details, this design has 64 kB of on-chip memory, which can be partitioned into shared memory and first-level cache. A convenient configuration for our tridiagonal solver would thus be 16 kB L1 and 48 kB shared memory, enabling the computations on refinement level 10 and 9 in single and double precision, respectively.

We compare the performance of our CR implementation with the best algorithm from the work by Zhang et al., which clearly outperforms previous GPU implementations. Both Zhang et al. and we deal with the simultaneous solution of M tridiagonal systems with M unknowns each. In our case (due to the finite element discretisation), M is always a power of two plus one, whereas Zhang et al. support powers of two, but the additional 'plus one' does not make the code much more difficult. They report that the standard, in-place CR implementation on a GeForce GTX 280 GPU for 512 systems of 512 unknowns executes in 1.066 ms. Their fastest solver, a hybrid combination of CR and PCR, needs less than half the time: 0.422 ms. In comparison our memory friendly CR implementation executes in 0.446 ms for 513 systems of 513-unknowns on the same GPU, see Section 5.4.7. In other words, the additional memory requirements of our approach are well-invested.

5.4.5. Combination of Gauß-Seidel and Tridiagonal Preconditioners

The TRIGSROW preconditioner combines the TRIDIROW and GSROW preconditioners:

$$\mathbf{C}^{\text{TRIGSROW}} := (\mathbf{LL} + \mathbf{LC} + \mathbf{LU} + \mathbf{CL} + \mathbf{CC} + \mathbf{CU})$$

$\mathbf{C}^{\text{TRIDIROW}}$ couples each grid point with its immediate left, bottom and right neighbours, see Figure 5.14 on page 142. To apply this preconditioner, we iterate over the rows of the mesh, shift the (already known) values from the previous row to the right hand side (Gauß-Seidel) and solve the remaining tridiagonal system. This solve can be performed efficiently in parallel with our cyclic reduction implementation, but in contrast to the TRIDIROW preconditioner, the rows depend sequentially on each other. To recover parallelism, we decouple the lines by multicolouring. The minimum number of colours is two, first treating all even-indexed lines with the TRIDIROW preconditioner, and then all odd-indexed ones with the full TRIGSROW operation.

Let $I_k = \{kM, \ldots, kM + M - 1\}$ denote the index set corresponding to the k-th line in the grid, $k = 0, \ldots, M$.

1. For all even k in parallel:
   $\mathbf{x}_i = (\mathbf{C}^{\text{TRIDIROW}})^{-1}\mathbf{d}_i \qquad i \in I_k$
2. For all odd k in parallel:
   $\mathbf{x}_i = \mathbf{d}_i - \mathbf{LU}_i\mathbf{x}_{i-M+1} - \mathbf{LC}_i\mathbf{x}_{i-M} - \mathbf{LL}_i\mathbf{x}_{i-M-1} \qquad i \in I_k$
   $\mathbf{x}_i = (\mathbf{C}^{\text{TRIDIROW}})^{-1}\mathbf{d}_i \qquad i \in I_k$

Figure 5.21: Multicoloured TRIGSROW preconditioner.

The implementation of this preconditioner (see Figure 5.21) is straightforward based on the ideas presented in Section 5.4.3 and in Section 5.4.4. The colouring scheme relaxes the coupling compared to the sequential implementation (cf. Section 5.4.3), and we could implement a stronger coupling similar to the GSFullRow preconditioner, incorporating not only the line below, but also the one above in the second sweep over the domain. We did not implement this feature yet, because stronger coupling can also be achieved by partitioning the computation with more colours: With $k = 4$ colours, it suffices to treat every fourth row with the TriDiRow preconditioner only, resulting in four parallel sweeps: First, the rows $0, 4, 8, \ldots$ are treated in parallel with the TriDiRow preconditioner, then the rows $1, 5, 9 \ldots$ can incorporate the previous row, followed by the rows $2, 6, 10, \ldots$ and finally the remaining rows $3, 7, 11, \ldots$. The same idea can also be applied using more colours, and in the limit $k = M$, the sequential variant is recovered. The actual choice of the number of colours k depends on the underlying hardware, i. e., how much independent parallelism is needed to sufficiently saturate the device. This parallelisation technique thus allows to balance the amount of coupling (better convergence rates) with the amount of decoupling (more independent work). We evaluate the variants with two and four colours in Section 5.5.2.

5.4.6. Alternating Direction Implicit Preconditioners

Due to the rowwise numbering, the dependencies within each row (TriDiRow) and to the left, right and bottom (TriGSRow) of the underlying mesh are implicit (i. e., the application of the preconditioner involves formally inverting a system) but other dependencies are explicit. The same idea can be formulated for a columnwise numbering, so that the dependencies along the columns are implicit, resulting in the elementary preconditioners TriDiCol and TriGSCol. Alternating the direction that is treated implicitly leads to the class of so-called *alternating direction implicit (ADI)* methods [179], and we denote the resulting elementary preconditioners by AdiTriDi and AdiTriGS. An analogous idea, even if the coupling is not implicit, can be formulated for any preconditioner. Jacobi acts entirely pointwise, so the row- and columnwise variants are identical. GSCol corresponds to a renumbering of the unknowns compared to GSRow. To apply the ADI variant of a preconditioner $\mathbf{C}$ to the vector $\mathbf{d}$, three steps are performed: First $\mathbf{C}^{\mathrm{ROW}}$ is applied, then the new defect is computed, and finally $\mathbf{C}^{\mathrm{COL}}$ is applied. Consequently, in terms of work performed, one ADI step corresponds to two steps of an elementary preconditioner.

Our realisation of the columnwise and ADI preconditioners follows the implementation in Feast: Instead of re-assembling the local matrices per subdomain based on a columnwise numbering of the mesh points, we rearrange the already assembled bands. We start with Figure 5.22 that illustrates the row- and columnwise numbering of a mesh with 25 grid points.

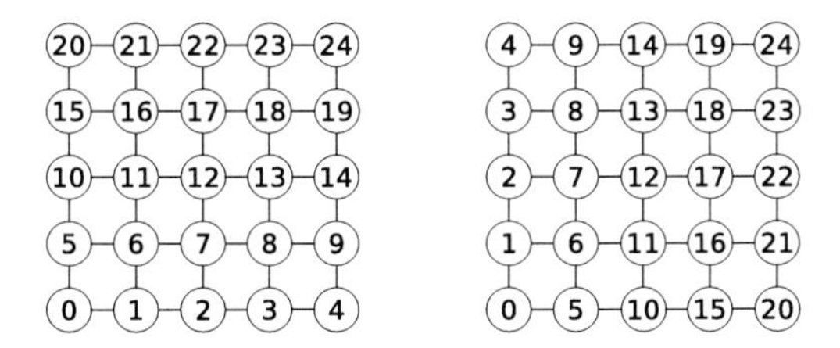

Figure 5.22: Row- and columnwise numbering of the unknowns.

Each matrix band corresponds to the coupling in exactly one direction, and we can thus write equivalently:

$$\begin{matrix} \mathrm{UL} & \mathrm{UC} & \mathrm{UU} \\ \mathrm{CL} & \mathrm{CC} & \mathrm{CU} \\ \mathrm{LL} & \mathrm{LC} & \mathrm{LU} \end{matrix} \quad \mapsto \quad \begin{matrix} \mathrm{LU} & \mathrm{CU} & \mathrm{UU} \\ \mathrm{LC} & \mathrm{CC} & \mathrm{UC} \\ \mathrm{LL} & \mathrm{CL} & \mathrm{UL} \end{matrix}$$

The actual implementation cannot simply copy all bands to their new counterparts, but has to respect the zero entries at locations in the off-diagonals corresponding to the mesh boundaries. This operation is performed in a preprocessing stage immediately after assembling the matrices, because each preconditioner is typically applied to many different vectors. Instead of recomputing this mapping on the fly, we introduce additional storage requirements of four bands (GSCOL), three bands (TRIDICOL) and six bands (TRIGSCOL), respectively. The preconditioners $\mathbf{C}^{\mathrm{ROW}}$ and $\mathbf{C}^{\mathrm{COL}}$ have exactly the same structure, and we can use the implementations for the rowwise variants developed throughout this section without any modifications for the columnwise and alternating direction implicit cases, we just have to provide the corresponding sets of precomputed matrix bands in the parameter lists.

The conversion of a vector from its row- to columnwise representation must be performed on the device, before and after the application of $\mathbf{C}^{\mathrm{COL}}$. The CUDA implementation of this operation is surprisingly challenging, because a naïve code has either coalesced reads and stride-M writes or vice versa. We instead use a tile-based approach where each block of threads rearranges one tile: Each thread reads four sequential values of a 32×32 tile, using two-dimensional addressing of a block with 256 threads. All reads are fully coalesced, and the values are stored in shared memory, directly at their permuted locations in the tile. Because of the sequential addressing, these writes are also free of bank conflicts. In the same way, each thread stores the rearranged tile back to off-chip memory, directly at its new location; these writes are fully coalesced. Our implementation does not run at peak off-chip bandwidth however because of the special cases of tiles that are not 'full', as the number of mesh points per direction, M, is not a multiple of 32.

5.4.7. Performance Microbenchmarks

Figure 5.23 presents timing results for the JACOBI, GSROW, GSFULLROW, TRIDIROW and TRIGSROW smoothers, and additionally for the conversion kernel from a row- to a columnwise representation of a vector. All timings are obtained on the 'high-end' test system which has been used for the CUDA microbenchmarks in Section 5.3.3, i. e., a Core2Duo CPU, a GeForce GTX 280 GPU and fast PC800 DDR2 memory. As the implementations of the CPU and GPU variants differ significantly, we measure performance in milliseconds and not MFLOP/s, this approach gives us a more meaningful, 'honest' comparison.

Table 5.1 highlights the speedup factors we obtain with our GPU implementation. None of the timings include any damping (as required, e. g., within the smoothing iteration of a multigrid solver), we only measure the application of each preconditioner, i. e., the operation $\mathbf{y} = \mathbf{C}^{-1}\mathbf{x}$.

Looking at the performance of the JACOBI kernel first, we observe a significant drop in performance when going from single to double precision on the GPU. The reason for this is that the only arithmetic operation performed in this kernel is a division, which is less efficient in double than in single precision: There is only one double precision ALU per eight single precision ones.[15] Nonetheless, the JACOBI smoother achieves a speedup of 26 and 11 for the largest problem size in single and double precision, respectively; corresponding to a memory throughput of 120 and 43 GB/s. On the other hand, this operation requires most unknowns compared to the other smoothers to achieve a noteworthy speedup, for the smaller problem sizes that do not saturate the device, we observe significant slowdowns on the GPU.

[15] As GPU memory is in general a more precious resource than CPU memory, we do not precompute the inverse of the matrix diagonal for any of our smoother implementations, to trade more memory for faster performance.

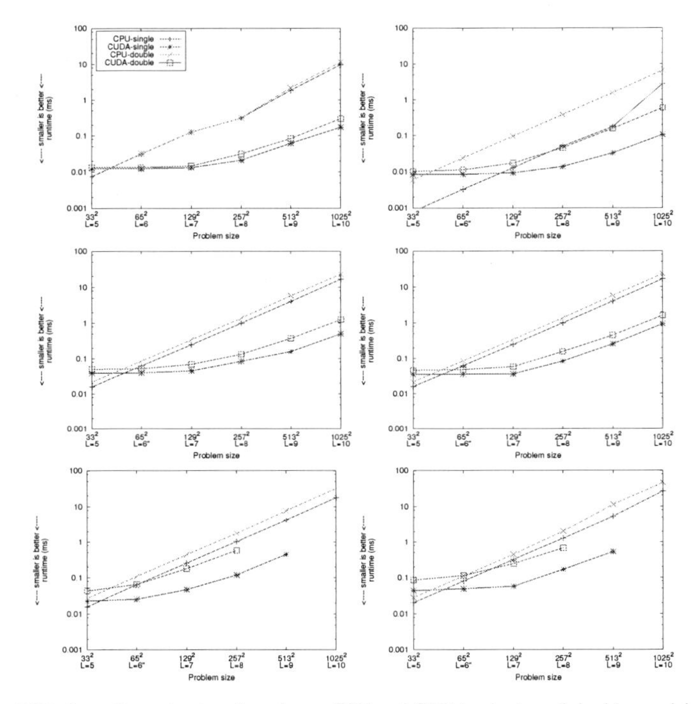

Figure 5.23: Smoother microbenchmarks on CPU and GPU in single and double precision (timings in milliseconds, logscale plot to emphasise differences). Top row: Vector permutation and JACOBI. Middle row: GSROW and GSFULLROW (the CPU measurements are the same in the two plots). Bottom row: TRIDIROW and TRIGSROW. Missing data points indicate insufficient shared memory resources for the tridiagonal solver. All plots share the same legend, and the y-axis uses a logarithmic scale.

Our multicoloured implementation of the GSROW preconditioner achieves the largest speedup overall, because the CPU compiler only partly vectorises the inherently sequential forward substitution.[16] On the GPU, its application executes only four and two times longer than the application of the JACOBI kernel in single and double precision, respectively. While these measurements are at first surprising because the kernel loads much more data and executes much more arithmetic operations, they underline that modern GPUs are throughput-oriented: This preconditioner benefits to a great extent from the ability of the hardware to hide stalls. Furthermore, all additional arithmetic operations compared to the JACOBI kernel are multiplications and additions rather than divisions, which all execute in four cycles per entire warp. GSROW achieves a speedup already for the small problem size of 4 225 degrees of freedom ($L = 6$). For the GSFULLROW preconditioner, the speedup numbers are smaller, since one additional mesh row has to be loaded for the second half of the unknowns. The drop in performance is however surprisingly small, expecially

[16] This has been determined by studying the output of the Intel compiler's optimisation report, but can also be seen in the plots because there is no big difference between single and double precision execution time.

in double precision.

The GeForce GTX 280 GPU only has 16 kB of shared memory, and therefore, the largest problem size we can execute our implementation of a tridiagonal solver for is limited to refinement level 9 and 8 in single and double precision, respectively. Consequently, the speedup compared to a CPU implementation, which furthermore implements a more efficient algorithm, is comparatively small, reaching more than a factor of eight for the largest two possible problem sizes and the TRIDIROW preconditioner, and a factor of three in double precision. On the CPU, GSROW and TRIDIROW take approximately the same time; on the GPU, GSROW is roughly 2.5 times faster.

TRIGSROW is 25 % and 18 % slower than TRIDIROW on the CPU and GPU, respectively. Consequently, the speedups we obtain are slightly larger, reaching more than a factor of ten in single and still a factor of three for the largest admissible problem sizes.

The 'vector permute' operation, which rearranges data from a row- to a columnwise representation, achieves the largest speedup. This is mostly due to the fact that this kernel has not received exhaustive tuning in FEAST yet: Its absolute runtime is very small compared to that of the actual preconditioners and the defect calculations, which together form the application of an alternate direction implicit preconditioning.

	vector permute		JACOBI		GSROW		GSFULLROW		TRIDIROW		TRIGSROW	
L	single	double	single	double	single	double	single	double	single	double	single	double
5	0.61	0.58	0.09	0.59	0.41	0.43	0.46	0.47	0.69	0.63	0.47	0.33
6	2.54	2.38	0.38	2.12	1.58	1.65	1.75	1.77	2.52	1.67	1.62	0.97
7	9.59	8.62	1.38	5.55	5.51	4.94	6.82	5.85	5.42	2.44	5.52	1.81
8	14.87	9.93	3.66	8.40	11.77	10.28	11.81	8.77	8.68	3.08	7.60	3.05
9	30.15	26.07	5.36	9.98	25.23	15.45	15.67	12.84	9.62		10.28	
10	56.19	37.84	26.24	11.12	33.12	18.48	17.83	14.29				

Table 5.1: Speedups obtained by the GPU in microbenchmarks of preconditioners, for increasing level of refinement. Missing numbers indicate problem sizes that cannot be tested due to shared memory limitations.

5.4.8. Summary

We have presented efficient GPU implementations of the preconditioners available in FEAST to date, which are specifically tailored to the underlying generalised tensor product property of the mesh. All these preconditioners are inherently sequential, and we have applied two different techniques to decouple the dependencies and recover parallelism. Preconditioners of Gauß-Seidel type are parallelised via multicolouring techniques. The resulting coupling between the unknowns is weaker than in the sequential case. Tridiagonal solvers cannot be parallelised with this approach. Our novel implementation (co-developed with Robert Strzodka) of the cyclic reduction procedure alleviates most of the drawbacks of previous implementations, outperforms previously published implementations significantly, and performs on par with the fastest published tridiagonal solver known to the author of this thesis.

Our performance microbenchmarks indicate substantial speedups over tuned CPU implementations. In the next section, we analyse how these preconditioners perform in Krylov subspace and multigrid solvers, both in terms of convergence rates and runtime. We are particularly interested in how much numerical efficiency is lost by the decoupling of dependencies, and to which extent the performance microbenchmark results carry over to full solvers.

5.5. Numerical Evaluation of Parallelised Preconditioners in Iterative Solvers

In this section we numerically analyse the (parallelised) preconditioners developed in Section 5.4 in more detail. We are interested in the numerical performance (in terms of convergence rates), their robustness and applicability, and ultimately, their runtime efficiency. Special focus is placed on differences between the CPU and the GPU versions, in particular for the multicolouring variants that effectively reduce the degree of coupling between the values taken into account by the preconditioner. We focus on multigrid solvers, but also briefly test the applicability of these preconditioners in Krylov subspace methods. None of the results in this section have been published previously.

5.5.1. Overview and Test Procedure

To avoid convergence effects that may occur on purely isotropic unitsquare configurations, we use the distorted geometry depicted in Figure 5.24. The geometry is covered with one subdomain, and either regular or anisotropic refinement is applied, the latter only in the first refinement step. Dirichlet boundary conditions on the bottom and left side of the domain are prescribed, and Neumann boundary conditions otherwise. On these domains, we solve the Poisson problem $-\Delta \mathbf{u} = 1$.

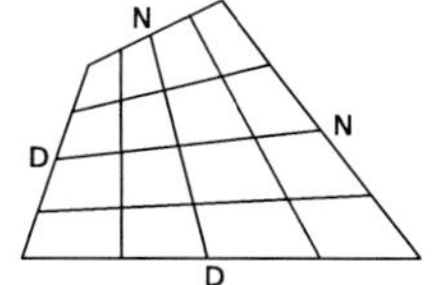

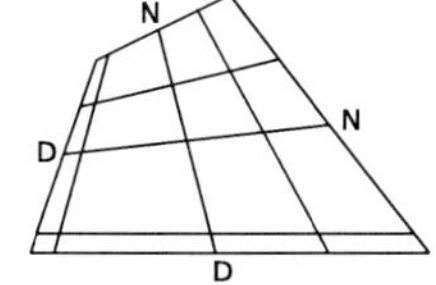

Figure 5.24: Benchmark grids used to evaluate the simple (left) and advanced (right) smoothers and preconditioners (refinement level $L = 2$).

All computations are performed entirely in double precision, in particular we do not employ our mixed precision iterative refinement scheme in order to clearly separate the effects of the parallelised smoothers and preconditioners from precision issues that may arise in the mixed precision setting (cf. Section 5.6.2 and Section 5.7).

We emphasise that in our implementation, smoothing is always realised by a damped, preconditioned Richardson iteration. This means that the smoother is always applied first, and the damping is applied afterwards. If the damping parameter is larger than 1.0, we speak of overrelaxation. Gauß-Seidel with overrelaxation is thus not equivalent to an SOR (successive overrelaxation) smoother, as it is the case in other formulations of the scheme. We analyse the JACOBI, GSROW, GSFULLROW, ADIGS, ADITRIDI and ADITRIGS operators, and refer to Altieri [1] for a similar analysis of multigrid smoothers specifically tailored to generalised tensor product meshes. Since the first four operators are less robust, they are evaluated on the isotropic domain, and the latter two are employed on the more challenging anisotropic case. For the isotropic test case, the GSCOL operator behaves almost identically to the GSROW variant, because it provides the same degree of coupling. In practice, this may not always be the case, for instance for problems involving convection given an appropriate numbering of the mesh nodes. The TRIDIROW and TRIDICOL preconditioners are very well suited to capture anisotropies that only occur in one, 'matching' mesh direction, for instance when boundary layers are resolved in the simulation of highly viscous fluids. To keep the number of tests manageable, we immediately evaluate the alternating direction implicit variants of the ADITRIDI and ADITRIGS preconditioners, the latter is the most robust one available in FEAST.

As explained in Section 5.4.4, the maximum admissible problem size (in double precision) for the two operators that involve solving tridiagonal systems is limited to refinement level $L = 8$ (66 049 unknowns) in our current implementation. All other tests are performed for up to one million grid points (refinement level $L = 10$).

All tests in this section are executed on the 'high-end' test system which is conformant to IEEE-754 in double precision on both the CPU and the GPU, see Section 5.3.3 for the exact machine specifications.

Solver Configurations

All solvers are configured to reduce the initial residuals by eight digits. Our main focus is placed on using the (parallelised) operators as smoothers in a multigrid scheme. The multigrid solver is always configured to perform four pre- and postsmoothing steps each with an elementary smoother, and two with an ADI smoother (we count one application of an alternating direction implicit smoother as two elementary smoothing steps, e. g., four JACOBI steps count as two TRIDIROW and two TRIDICOL steps). We employ a V cycle, and vary the damping factor in steps of 0.1. A conjugate gradient solver is used for the coarse grid problems. As not all preconditioners are symmetric, we use the JACOBI preconditioner for the coarse grid solver when the multigrid solver employs JACOBI or a Gauß-Seidel smoother, and ADITRIDI when the multigrid solver employs ADITRIDI or ADITRIGS.

In Section 5.5.4 we briefly comment on the suitability of these operators as preconditioners in Krylov subspace schemes.

Target Quantities

We measure the 'pure' numerical efficiency in terms of the *convergence rate* κ, which quantifies the reduction of the residuals after k iterations:

$$\kappa := \left(\frac{\|\mathbf{A}\mathbf{x}^{(k)} - \mathbf{b}\|_2}{\|\mathbf{A}\mathbf{x}^{(0)} - \mathbf{b}\|_2} \right)^{1/k}$$

Here, superscript k denotes the iteration number, and $\mathbf{A}\mathbf{x} = \mathbf{b}$ is the sparse linear system to be solved. For the solution of finite element problems, an ideal multigrid solver converges independently of the mesh width h, while the number of iterations increases proportionally to h for the conjugate gradient and BiCGStab solvers, see Section 2.2. A value of $\kappa = 0.1$ means that the given solver gains one digit per iteration.

The ultimate metric from a user perspective is of course the time to solution. However, due to granularity effects, this metric can be too coarse-grained to allow a meaningful comparison. Such granularity effects may occur for instance when a solver just barely misses the prescribed stopping criterion and performs an additional iteration, which leads to the final residual being (in absolute numbers) possibly significantly smaller than the prescribed stopping criterion. Another solver may just 'be lucky' and just barely meet a prescribed criterion. The situation may very well be the other way around for a different problem size, or for a different stopping criterion. The convergence rate alone does not capture the runtime of a given iterative solver at all.

For these reasons, we also compute the *total (runtime) efficiency* E_{total}, which measures the time in µs per unknown necessary to gain one digit in the residual for all tests. It allows a better comparison, independent of the problem size (refinement level, mesh width), the solver configuration, the stopping criterion and the problem under consideration. The total efficiency is defined as follows,

$$E_{\text{total}} := -\frac{T \cdot 10^6}{\log_{10}(\kappa) \cdot N \cdot k},$$

where T denotes the time to solution in seconds and N the number of unknowns. The total runtime efficiency thus reduces the performance and (numerical) efficiency of a given iterative solver to a single number, smaller values are better.

All timings of the GPU solver include the initial transfer of the right hand side to the device (we always start with the zero vector as initial iterate) and the final transfer of the computed solution back to the host, but do not include transfers of matrix data, because this is the relevant setting in our large-scale tests in Chapter 6.

5.5.2. Pure Numerical Efficiency of Multigrid Solvers

JACOBI Smoother

Figure 5.25 depicts convergence rates of the multigrid solver equipped with the simple JACOBI smoother. This operation acts entirely pointwise and is thus trivially parallelisable. As expected, the CPU and GPU variants perform identically. The convergence rate is highly dependent on the (manual) choice of an 'optimal' damping parameter, in fact, the multigrid scheme diverges for damping factors larger than 0.8. For this test case, a damping parameter of 0.7 is the best choice, further underrelaxation increases the convergence rate.

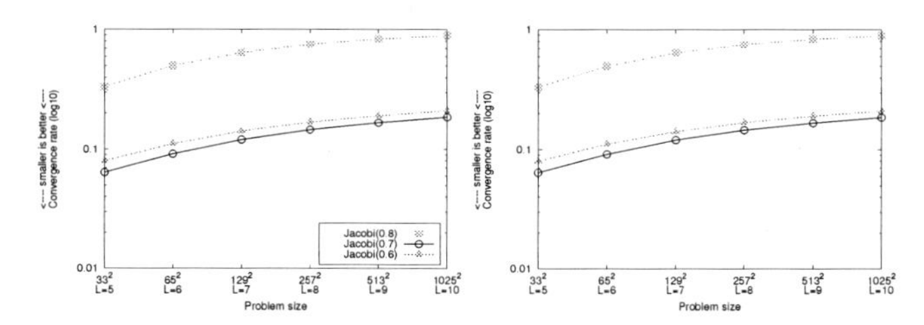

Figure 5.25: Convergence rates depending on the damping factor for the JACOBI smoother on the CPU (left) and the GPU (right). The y-axis uses a logarithmic scale and both plots share the same legend.

GSRow Smoother

On the CPU, the convergence rate of the multigrid solver with the damped GSROW smoother slightly more than halves compared to the best JACOBI case when the number of smoothing steps is kept constant, see Figure 5.26 (left). Equally important, the GSROW smoother is almost independent of the choice of the damping parameter and only starts to degrade for underrelaxation with more than 0.8 or overrelaxation with more than 1.4 (not shown in the figure). For this configuration, small overrelaxation with a factor of 1.1 gives optimal results.

The multicoloured GPU implementation (see Section 5.4.3) provides looser coupling as it takes less values into account. Consequently, the convergence rate is higher, i. e., worse, between 66 % on small levels of refinement and 45 % on larger problem sizes, respectively. For this particular test problem, the effect manifests itself as one additional multigrid iteration performed by the GPU variant. Furthermore, overrelaxation leads to worse convergence or even divergence, and optimal results are obtained without damping. As explained in Section 5.4.3, this is the expected behaviour: The multicoloured variant treats only asymptotically 1/4 of all unknowns as strong as the natural order Gauß-Seidel variant available on the CPU. The experiment confirms that the decoupled variant still leads to faster convergence than the simple JACOBI smoother, and exhibits

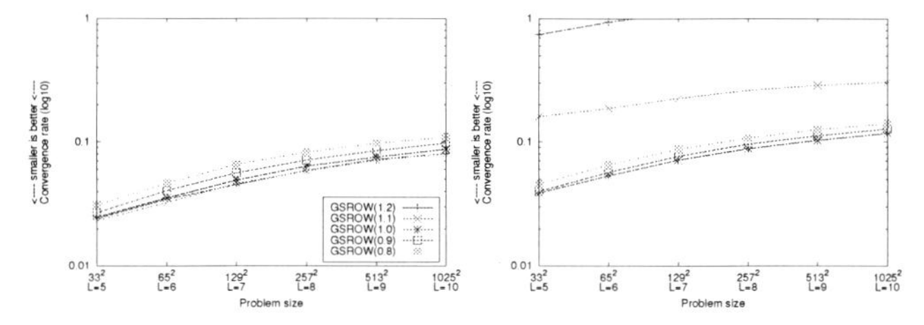

Figure 5.26: Convergence rates depending on the damping factor for the GSRow smoother on the CPU (left) and the GPU (right). The y-axis uses a logarithmic scale and both plots share the same legend.

much less sensitivity to the optimal choice of the damping parameter as long as overrelaxation is not applied.

We experimented with three different ideas to obtain faster convergence with the less powerful GPU-based smoother. Performing six pre- and postsmoothing steps instead of four as on the CPU yields (almost) identical convergence rates as on the CPU, thus alleviating the disadvantage of the multicoloured algorithm. The sensitivity to overrelaxation is not removed. However, this makes the application of the smoother more expensive, and we discuss if the additional work is invested well in terms of total runtime efficiency in Section 5.5.3. In the second and third experiment, we alternate the colouring of the mesh points, effectively varying the degree of coupling that a given mesh point receives inbetween applications of the smoother. For simplicity, we only swap the role of two pairs of colours: In the variant labeled `alternate-in-smoothing`, in all even smoothing steps we treat the unknowns in even-indexed rows of the grid with the first two colours and the unknowns in odd-indexed rows with the remaining two colours. In odd-indexed steps, odd-indexed rows are treated with the first two colours and even-indexed rows with the other two. In the variant labeled `alternate-in-cycles`, all pre- and postsmoothing smoothing steps in one multigrid cycle use the same colouring, but we alternate it analogously in subsequent multigrid cycles. As the results in Figure 5.27 show, the first strategy leads to (slightly) slower convergence, and the second achieves the same convergence rate as leaving the colouring fixed. We therefore do not alternate the colouring of Gauß-Seidel based smoothers in the remainder of this thesis.

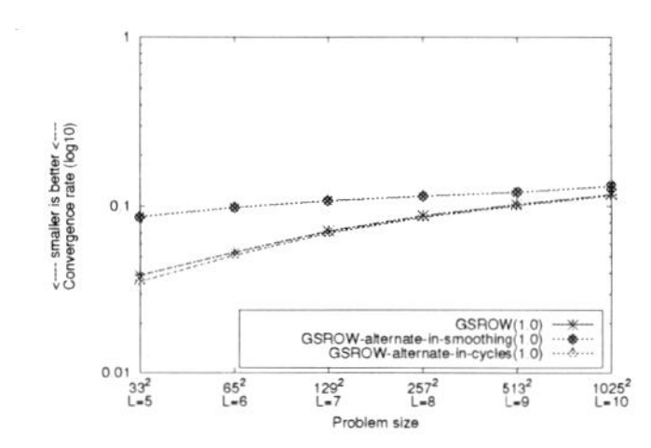

Figure 5.27: Influence of alternating colours in the GPU implementation of the GSRow smoother. The y-axis uses a logarithmic scale.

GSFullRow Smoother

The GSFullRow smoother is an alternative idea to improve the numerical properties of the parallel GSRow preconditioner, which includes all previously updated values in the decoupled sweeps through the domain and is thus equivalent to a renumbering of the unknowns. As depicted in Figure 5.28, it gives best results for this test problem when overrelaxation is applied. Even in the case of damping with 0.9, the convergence rates are better than of the sequential CPU variant. For damping by 1.3 (more overrelaxation reduces performance again), the best variant exhibits between 70 % (small problem sizes) and 40 % ($L = 10$) better convergence rates than the sequential variant. These results underline the effectiveness of our idea to include all already available intermediate values. In Section 5.5.3 we investigate if the slower execution of this variant compared to the straightforward multicoloured GSRow smoother leads to better total efficiency.

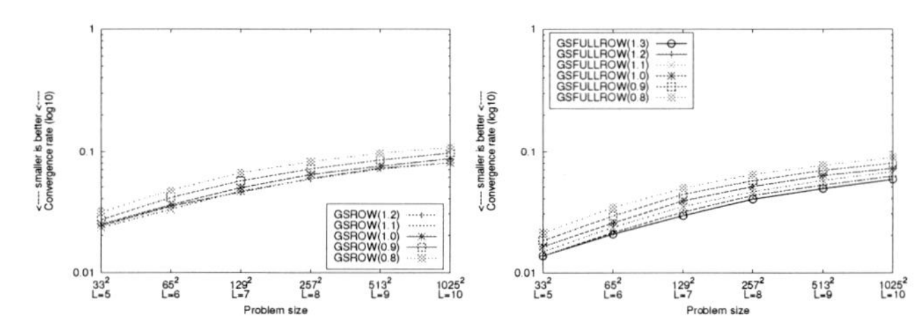

Figure 5.28: Convergence rates depending on the damping factor for the GSRow smoother on the CPU (left, repeated from Figure 5.26) and the GSFullRow smoother on the GPU (right). The y-axis uses a logarithmic scale.

ADIGS Smoother

Figure 5.29 depicts the convergence rates obtained when using the ADIGS smoother.

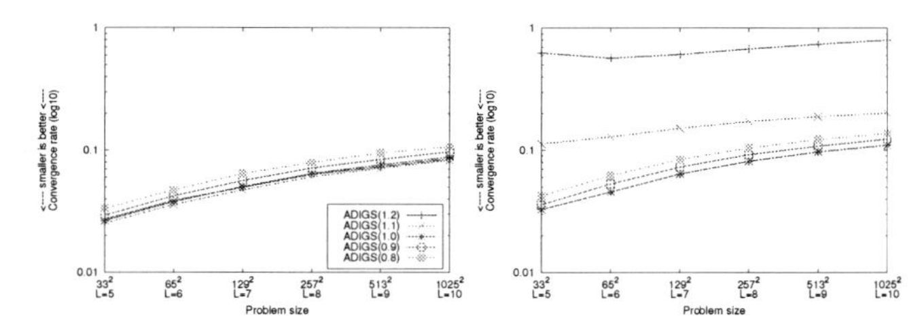

Figure 5.29: Convergence rates depending on the damping factor for the ADIGS smoother on the CPU (left) and the GPU (right). The y-axis uses a logarithmic scale and both plots share the same legend.

On the CPU, we observe convergence rates that are up to 5 % worse than for the standard Gauß-Seidel algorithm, in practice this means that two applications of the GSRow smoother are equivalent to one application of the ADIGS one. This is the expected behaviour, as we see no difference between the GSRow and GSCol variants for this configuration (cf. Section 5.5.1). On the GPU however, employing the alternating direction variant is beneficial in terms of convergence rates and robustness: Overrelaxation does not lead to divergence any more, although we observe a dependency on the choice of the damping parameter. The disadvantage of the multicolouring slightly reduces from 45 % (GSRow) to 33 % (ADIGS). We investigate the influence on the total efficiency in Section 5.5.3.

ADITRIDI Smoother

The results in the next paragraphs are obtained on the anisotropic domain (cf. Figure 5.24 (right)). Due to limitations of the current GPU implementation, we can only compute problem sizes up to refinement level $L = 8$.

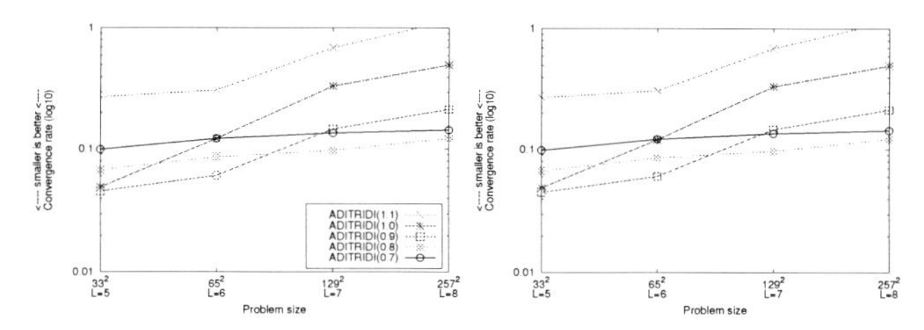

Figure 5.30: Convergence rates depending on the damping factor for the ADITRIDI smoother on the CPU (left) and the GPU (right). The y-axis uses a logarithmic scale and both plots share the same legend.

The CPU- and the GPU-based solvers equipped with the ADITRIDI smoother exhibit identical convergence rates (see Figure 5.30), underlining that the inversion of tridiagonal systems with the cyclic reduction algorithm is as stable as the standard Thomas algorithm: The matrices stemming from the discretisation of all problems in this thesis are diagonally dominant and thus, no pivoting needs to be applied which would impact stability. The performance of the ADITRIDI smoother depends significantly on an optimal choice of the damping parameter: Overrelaxation leads to divergence, and for this configuration, optimal results are obtained for a damping factor of 0.7–0.8. Further underrelaxation increases the convergence rates again.

ADITRIGS Smoother

The ADITRIGS smoother is the most robust smoother available in FEAST. We discuss the CPU variant first. As depicted in Figure 5.31 (left), the resulting convergence rates of the multigrid solver are more or less independent of the damping parameter only too strong underrelaxation starts to have a deteriorating effect. Best results for this test case are obtained without any damping. Compared to the ADITRIDI smoother, we observe more than a halving of the convergence rates.

The implementation on the GPU uses a bicolouring of the mesh rows to decouple the dependencies of the inherently sequential Gauß-Seidel part, see Section 5.4.5. Similar to the observations for the GSRow smoother, we see that the parallelised variant is no longer independent of the choice of the damping parameter. Moderate underrelaxation by a factor of 0.9 yields best results, and overrelaxation is generally bad. The convergence rates of the GPU solver are approximately 60–80 % higher than on the CPU, which manifests itself as one additional multigrid cycles compared to the CPU version. It would be interesting to alternate the colouring for this smoother as well, but discouraged by the limited success we had with this approach for the GSRow smoother, we did not implement this technique for ADITRIGS. In future work, we will implement a stronger coupling in every second row of the mesh as well, analogously to the GSFULLROW smoother.

Figure 5.32 depicts the convergence rates for varying damping parameters when using four

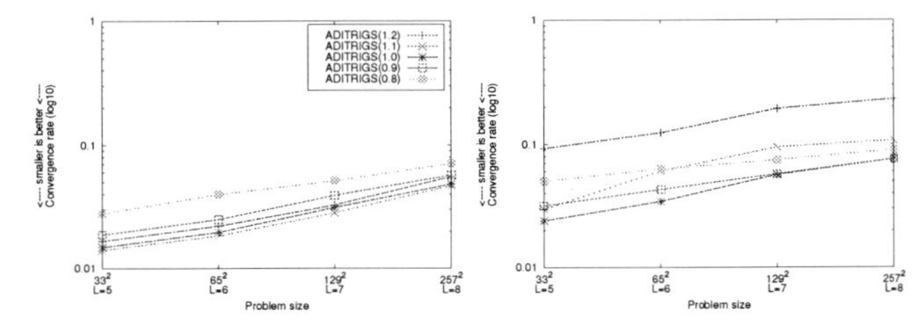

Figure 5.31: Convergence rates depending on the damping factor for the ADITRIGS smoother on the CPU (left) and the bicoloured variant on the GPU (right). The y-axis uses a logarithmic scale and both plots share the same legend.

instead of two colours, which improves the coupling because only one quarter instead of half the rows in the mesh are treated with a tridiagonal preconditioner only. Consequently, the convergence rates improves as expected, and are now only 40 % higher than those of the sequential variant on the CPU.

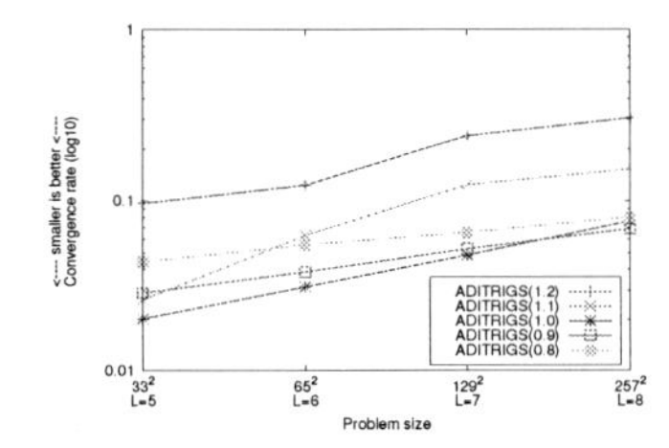

Figure 5.32: Convergence rates depending on the damping factor for the ADITRIGS smoother on the GPU when using four instead of two colours. The y-axis uses a logarithmic scale.

5.5.3. Total Runtime Efficiency of Multigrid Solvers

In the following, we compare the total efficiency of multigrid solvers on the CPU and the GPU for the six different smoothers under consideration in this section. We restrict ourselves to the configurations and parameter settings identified as optimal in terms of convergence rates for the test problems under consideration in Section 5.5.2.

Mostly Isotropic Case: JACOBI, GSROW, GSFULLROW and ADIGS

Figure 5.33 depicts the total runtime efficiency (measured in μs) of multigrid solvers equipped with the JACOBI, GSROW, GSFULLROW and ADIGS smoothers for a Poisson problem on the mostly isotropic test domain. Looking at the CPU results first, we observe that for all problem sizes the GSROW smoother yields exactly the same efficiency as the JACOBI smoother, which requires the tedious, manual search for an 'optimal' damping parameter. In terms of runtime efficiency, the independence of the GSROW smoother of the damping parameter is practically for free: For the largest problem size, we measure a total runtime efficiency of 0.654 μs and 0.648 μs for JACOBI and GSROW, respectively. Since the JACOBI and GSROW smoothers are applicable in the same situations, GSROW should thus always be used instead of JACOBI. In other words, the additional

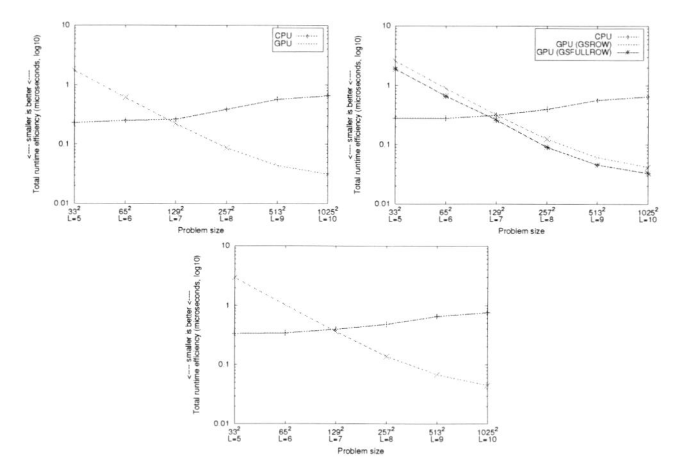

Figure 5.33: Total runtime efficiency on the CPU and GPU, JACOBI (top left), GSROW and GSFULLROW (top right) and ADIGS (bottom). The y-axis uses a logarithmic scale and all plots share the same legend.

computations performed by the GSROW smoother are well invested. In terms of convergence rates, GSROW and ADIGS are practically equally efficient. The conversion operations from a row- to a columnwise representation and vice versa of the defect vector however have a significant effect in terms of total efficiency, leading to an approximately 20 % higher value of E_{total} for the latter. The additional work thus does not yield returns in this test case.

On the GPU, the situation is different, as the multicoloured parallelisation of Gauß-Seidel type smoothers reduces the 'pure' numerical performance. We observe between 66 % and 45 % worse convergence rates for the GSROW smoother (cf. Figure 5.26) as compared to the natural ordering sequential CPU implementation. These factors are identical when measuring the impact in terms of total efficiency. The total efficiency of multigrid with GSROW is 33 % worse than that of JACOBI for large problem sizes, while its convergence rate is approximately 60 % better (compared to more than 200 % on the CPU). The conclusion is thus not as straightforward as on the CPU, as the better numerical efficiency and the worse runtime do not cancel out as favourably. From a practical point of view we nonetheless suggest to use GSROW instead of JACOBI, because the multicoloured variant is still almost independent of the choice of the damping parameter as long as overrelaxation is avoided. The GSFULLROW variant is however always more efficient than the parallel GSROW smoother. Its more favourable numerical properties are not offset by the higher runtime due to higher bandwidth demands and more computations. In the remainder of this thesis, we therefore always use this variant of the parallel Gauß-Seidel smoother. ADIGS leads to almost exactly the same total efficiency as GSROW on the GPU while reducing the difference (compared to the natural order CPU algorithm) in terms of convergence rates from 45 % to 33 % ($L = 10$, similar improvements for smaller problem sizes), so in contrast to the CPU, the alternating direction implicit variant is more favourable.

Finally, we want to highlight the speedup obtained by the GPU. We measure this speedup only in terms of total efficiency to abstract from purely numerical performance. This approach is justified by the fact that the parallelised variants are in general applicable for the same configurations as their CPU variants. We first note that—as expected from our microbenchmark results in Section 5.3.3 and Section 5.4.7—total efficiency on the CPU degrades significantly as soon as the problem size does not fit entirely into on-chip cache memories any more. In contrast, the GPU is always most efficient for the largest problem size, as the device needs to be saturated to be able to fully hide latencies of accesses to off-chip memory. More importantly, the latency[17] of the PCIe transfers (which we always include in our timings of entire subdomain solvers) impacts efficiency for smaller problem sizes to a greater amount: The break-even point at which the GPU reaches the same performance as the CPU implementation is reached for a problem size of 16 641 unknowns (refinement level 7). For larger problem sizes, we observe speedup factors of 4.5, 13.1 and 20.9 ($L = 8, 9, 10$, JACOBI smoother), of 3.2, 9.3 and 15.5 (GSROW smoother), of 4.4, 12.3 and 19.6 (GSFULLROW smoother) and of 3.5, 9.6 and 16.8 (ADIGS smoother). Roughly speaking we can say that in order to achieve (close to) an order of magnitude speedup for the entire multigrid solver, we need at least a problem size of 263 169 degrees of freedom (corresponding to refinement level $L = 9$). For the sake of completeness, we also state the absolute time to solution for the largest problem sizes, which amount to 0.27 s, 0.37 s, 0.30 s and 0.41 s for the four different smoothers, respectively.

Anisotropic Case: ADITRIDI and ADITRIGS

For these two more robust smoothers which are evaluated for the anisotropic test problem, the maximum problem size is currently limited to refinement level $L = 8$ in double precision on the GPU. Figure 5.34 depicts the total runtime efficiency of multigrid solvers using these two smoothers. On the CPU, ADITRIGS is always more efficient than ADITRIDI, the superior numerical properties translate directly to efficiency, and the additional computations are well invested. The situation is symmetric to the comparison between JACOBI and GSROW in the previous paragraph. Consequently, ADITRIGS should always be used, since it is more robust, it is applicable to a wider range of problems, and the additional computations lead to good returns in efficiency. However, in certain situations, there may not be enough memory available to store the additional vectors required by the alternating direction implicit implementation: ADITRIDI only needs three additional vectors while ADITRIGS needs six.

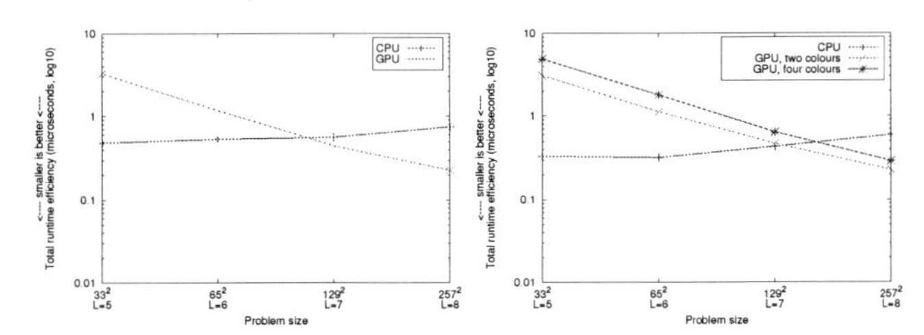

Figure 5.34: Total runtime efficiency on the CPU and GPU, ADITRIDI (left) and ADITRIGS (right). The y-axis uses a logarithmic scale and both plots share the same legend.

On the GPU, the decoupling introduced by the bicolouring or quadcolouring of the parallelised ADITRIGS smoother results in a deterioration of convergence rates compared to the CPU variant, while the ADITRIDI smoothers are identical on the two platforms. ADITRIGS improves convergence rates by a smaller amount on the GPU, leading to an increase of E_{total}. The exact numbers

[17] Latency is more important than bandwidth for small problem sizes, as the transferred amount of data is rather small and only increases by a factor of four for increasing level of refinement. In essence, the penalty to transfer 16 641 values via PCIe is the same as for any smaller data size, see also the measurements by Kindratenko et al. [132].

are 0.7674 µs vs. 0.2295 µs for multigrid with ADITRIDI compared to 0.5967 µs vs. 0.2282 µs for ADITRIGS (with two colours) on the CPU and GPU for the largest possible refinement level $L = 8$. Using four colours in the parallelised ADITRIGS smoother improves convergence rates slightly, but this advantage does not pay off in terms of total runtime efficiency. For this test problem, the efficiency is always worse than for the bicoloured variant, approximately 28 % for the largest admissible problem size compared to a 11 % better convergence rate. This result is surprising, because for the largest admissible problem size, only one mesh row is resident on each multiprocessor at a given time, independent of the amount of colours, and one would expect that the computations on the finest level of refinement dominate the absolute runtime. However, computations on smaller levels of refinement, where the device is not saturated, suffer disproportionately: For smaller problem sizes, the improvements in terms of convergence rates are almost identical (cf. Figure 5.31), but the total efficiency of the four-coloured variant is 58, 59, 37 and 28 % higher than of the two-coloured variant, for refinement levels $L = 5, \dots, 8$, respectively. The GPU used for these tests has 30 multiprocessors, which means that the device is severely underutilised for small problem sizes: For 1 089 unknowns ($L = 5$), only 16 and eight multiprocessors are actually active (33 mesh rows partitioned into two and four kernel calls, respectively). Underutilisation is even worse for coarser mesh resolutions that are visited in the course of the multigrid cycle: Each block of threads comprises only one warp for problem sizes below 1 089 unknowns (each row/block has only 16 unknowns), and consequently, the multiprocessors cannot benefit from latency hiding to off-chip memory any more: In the limit case of the coarsest level of refinement, the GPU behaves more like a 1.3 GHz serial processor with a high latency to access uncached off-chip memory. Alternatively, one could argue that coarser levels in multigrid suffer to a great extent from serialisation effects (that prevent strong scaling improvements with an increasing amount of multiprocessors) on GPUs than on CPUs.

As all these effects on coarser levels of refinement build up in the course of the computation, their individually negligible impact is no longer negligible for the entire multigrid solver. We discuss this aspect further in Section 5.8 in the context of F and W cycle multigrid, which spend much more time on coarser resolutions than the V cycle we employ here. A final conclusion which of the two parallel ADITRIGS implementations is better cannot be made yet, and we revisit this question during our evaluation of mixed precision methods in Section 5.7.3, because we can apply this smoother on refinement level $L = 9$ in single precision.

Even better convergence rates can be obtained when decoupling the treatment of rows to a lesser degree, i. e., by introducing more than four colours (see Section 5.4.5). However, with eight colours, the device is already underutilised for refinement level $L = 7$ (129 mesh rows lead to eight sweeps with 17 and 16 active multiprocessors). Given the discouraging analysis of the slowdown introduced by computations on smaller levels of refinement, we did not implement this variant, and always employ the bicoloured variant of ADITRIGS in double precision in the following, unless explicitly mentioned otherwise.

Performing more smoothing steps does improve the convergence rates of multigrid with ADITRIGS smoother, but decreases its efficiency, so the extra work is not invested well (results not shown in the plots).

In terms of speedup, the break-even point from which on the GPU implementation is faster than the CPU (including transfers via PCIe) is the same as for the 'simple' smoothers, except for multigrid with the ADITRIGS smoother which only achieves speedups for the largest problem size under consideration. Nonetheless, we observe speedup factors of 3.3× and 2.6× (2.1× for the four-coloured variant) for E_{total} on refinement level $L = 8$, in line with speedups measured for simple smoothers and this problem size. Future hardware will enable computations on larger problem sizes with more shared memory [172], so we expect larger speedups for the more advanced smoothers as well, see also Section 5.7.4 for the single precision case which allows computations up to $L = 9$.

5.5.4. Krylov Subspace Solvers

We want to briefly examine if our findings carry over to the case of preconditioned Krylov subspace methods. We only consider the largest admissible problem size of $L = 8$, because we only employ Krylov subspace methods as coarse grid solvers in the multigrid algorithm, which is much more efficient than Krylov subspace methods for the problems under consideration in this thesis. At the time of writing, our implementation provides the conjugate gradients and BiCGStab algorithms. Only the latter can use unsymmetric preconditioners of Gauß-Seidel type, therefore we limit our tests to this solver. BiCGStab is known to react sensitively to floating point noise, i. e., very small variations can influence the convergence behaviour dramatically. Table 5.2 lists the number of iterations required until convergence when using different preconditioners for a BiCGStab solver, the more powerful ADITRIDI and ADITRIGS preconditioners are evaluated on the anisotropic domain, and the remaining four on the isotropic one.

	JACOBI		GSROW / GSFULLROW			ADIGS		ADITRIDI		ADITRIGS (2,4 colours)		
Level	CPU	GPU	CPU	GPU	GPU	CPU	GPU	CPU	GPU	CPU	GPU	GPU
5	91	94	58	73	54	33	45	24	24	17	22	20
6	199	190	118	166	128	69	87	48	48	35	45	40
7	387	385	296	322	258	147	283	106	113	75	91	84
8	795	755	739	826	562	358	624	206	222	167	221	209

Table 5.2: Number of iterations until convergence of a BiCGStab solver with different preconditioners.

The GPU solver is, compared to the CPU case, particularly bad for ADIGS. For JACOBI and ADITRIDI, the number of iterations until convergence is almost identical on the CPU and on the GPU except for the largest problem size where floating point noise starts to deteriorate the convergence. This is the expected behaviour, because the two implementations are equivalent and do not involve weaker coupling due to multicolouring techniques. The beneficial effects of the GSFULLROW preconditioner, and of using four rather than two colours in the ADITRIGS preconditioner carry over from the multigrid solver. For instance, the GSFULLROW preconditioner leads to a reduction from 739 to 562 iterations compared to the standard, sequential Gauß-Seidel preconditioner on the CPU.

5.5.5. Summary

The numerical evaluation of the preconditioners developed in Section 5.4 yields a number of interesting findings. The following conclusions are valid for both multigrid and Krylov subspace solvers.

On the CPU, the GSROW smoother is as efficient (in terms of total runtime efficiency) as the JACOBI smoother, its better convergence rates are perfectly balanced by the higher amount of computations and memory accesses. Since JACOBI is less robust and depends to a high degree on the manual, 'optimal' choice of the damping parameter, GSROW is always more favourable than JACOBI. The alternating direction implicit variant of the Gauß-Seidel smoother is less efficient, which is surprising because it only corresponds to a renumbering of the mesh in its columnwise sweeps. Analogously, the ADITRIGS smoother, the most robust smoother available in FEAST, is more (runtime-) efficient than using only the tridiagonal part as with the ADITRIDI smoother.

On GPUs, the analysis is not as straightforward. JACOBI and ADITRIDI yield identical convergence rates on CPUs and GPUs, and both depend on the 'optimal' manual choice of the damping parameter. With moderate damping, we achieve best results. The multicolouring to recover parallelism from the inherently sequential Gauß-Seidel type smoothers relaxes dependencies and thus provides weaker coupling. In our experience, the parallelised variants still provide enough coupling so that they are applicable in the same situations the natural ordering variant on the CPU. GSROW on the GPU is mostly invariant of the relaxation parameter unless overrelaxation is applied. The convergence rate is worse than on the CPU; in this particular test case, the GPU performs one more multigrid cycle. In terms of runtime efficiency, the better convergence rate and slower execution compared to JACOBI do not cancel out as favourably as on the CPU. We analysed two techniques that alternate the colouring to improve numerical performance of the parallelised Gauß-Seidel smoothers, but none proved to be beneficial. The alternating direction implicit variant of the GSROW smoother is advantageous both in terms of convergence rates and runtime efficiency on the GPU. The decoupling makes it possible to interpret the Gauß-Seidel smoother in a different way, by always incorporating all previously computed intermediate results in later sweeps over the domain. This idea has been co-developed with Robert Strzodka, and we termed this stronger coupling parallel variant GSFULLROW. Our numerical evaluation of this preconditioner leads to very promising results, in our experiments, we observe that with moderate overrelaxation, we are able to outperform the sequential, natural ordering Gauß-Seidel operation both in terms of convergence rates of Krylov subspace and multigrid solvers equipped with this preconditioner, and, more importantly, also in terms of total runtime efficiency. Finally, the parallelised ADITRIGS variants using two and four colours to decouple the sequential dependencies of mesh rows on their predecessors in the numbering lose numerical efficiency compared to the sequential algorithm on the CPU. Using four colours rather than just two, i. e., treating only every fourth row in the mesh with the tridiagonal preconditioner only, is advantageous in terms of convergence rates. Due to device underutilisation (only refinement level $L = 8$ can be treated in double precision), we have not been able to determine the better of the two choices yet.

In terms of speedup, we observe more than one order of magnitude (in fact, a factor of 20) for large problem sizes and the simpler smoothers, and factors of 3.3 and 2.6 for more robust smoothers and medium problem sizes. Due to limitations of the current implementation, we cannot apply the more robust smoothers to large problem sizes where a full saturation of the device and hence an even larger speedup is achieved.

We analyse the smoothers again in the context of mixed precision iterative refinement solvers on a single subdomain (Section 5.7), and in our evaluation of accelerating applications built on top of FEAST, in particular for the Poisson problem (Section 6.3), and for the Navier-Stokes application in Section 6.5, where the difficulty to solve the linearised problems does not result from (artificially introduced) anisotropies but from nonlinearities in the governing equations.

In future work, we plan to add the stronger coupling which incorporates all already known values during the sweeps of a multicolour parallelisation also to the ADITRIGS preconditioner which uses two colours, because this technique has proven to be highly beneficial in the GS-FULLROW case. Furthermore, MG-Krylov techniques can be implemented and analysed, i. e., techniques that use few iterations of for instance the preconditioned conjugate gradient algorithm as a smoother within the multigrid solver. The advantage of this approach is that due to the underlying **A**-orthogonalisation, the smoother computes optimal steplength parameters, i. e., optimal damping factors.

The best known preconditioners in practice—for these kinds of problems—are based on an *incomplete LU decomposition (ILU)*, see for instance the textbook by Barrett et al. [13]. These techniques generally change the sparsity pattern of the matrices, and are thus not applicable in our context, but rather in the context of completely unstructured grids. The *sparse approximate inverse (SPAI)* algorithm promises to bridge this gap [101]: It computes an explicit representation—with a fixed sparsity pattern—of the inverse of a given matrix by minimising $||\mathbf{AC}-\mathbf{I}||$ in the Frobenius norm. The SPAI algorithm has been implemented by Marcus J. Grothe, and the code is freely available.[18] According to the developers, it is possible to configure the SPAI package to compute an approximate inverse that exactly matches our banded matrix structure. Once such an approximate inverse is available, preconditioning reduces to matrix-vector multiplication, and it would be interesting to analyse such approaches, both from a numerical point of view, and in the context of Chapter 6 and nonlinear problems in general where we cannot neglect matrix assembly and the associated 'preprocessing' steps of assembling preconditioners any more.

[18] `http://www.computational.unibas.ch/software/spai/`

5.6. Accuracy of Emulated and Mixed Precision Solvers

In Section 4.5 on page 95 we have demonstrated numerically—for the prototypical yet fundamental Poisson problem—that single precision is insufficient to solve many ill-conditioned problems stemming from finite element discretisations. The conventional solution to this problem, applied by the vast majority of practitioners in the field of numerical mathematics, is to use double precision exclusively, simply because double precision as defined by the IEEE-754 standard [122] is the highest precision natively available on commodity hardware. However, this is not necessarily the best approach in terms of performance, or might even be impossible on certain hardware. The ideas to emulate high precision formats based on unevaluated sums of native low precision values (Section 4.2), and to use different precision formats in the same algorithm (Section 4.3), have many potential benefits, as outlined in Section 4.1.2. Before evaluating performance aspects in Section 5.7, the goal of this section is to answer the following simple, yet important question: *Are emulation or iterative refinement techniques able to deliver the same accuracy as computing entirely in high native precision for the problems in the scope of this thesis?* If this is not the case, then there is no alternative to using high precision exclusively.

Section 5.6.1 describes our test environment. In Section 5.6.2 and Section 5.6.3 we examine the iterative refinement procedure (using standard double and single precision) and the emulated high precision format, respectively. We conclude with a summary in Section 5.6.4. Although we only focus on double, single and emulated double-single precision, some of our findings are also relevant for emulated quadruple precision formats, which may become important if for some applications double precision is not sufficiently accurate any more. All tests in this section are based on the Poisson problem, and constitute an updated and extended version of previously published joint work with Robert Strzodka [81], and with Robert Strzodka and Stefan Turek [85].

5.6.1. Overview and Test Procedure

Three different floating point formats are available for our tests: Single precision (s23e8), double precision (s52e11) and the emulated double-single format (s46e8, see Section 4.2). Single and double precision are available on the CPU. Single precision (though not entirely IEEE-754 conformant) is available on all GPUs, both in our legacy OpenGL implementation and in CUDA. Emulated double-single precision is only available in our OpenGL implementation, and finally, the GeForce GTX 280 (included in the 'high-end system', see Section 5.3.3) is the only GPU available to us that supports native double precision, compliant with the IEEE-754 specification. All tests below are executed on this 'high-end' test system.

As explained in Section 4.5.1, the condition number of the system matrix determines to what extent errors in the input data and errors accumulating in the course of the solution are amplified. The mixed precision iterative refinement procedure is only guaranteed to converge to working accuracy if the system to be solved is 'not too ill-conditioned' in the low precision format, see Section 4.3. Instead of using arbitrary test matrices with known estimates of the condition number (e. g., Hilbert matrices of increasing dimension), we employ *mesh anisotropies*. Such anisotropies are very common in practice, and we have the additional benefit that all anisotropies we consider do not break the logical tensor product property of the mesh, and thus the banded structure of the system matrices. In the following, we use two different methods to introduce anisotropies and thus strongly varying coefficients to the system matrices:

- **Uniform anisotropies:** Using rectangular instead of square elements allows us to specify the aspect ratio of each element in the refined discretisation. We refine the mesh uniformly (cf. Figure 5.35 on the left) which leads to uniformly distributed varying coefficients in the matrix.

- **Anisotropic refinement:** In this strategy, we start with the unitsquare and in each refinement step, we subdivide the bottommost and rightmost layer of elements anisotropically (cf. Figure 5.35 on the right). This refinement scheme is often used to accurately resolve boundary layers in fluid dynamics, or discontinuities in solid mechanics. For each refined element in these layers, the new midpoint x_c is calculated by recursively applying the formula $x_c = x_l + \nu \cdot \frac{x_r - x_l}{2}$ with a given *anisotropy factor* ν ($\nu = 1.0$ yields uniform refinement) and x_l and x_r denoting the x-coordinates of the left and right edge of a element before subdividing (analogously for the y-component). All other elements are refined uniformly. This leads to matrices with locally condensed, non-uniformly distributed anisotropies instead of uniform ones.

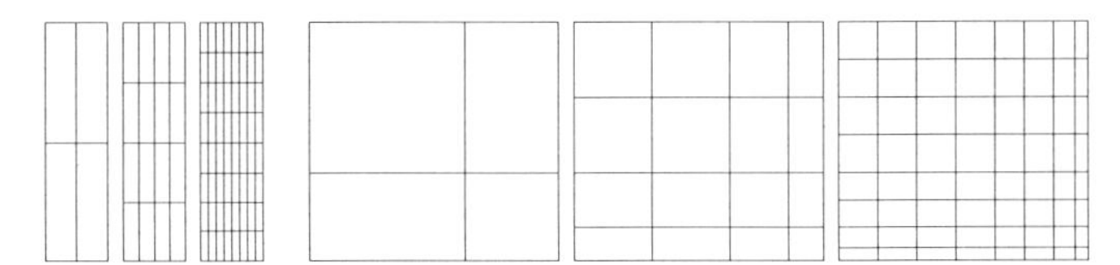

Figure 5.35: Example of different mesh anisotropies on refinement levels 1–3. Left: Uniform subdivision of an anisotropic coarse mesh. Right: Anisotropic subdivision of the rightmost and bottommost element layer in each refinement step.

We note that the above cases generate matrices which also occur when discretising anisotropic operators. For example, operators of the type $\text{div}(\mathbf{G}\nabla\mathbf{u})$ for a constant (i. e., independent of the spatial location) diagonal matrix $\mathbf{G}$ lead to the same matrix entries as introducing a corresponding degree of anisotropy directly on the coarse mesh level. Consequently, our test results also apply directly to these problems of operator anisotropy (cf. also the description of the discrete systems arising in linearised elasticity in Section 2.3.5).

Detailed Test Cases

Our tests comprise the following eight configurations, which have already been used in previously published work [81, 85]. In addition to the dimensions of the domain Ω, we also list the maximum aspect ratio on the finest level of refinement (AR_{max}) and the smallest mesh width h_{min}. Furthermore, we compute the condition numbers of the system matrix for refinement levels $L \leq 6$ by brute force, see Section 4.5.1. This allows us to estimate an upper bound of the condition number for the larger problems, because it scales with the square of the mesh width, and hence, with the square of h_{min}.

UNI1 No anisotropies: $\Omega = [0,1]^2$, uniform refinement, $\text{AR}_{\text{max}} = 1.0$, $h_{\text{min}} = 9.77\text{E-}4$. Condition numbers: 1.33E4 ($L = 8$), 5.31E4 ($L = 9$) and 2.12E5 ($L = 10$).

UNI2 Weak uniform anisotropy: $\Omega = [0,0.25] \times [0,1]$, uniform refinement, $\text{AR}_{\text{max}} = 4.0$, $h_{\text{min}} = 2.44\text{E-}4$. Condition numbers: 2.50E4 ($L = 8$), 1.00E5 ($L = 9$) and 4.00E5 ($L = 10$).

UNI3 Medium uniform anisotropy: $\Omega = [0,0.0625] \times [0,1]$, uniform refinement, $\text{AR}_{\text{max}} = 16.0$, $h_{\text{min}} = 6.10\text{E-}5$. Condition numbers: 2.70E4 ($L = 8$), 1.08E5 ($L = 9$) and 4.31E5 ($L = 10$).

ANISOREF1 Weak anisotropic refinement: $\Omega = [0,1]^2$, $\nu = 0.75$, $\text{AR}_{\text{max}} = 22.20$, $h_{\text{min}} = 5.50\text{E-}5$. Condition numbers: 7.20E4 ($L = 8$), 3.84E5 ($L = 9$) and 2.05E6 ($L = 10$).

ANISOREF2 Medium anisotropic refinement: $\Omega = [0,1]^2$, $\nu = 0.5$, $\text{AR}_{\text{max}} = 1.54\text{E}3$, $h_{\text{min}} = 9.54\text{E-}7$. Condition numbers: 1.25E6 ($L = 8$), 1.00E7 ($L = 9$) and 8.00E7 ($L = 10$).

ANISOREF3 Hard anisotropic refinement: $\Omega = [0,1]^2$, $\nu = 0.25$, $\text{AR}_{\text{max}} = 1.84\text{E}6$, $h_{\text{min}} = 9.31\text{E-}10$. Condition numbers: 2.11E8 ($L = 8$), 3.38E9 ($L = 9$) and 5.41E10 ($L = 10$). This test case yields a problem which cannot be assembled in single precision, because the Jacobi determinant of the mapping from the real long and thin elements to the reference element evaluates to zero.

ANISOREF4 Hard anisotropic refinement: $\Omega = [0,1]^2$, $\nu = 0.0625$, $\text{AR}_{\text{max}} = 2.13\text{E}12$, $h_{\text{min}} = 8.88\text{E-}16$. Condition numbers: 1.07E13 ($L = 8$), 6.83E14 ($L = 9$) and 4.37E16 ($L = 10$). This test case also yields a problem which cannot be assembled in single precision.

ANISOREF5 Extremely hard anisotropic refinement: $\Omega = [0,1]^2$, $\nu = 0.03125$, $\text{AR}_{\text{max}} = 2.08\text{E}15$, $h_{\text{min}} = 9.24\text{E-}19$. Condition numbers: 2.57E15 ($L = 8$), 3.29E17 ($L = 9$) and 4.21E19 ($L = 10$). This test case can barely be assembled in double precision for refinement level 9, cannot be assembled for refinement level 10 in double precision, and is therefore extremely challenging for our mixed precision solver.

Looking at the condition numbers for uniform anisotropies, we see that they increase very slowly with the degree of anisotropy. Our tests thus concentrate on the ANISOREF configurations, because they lead to ill-conditioned problems much more quickly.

For a domain $\Omega := [0,A] \times [0,B]$, we define the function $u(x,y) := x(A-x)y(B-y)$ and use its analytical Laplacian to prescribe a right hand side so that the Poisson problem has a known analytical solution. This is a straightforward generalisation of the 'hard' function which we already analysed in Section 4.5. Note that configuration 'UNI1' is therefore not identical to the previous tests in Section 4.5. Homogeneous Dirichlet boundary conditions are prescribed along the whole boundary. We measure the relative L_2 error of the computed solution against the analytical reference solution, as explained in Section 2.1.3. A computed solution is considered accurate if the L_2 errors are reduced by a factor of four per refinement step up to $L = 10$, the finest discretisation we apply for these tests.

The L_2 error is computed in double precision; and to minimise errors in the input data, the matrix and right hand side are always assembled in double precision and casted to single or emulated precision as needed.

Solver Configuration

We employ a multigrid solver with the ADITRIDI smoother unless stated otherwise, either to solve the entire system directly, or to approximately solve the (inner) defect problems in the mixed precision scheme (cf. Figure 4.4 on page 92). We emphasise that the results are quantitatively and qualitatively identical independent of the actual solver configuration. We use ADITRIDI because it is particularly well suited for these kinds of anisotropies. The (outer) stopping criterion is set to reduce the initial residual by eight digits, and the solver performs four pre- and postsmoothing steps, in a V cycle. We count one application of the alternating direction implicit smoother as two smoothing steps. As explained in Section 5.4.4, our CUDA implementation of the ADITRIDI smoother does not support the largest problem size (refinement level $L = 10$) in single precision due to the limited size of the shared memory per multiprocessor. Therefore, we execute all mixed precision solvers also on the CPU to be able to obtain results for the largest problem size. As we demonstrate in Section 5.6.2, the results are quantitatively and qualitatively independent of the hardware architecture they are computed on, in particular, small differences in compliance to the IEEE-754 standard (lack of denormalisation, flush-to-zero, slightly higher error bounds for some operations) in single precision do not have any effect on the achieved accuracy.

5.6.2. Accuracy of Mixed Precision Iterative Refinement Solvers

In this test series, we execute all eight configurations first entirely in double precision, and then in a mixed precision setting using double precision for the outer loop and single precision for the inner solver iterations. As the exact numbers in Table 5.3 demonstrate, the mixed precision iterative refinement scheme computes identical results compared to the standard double precision case, even for extremely ill-conditioned problems. We also observe that the differences between the CPU and the GPU, and between double and mixed precision are always in the noise. This is remarkable for the hardest cases ANISOREF4 and ANISOREF5, because the condition numbers on refinement levels 10 and 9 are close to the limits of double precision already. Based on the mixed precision iterative refinement results on the CPU, it is safe to assume that the differences between the CPU and the GPU will continue to be in the noise, as soon as sufficient resources are available on future hardware to compute on the finest level of refinement [172].

		Double precision CPU		Mixed precision CPU		Mixed precision GPU	
Configuration	Level	L_2 error	Reduction	L_2 error	Reduction	L_2 error	Reduction
UNI1	8	1.7344895E-5	4.00	1.7344902E-5	4.00	1.7344902E-5	4.00
UNI1	9	4.3362264E-6	4.00	4.3362356E-6	4.00	4.3362292E-6	4.00
UNI1	10	1.0841185E-6	4.00	1.0841276E-6	4.00		
UNI2	8	1.6946217E-5	4.00	1.6946284E-5	4.00	1.6946281E-5	4.00
UNI2	9	4.2365330E-6	4.00	4.2365355E-6	4.00	4.2363011E-6	4.00
UNI2	10	1.0590902E-6	4.00	1.0591051E-6	4.00		
UNI3	8	1.6603963E-5	4.00	1.6603658E-5	4.00	1.6603650E-5	4.00
UNI3	9	4.1508011E-6	4.00	4.1508177E-6	4.00	4.1504573E-6	4.00
UNI3	10	1.0377274E-6	4.00	1.0377500E-6	4.00		
ANISOREF1	8	2.2559231E-5	4.00	2.2559231E-5	4.00	2.2559230E-5	4.00
ANISOREF1	9	5.6398002E-6	4.00	5.6397700E-6	4.00	5.6397099E-6	4.00
ANISOREF1	10	1.4099726E-6	4.00	1.4099751E-6	4.00		
ANISOREF2	8	3.3671244E-5	4.00	3.3671243E-5	4.00	3.3671243E-5	4.00
ANISOREF2	9	8.4177915E-6	4.00	8.4177934E-6	4.00	8.4177905E-6	4.00
ANISOREF2	10	2.1044773E-6	4.00	2.1047529E-6	4.00		
ANISOREF3	8	4.9063089E-5	4.00	4.9063092E-5	4.00	4.9063092E-5	4.00
ANISOREF3	9	1.2265724E-5	4.00	1.2265729E-5	4.00	1.2265727E-5	4.00
ANISOREF3	10	3.0664399E-6	4.00	3.0664452E-6	4.00		
ANISOREF4	8	6.3654794E-5	4.00	6.3654923E-5	4.00	6.3654654E-5	4.00
ANISOREF4	9	1.5913491E-5	4.00	1.5913579E-5	4.00	1.5913506E-5	4.00
ANISOREF4	10	3.9782878E-6	4.00	3.9784532E-6	4.00		
ANISOREF5	8	6.6448219E-5	4.00	6.6447627E-5	4.00	6.6447308E-5	4.00
ANISOREF5	9	1.6612151E-5	4.00	1.6611850E-5	4.00	1.6612856E-5	4.00
ANISOREF5	10						

Table 5.3: Solution of a Poisson problem directly in double precision and with the mixed precision iterative refinement solver (relative L_2 errors). Missing GPU results are due to insufficient resources on the device.

5.6.3. Accuracy of the Emulated Double-Single Format

The emulated double-single format using the unevaluated sum of two single precision values (also called 'double-single') is only available in our legacy OpenGL-based implementation and has not been implemented in the CUDA backend yet. Hence, this experiment can only be performed for the isotropic test case (labeled 'UNI1'), as our legacy OpenGL implementation does not provide stronger smoothers other than JACOBI.

Emulated Double Precision Solvers

Table 5.4 compares the accuracy of the computed results using double precision on the CPU with that of a GPU solver executing in emulated double-single precision; single precision data is included for reference. We observe that the double-single format is not sufficient for this numerical test: While the results are better than in single precision (differences only occur on refinement level 9 rather than level 6), our accuracy demands are not met for the largest problem size. Similar to the single precision case (cf. Section 4.5.3), convergence control fails already on smaller levels of refinement, because the residuals of intermediate solutions cannot be reduced further. This behaviour can be explained by the fact that the double-single format does not increase the dynamic range beyond that of single precision, despite twice the bit length of the combined mantissa: The emulated format is not equivalent to native double precision (s52e11), but rather to a s46e8 format.

	Double precision CPU		Single precision CPU		Emulated double-single precision GPU	
Level	L_2 error	Error reduction	L_2 error	Error reduction	L_2 error	Error reduction
5	1.1102361E-3	4.00	1.1110228E-3	4.00	1.1102652E-3	4.00
6	2.7752813E-4	4.00	2.8084065E-4	3.96	2.7756658E-4	4.00
7	6.9380090E-5	4.00	6.1993631E-5	4.53	6.9414942E-5	4.00
8	1.7344930E-5	4.00	4.4250309E-5	1.40	1.7364390E-5	4.00
9	4.3362698E-6	4.00	2.9111338E-5	1.52	4.4038127E-6	3.94
10	1.0841660E-6	4.00	5.7460920E-5	0.51	1.4652436E-6	3.01

Table 5.4: Solution of a Poisson problem in double, single and emulated double-single precision (relative L_2 errors, UNI1 configuration).

Mixed Precision Solvers Using Emulated Double Precision

As the emulated double-single format does not give sufficient accuracy when used as a replacement of native double precision, the question arises if the emulated format is precise enough to execute the high precision correction loop within a mixed precision iterative refinement scheme. This experiment also determines if the problems observed in the previous paragraph are due to accumulated roundoff and cancellation errors in the course of the computation, or due to the general inappropriateness of the emulated precision format. The benefit of such a configuration is that it can be executed entirely on the device when native double precision is not available, such as on older GPUs. Without expensive data transfers over the narrow AGP and PCIe buses (cf. Section 3.1.3), the GPU is almost entirely decoupled from the CPU, leaving the CPU mostly idle for other tasks. Thus, it would be very beneficial in terms of performance if the emulated format were accurate enough.

	Double precision CPU		Mixed precision (double-single/single) GPU	
Level	L_2 error	Error reduction	L_2 error	Error reduction
5	1.1102361E-3	4.00	1.1102578E-3	4.00
6	2.7752813E-4	4.00	2.7755694E-4	4.00
7	6.9380090E-5	4.00	6.9400158E-5	4.00
8	1.7344930E-5	4.00	1.7373998E-5	3.99
9	4.3362698E-6	4.00	4.4217465E-6	3.93
10	1.0841660E-6	4.00	1.5605046E-6	2.83

Table 5.5: Solution of a Poisson problem directly in double precision and with a mixed precision iterative refinement solver using emulated double precision in the outer loop (relative L_2 errors, UNI1 configuration).

Table 5.5 shows that this is not the case. The L_2 errors of the computed solutions are similarly unacceptable for this configuration. This is in some sense the expected behaviour, as the solver is indeed converging to working precision, i. e., emulated double-single precision. Consequently, the emulated format does not provide sufficient accuracy.

5.6.4. Summary

The experiments in this section demonstrate that emulated high precision in a double-single combined format does not meet the accuracy demands of this thesis. In addition, the microbenchmarks of individual operations in this format (see Section 5.2.4) show that the performance of the emulated format is significantly slower than native single precision on the device, and we conclude that emulated precision is not a feasible approach for the applications in the scope of this thesis. Consequently, further performance benchmarks and parameter studies for solvers using this format are not presented in the remainder of this thesis.

On the other hand, the mixed precision iterative refinement technique always converges to double precision accuracy if the outer loop is executed in double precision. We have verified this claim even for very ill-conditioned problems, in particular for problems that cannot be assembled in single precision, and will return to this question in the context of the large-scale solution on GPU-accelerated clusters of problems from linearised elasticity in Section 6.4.2 and laminar flow in Section 6.5.3. We emphasise that we always obtain double precision accuracy, independent of the actual solver configuration (smoother/preconditioner, hardware, etc.).

For older GPUs that only support single precision natively in hardware, our results imply that *the mixed precision approach that performs the correction loop in native double precision on the CPU is the only viable solution scheme.* This approach is potentially slower due to the AGP and PCIe bus transfers of right hand sides and solution updates once per outer iteration, but this is a price we cannot avoid to pay without sacrificing accuracy. In other words, single-precision GPUs can only be used as co-processors to the general-purpose CPU, and can never execute self-contained solvers. We return to the performance and efficiency aspects of mixed precision iterative refinement in Section 5.7, for GPUs with and without native double precision support.

Interestingly enough, the failure of the emulated precision format also indicates a potential problem for extremely ill-conditioned linear systems for which native double precision may not be sufficient. No current CPU supports quadruple precision natively in hardware, so an emulated quad-double format using the unevaluated sum of two double precision values to construct an s104e11 number representation is a feasible approach. However, the emulation is potentially bound to fail, and a more expensive form of emulation has to be applied. We will examine this aspect in future work.

5.7. Numerical Evaluation of Mixed Precision Iterative Solvers

In Section 5.6 we have experimentally demonstrated that mixed precision iterative refinement solvers using double and single precision converge to double precision accuracy even for extremely ill-conditioned problems. In this section we are interested in the numerical performance, their robustness and their runtime efficiency compared to solvers that execute entirely in double precision. Special focus is placed on the speedups of mixed precision solvers over executing in double precision alone, both on the CPU and on the GPU. We investigate the acceleration achieved by GPU-based solvers compared to conventional CPU versions, and of hybrid variants that execute the correction loop on the CPU but the inner solver on the GPU.

Section 5.7.1 motivates and summarises the goals of the following experiments in detail. Efficiency on the CPU and the GPU is evaluated in Section 5.7.2 and 5.7.3, respectively. We discuss the speedups achieved by the GPU over the CPU in Section 5.7.4 and conclude with a summary of the most important findings in Section 5.7.5.

The presented results are significantly updated and extended from previously published joint work with Robert Strzodka and Stefan Turek [81, 85].

5.7.1. Overview and Test Procedure

For dense systems, mixed precision iterative refinement (cf. Section 4.3) is typically applied by computing an LU decomposition in low precision and using it to *solve* the auxiliary systems by forward- and backward substitution, to low precision accuracy. The speedup in this case stems from the fact that asymptotically, the computation of the decomposition dominates with its cubic runtime. Provided that the system matrix $\mathbf{A}$ is 'not too ill-conditioned' in low precision, the number of iterations required until convergence can be bounded from above by (see for instance Langou et al. [142]):

$$\left\lceil \frac{\log_{10} \varepsilon_{\text{high}}}{\log_{10} \varepsilon_{\text{low}} - \log_{10} \operatorname{cond}(\mathbf{A})} \right\rceil$$

Here, $\varepsilon_{\text{high}}$ and ε_{low} denote the machine epsilon of the high and low precision floating point formats. In the case of single and double precision, $\log_{10} \varepsilon_{\text{double}} \approx -16$ and $\log_{10} \varepsilon_{\text{single}} \approx -8$, and Langou et al. report that in practice this bound is reasonably exact and usually, 3–4 iterations suffice.

In our case of large, sparse systems, we cannot fully solve the auxiliary systems: As demonstrated in Section 4.5, iterative solvers are not guaranteed to converge in single precision, they may stall or even diverge. The decisive parameter, to be set manually by the user, is thus the stopping criterion of the inner iterative solver. Consequently, the above bound on iteration numbers does not hold in our case. We are not aware of any theoretical analysis in the literature that yields a similar asymptotic bound for the iteration number, so we perform numerical experiments to gain insight into the relation between the amount of work performed (the accuracy gained) by the inner solver and the convergence rate and total efficiency of the (outer) iterative refinement scheme.

Our primary motivation to investigate mixed precision iterative refinement methods is to improve performance and efficiency over standard solvers employing only double precision, without sacrificing accuracy. When using multigrid as the inner solver, we do not benefit from the asymptotics as in the case of dense systems and direct solvers. In our case, convergence rates alone do not convey enough information to assess the numerical performance: The single precision approximate solution is used as a preconditioner of the high precision Richardson iteration, and the more exact the approximate solution is, the better the convergence rate is. We therefore use the total runtime efficiency E_{total} as the only metric to quantify performance in the tests in this section, see Section 5.5.1 for its definition. This quantity alleviates all granularity effects incurred by varying numbers of iterations, and also captures differences in floating point performance, for instance in the case where high precision is only available at the penalty of a bus transfer between host and co-processor.

In Section 5.7.2 we numerically analyse the impact of several typical choices of the stopping criterion. We use the same ill-conditioned test cases as in the accuracy measurements and in previously published work [81, 85], see Section 5.6.1 for the exact definitions. However, we omit the tests for the uniformly refined anisotropic domains, because they do not exhibit challenging condition numbers. A second important question in this context is whether the accurate solution (Section 5.6.2) can only be achieved with an disproportionately increasing amount of iterations, or more generally, if our mixed precision iterative refinement schemes using only approximate inner solutions are efficient compared to using double precision exclusively, for practical purposes independent of the condition number.

On GPUs, performance characteristics of single and double precision arithmetic are potentially different to CPUs. This is especially important on older GPUs, where native double precision is not available and the refinement loop has to be performed on the CPU, involving comparatively expensive bus transfers of right hand sides and solution updates. This situation is symmetric to our large-scale FEAST solvers, where results computed by GPU-based subdomain solvers have to be incorporated in the global solution and communicated between the nodes in a cluster, see Section 2.3 and Chapter 6. Therefore, we analyse efficiency on the GPU separately in Section 5.7.3. The speedups we achieve with various GPU-based solvers over CPU implementations are discussed in Section 5.7.4.

All tests in this section are performed on the 'high-end system', see Section 5.3.3, and as before, we always include at least one pair of bus transfers into our timings (see Section 5.5.1 for our rationale for this test procedure).

Solver Configuration

The majority of our experiments is performed using a multigrid solver with ADITRIDI smoother. As pointed out in Section 5.6.1, this solver is particularly well-suited for the anisotropic problems that we use to rapidly increase the condition numbers in a controlled fashion. We demonstrate that the results are qualitatively and quantitatively independent of the choice of the smoother by performing selected experiments also on an isotropic configuration for which the JACOBI smoother suffices.[19] Finally, we also briefly evaluate the ADITRIGS smoother in the context of mixed precision methods. In practice, the same guidelines to choosing a suitable smoother apply to both solving entirely in double precision and to using mixed precision iterative refinement, at least for multigrid. In any case, the multigrid solver is configured to execute four pre- and postsmoothing steps in a V cycle. We count one application of this alternating direction implicit smoother as two elementary smoothing steps, see Section 5.5.1. The coarse grid problems are solved with a conjugate gradient algorithm, preconditioned with the same operator as used for smoothing. All solvers are configured to reduce the initial residual by eight digits, either the multigrid itself when solving entirely in double precision, or the Richardson solver when using iterative refinement. The stopping criterion of the inner solvers is varied.

At the end of Section 5.7.2, we briefly comment on using Krylov subspace solvers in a mixed precision iterative refinement context.

[19] We use the JACOBI smoother despite having established in Section 5.5.3 that at least on the CPU, one should always favour Gauß-Seidel type smoothers over JACOBI. Robust and strong smoothers are however a fairly recent addition to our code base, and many of the results we present in Chapter 6 and in previously published work [84, 85, 87, 90] have been obtained at a time when only JACOBI was available.

5.7.2. Efficiency on the CPU and Influence of the Stopping Criterion

Figure 5.36 compares the total efficiency of multigrid solvers executing entirely in double precision with that of mixed precision iterative refinement solvers using multigrid for the low-precision auxiliary problems, for all six increasingly ill-conditioned test cases.

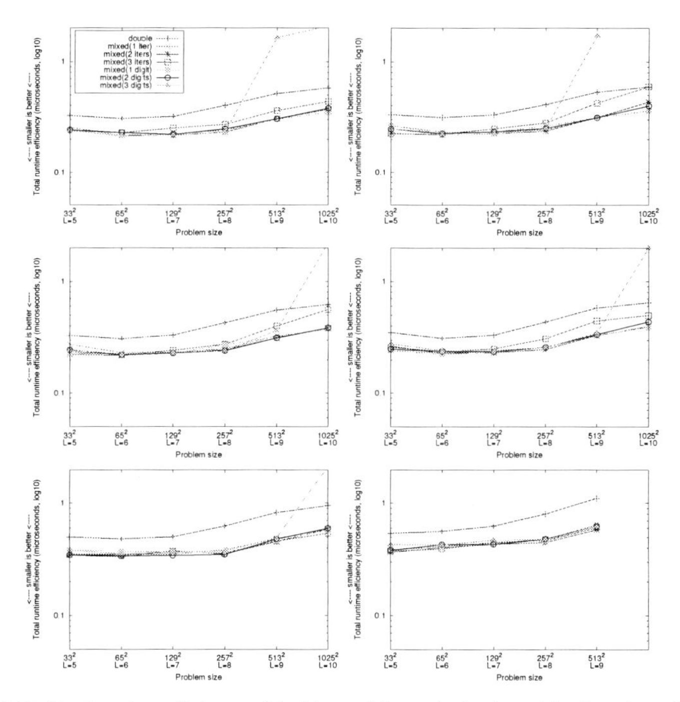

Figure 5.36: Total runtime efficiency of double precision and mixed precision iterative refinement multigrid solvers with varying stopping criteria, on the CPU. Top row: UNI1, ANISOREF1. Middle row: ANISOREF2, ANISOREF3. Bottom row: ANISOREF4, ANISOREF5. The y-axis uses a logarithmic scale, and all plots share the same legend.

We first observe, by looking exemplarily at the total efficiency of the double precision solver (red dashes) on the highest level of refinement, that there is a certain degree of dependence on the condition number: The efficiency of the multigrid solver degrades with increasing condition number, because the systems become harder to solve and the number of iterations needed until convergence increases. From the UNI1 configuration to the ANISOREF5 configuration, we lose between 50 and 100 % efficiency, for all levels of refinement. The more important observation is however: *Mixed precision iterative refinement using multigrid cycles with limited stopping criteria, i. e., approximate solutions, exhibits the same behaviour, even for very ill-conditioned systems and high levels of refinement.* For all six configurations, the plots of the double and best mixed precision solvers are apart by a constant offset. Consequently, we can conclude that the identical accuracy independent of the condition number (as demonstrated in Section 5.6.2) is not paid for by a disproportionate increase in iteration numbers compared to using double precision alone.

The best mixed precision scheme is, for the first four test cases, always between 1.4 (small problem sizes) and 1.7 times more efficient than using double precision exclusively. For the two hardest test problems, the factor even increases to 1.8–1.9. Based on bandwidth considerations, the maximum speedup we expect is a factor of two, because we can transfer twice the amount of values to and from off-chip memory in the same time (cf. Section 4.1). The measured speedups are in line with this upper bound, as the mixed precision solver actually performs more arithmetic work, namely a defect calculation and a vector update once per outer loop iteration.

We now discuss the efficiency of iterative refinement solvers as a function of the 'inner' stopping criterion. In other words, we want to establish a rule of thumb which setting of the user-defined inner stopping criterion yields some sort of black-box character. As our results indicate, reducing the residuals by three digits or even more is always suboptimal for the largest problem size under consideration. Depending on the actual test case, the inner solver mostly reaches its maximum prescribed number of iterations because at some point it stalls, and the total efficiency E_{total} loses its expressiveness. The stalling of the inner solvers for large problem sizes is in fact the expected behaviour: Single precision is not only insufficient to give accurate results, but can lead to stalls for moderately complex scenarios, as discussed in Section 4.5.3. Consequently, we omit the configuration that gains three digits in the following, three digits are not a suitable stopping criterion. We emphasise nonetheless that the solvers always converge to double precision accuracy in our tests. The inner solver stalls but computes a solution that still advances the global solution. To make the solvers more robust, in future work we will add some detection mechanism for this behaviour, e. g., by monitoring the reduction of residuals over three or more iterations.

For the less ill-conditioned test problems, the difference in efficiency is more pronounced, and the variant that performs three inner multigrid cycles per update step loses efficiency. This situation is symmetric to the variant that gains three digits. For all configurations, performing either one or two iterations, or reducing the residual by one digit gives optimal efficiency. In fact, in all tests, performing exactly one multigrid cycle in the inner solver yields a sum of all inner iterations which is identical to the number of cycles performed by the double precision solver.

JACOBI and ADITRIGS Smoothers

Figure 5.37 (left) depicts the results of using multigrid with a JACOBI smoother, configured to perform four pre- and postsmoothing steps. This experiment is only performed for the UNI1 configuration. The plots on the right illustrates the performance of ADITRIGS on the ANISOREF5 configuration. The results confirm that—as expected—the above findings are independent of the smoothing operator in the multigrid solver: The shapes of the curves are identical to the ones obtained when using ADITRIDI as smoother. The speedups of mixed precision iterative refinement solvers are maximised for the variants that either perform two iterations or reduce the initial residuals by two digits. We observe approximately a factor of 1.9–2.1 up to refinement level $L = 9$ for both smoothers, and still 1.7 for the largest problem size and JACOBI at which the doubled effective bandwidth of the mixed precision approach dominates.

Krylov Subspace Solvers

In this experiment, we use a conjugate gradient solver with JACOBI preconditioner to solve the UNI1 configuration. This solver does not converge independently of the mesh width, so prescribing a fixed number of inner iterations needs to incorporate some heuristics depending on the problem size. We thus only analyse the case where the conjugate gradient solver is configured to reduce the initial residuals by one, two or three digits.

Table 5.6 presents the results, we only discuss iteration numbers because the effects are visible more clearly than in terms of total efficiency. We first observe that the double precision con-

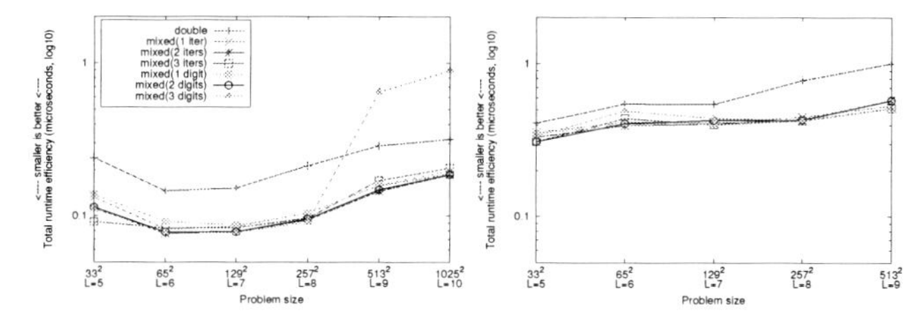

Figure 5.37: Total runtime efficiency of double precision and mixed precision iterative refinement multigrid solvers on the CPU, UNI1 configuration, JACOBI smoother (left); and ANISOREF5 configuration, ADITRIGS smoother (right). The y-axis uses a logarithmic scale and both plots share the same legend.

jugate gradient solver exhibits the expected behaviour: The number of iterations required until convergence doubles with each refinement step, i. e., proportionally to the reciprocal of the mesh width. This is not the case when using conjugate gradients in a mixed precision iterative refinement scheme, which requires significantly more iterations overall. The outer iteration numbers for the larger problem sizes also indicate that the solver does not converge smoothly: Gaining ten times one digit (or five times two digits etc.) only gives eight digits in total, this was not the case in multigrid. This effect is most pronounced for the first mixed precision configuration, but still clearly visible in the remaining two. In terms of total efficiency, only the latter two configurations perform on par with the double precision variant, but we only observe a speedup of approximately 10 % (total efficiency results not shown in detail).

Level	double	mixed (1 digit)	mixed (2 digits)	mixed (3 digits)	double (1 digit)	double (2 digits)	double (3 digits)
2	3	2:6	2:6	2:6	1:3	1:3	1:3
3	9	4:14	3:15	2:12	4:14	3:15	2:10
4	19	6:38	4:36	3:40	6:38	4:33	3:32
5	38	8:82	4:68	3:76	8:82	4:70	3:61
6	73	9:174	4:130	3:151	9:178	4:140	3:122
7	146	10:373	5:338	3:312	10:382	5:354	3:245
8	293	10:844	5:754	3:575	10:831	5:685	3:494
9	590	10:1723	5:1649	4:1762	10:1305	5:1360	3:991
10	1188	10:3878	5:3000	4:3458	11:3508	5:2603	3:2004

Table 5.6: Number of iterations required by a conjugate gradient solver on the UNI1 configuration in double precision (left), with mixed precision iterative refinement (middle columns) and with iterative refinement only in double precision (right columns). The shorthand notation $x : y$ abbreviates the number of outer iterations and the sum of all inner iterations.

In joint work with Robert Strzodka we have investigated the reason for this behaviour [211]. Looking at the algorithmic description of the conjugate gradient scheme, we see that it builds up a search space of **A**-orthogonal directions (see Figure 2.2 on page 25). One idea is that due to the reduced precision, these search directions are no longer **A**-orthogonal. To quantify this claim, we execute the iterative refinement solver again, but this time use double instead of single precision to solve the auxiliary problems (which means that we perform 'same precision iterative refinement'). For the two configurations that reduce the residuals by two and three digits respectively, the savings are significant but do not alleviate the initial problem. For the first stopping criterion, we see almost no difference, and even for the third one, we still perform almost twice the amount of conjugate gradient steps compared to solving directly in double precision. So the loss of **A**-orthogonality due to reduced precision is not causing the behaviour we observe. When comparing the results for varying stopping criteria, we observe that the conjugate gradient solver reacts very sensitively to restarts. In fact, when a conjugate gradient solver is restarted, the entire search space of **A**-orthogonal directions is lost. To alleviate this problem, we have developed a *residual-guided* conjugate gradient variant: The standard algorithm stores information about the solution process in the auxiliary vectors **p**, **q** and **r** (see Figure 2.2). In fact, **p** is the vector that stores the next orthogonal search direction. The idea of our variant is to maintain this search direction from one outer iteration to the next. It cannot be reused immediately, because in each outer iteration, the system is solved for a different right hand side. We thus have to add a correction step to the outer loop to orthogonalise the last search direction from the last solve prior to entering into inner iterations for the new problem. Due to space constraints (and as our focus is on efficient multigrid solvers), we do not present the improved solver in algorithmic form, and refer the interested reader to the original publication for details and a set of tests that demonstrate the feasibility of our modification [211].

5.7.3. Efficiency on the GPU

Figure 5.38 depicts the results of the same experiments as Figure 5.36, but this time on the GPU. We first confirm that there are no differences in the number of iterations and the relation of inner and outer iterations between the CPU and the GPU variants. As already observed in our microbenchmarks of basic operations (Section 5.3.3) and smoothing operators (Section 5.5), the GPU always achieves maximum efficiency for the largest problem sizes. In double precision, we can only solve problems up to refinement level $L = 8$ due to the limited size of shared memory per multiprocessor on current generation hardware; the mixed precision iterative refinement approach enables calculations corresponding to one additional level of refinement.

In double precision and for the largest possible problem size, we achieve a total efficiency of 0.12–0.13 µs per degree of freedom and digit gained for the first four test cases, and 0.2 and 0.26 µs for the last two, extremely ill-conditioned test cases. Setting the inner stopping criterion to reduce the initial residuals by three digits increases runtime for large levels of refinement and moderately ill-conditioned test cases, analogously to the CPU tests.

The speedup of the best mixed precision iterative refinement solver compared to using double precision exclusively is, for all test cases, approximately 1.2 for small levels of refinement and 1.9–2.1 for $L = 8$, the largest possible input size in double precision. For smaller problem sizes, this speedup is smaller than on the CPU, which further underlines that GPUs are only fully efficient if completely saturated, and that the two included bus transfers incur a performance penalty that cannot be ignored. With more shared memory available on the device, we expect the speedup to grow even larger for the problem on refinement level $L = 9$ (currently possible in single but not in double precision) and $L = 10$ (currently not possible at all). At first, it is surprising to see more than a factor of two, which on the GPU as well as on the CPU is the expected theoretical upper bound based on bandwidth considerations alone, at least for sufficiently large problem sizes. However, the parameter space on the GPU is higher-dimensional as on the CPU. The decisive performance factor is again the degree of saturation and consequently, the ability of the hardware thread scheduler to hide stalls by switching to different warps of execution. Our implementation of the tridiagonal solver at the core of this smoother maps each row in the mesh to one thread block. In single precision, the hardware can thus always switch between twice the amount of mesh rows per multiprocessor[20], except for the largest problem size which requires so much shared memory

[20] More correctly speaking, the hardware can switch between twice the amount of warps, because there are warps from at least one additional thread block of the same size. More active warps mean a higher probability that at a given

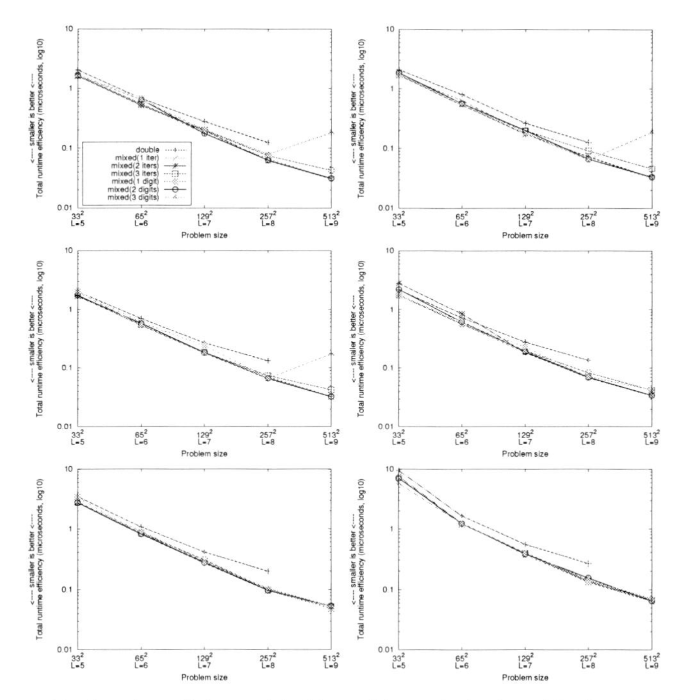

Figure 5.38: Total runtime efficiency of double precision and mixed precision iterative refinement multigrid solvers with varying stopping criteria on the GPU. Top row: UNI1, ANISOREF1. Middle row: ANISOREF2, ANISOREF3. Bottom row: ANISOREF4, ANISOREF5. The y-axis uses a logarithmic scale, and all plots share the same legend.

that only one thread block is active per multiprocessor.

This also explains why the difference between the various choices of stopping criteria is less pronounced on the GPU than on the CPU. Unless the number of inner iterations increases significantly because the solver cannot reduce the residuals to the prescribed threshold (i. e., gain three digits), the amount of work saved by performing fewer update steps in expensive double precision is balanced by the faster execution in single precision.

In summary, we can say that the 'rule of thumb' established for the CPU solvers, namely to perform either one or two inner multigrid cycles or to gain one or two digits in the inner solver, also holds on the GPU.

JACOBI and ADITRIGS Smoothers

Figure 5.39 (top) illustrates the results obtained on the UNI1 configuration when using the JACOBI smoother instead of ADITRIDI, the plots in the bottom row make the same comparison for ADITRIGS and the ANISOREF5 test case. The same conclusions can be drawn as in the CPU case. For the largest problem size and JACOBI, we achieve a total efficiency of 0.016 µs per unknown per

stall, another warp is instantaneously ready for execution.

digit gained in double precision, and 0.0085 µs for the mixed precision variants, corresponding to a speedup factor of 1.9. In terms of absolute timings, *the best mixed precision iterative refinement solver is able to solve a problem with one million unknowns in less that 0.08 seconds without sacrificing accuracy*. Admittedly, the 'barrier' of solving such problems accurately in less than 0.1 second is artificial and mostly of psychological nature, but the result in itself remains impressive. We discuss speedups achieved by the GPU further in Section 5.7.4. The ADITRIGS case is slightly less efficient than ADITRIDI when using four colours to parallelise the sequential Gauß-Seidel dependencies between subsequent mesh rows, and on par with ADITRIDI when only two colours are used. The gap between the (numerically stronger) four-coloured variant and the two-coloured smoother than exhibits more independent parallel work in each sweep over the domain is however narrower than in the case of solving entirely in double precision, see Figure 5.34 on page 166. The exact numbers for refinement level $L = 9$ are 0.0645 µs (two colours, inner stopping criterion two iterations) and 0.0692 µs (four colours, three iterations).

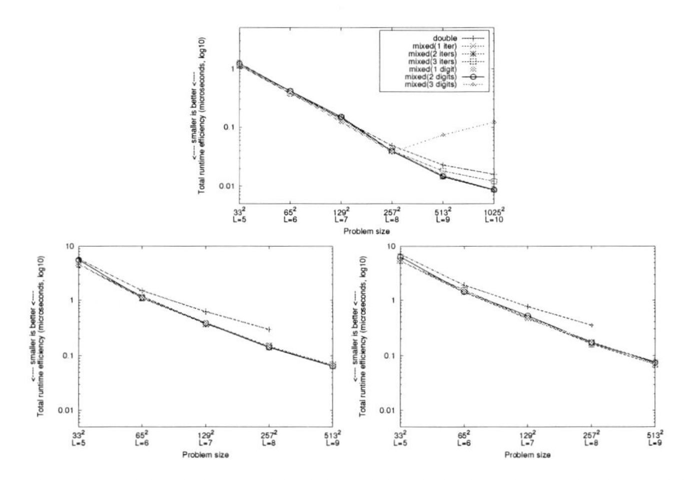

Figure 5.39: Total runtime efficiency of double precision and mixed precision iterative refinement multigrid solvers on the GPU, UNI1 configuration, JACOBI smoother (top); and ANISOREF5 configuration, ADITRIGS smoother (bottom left: two colours, bottom right: four colours). The y-axis uses a logarithmic scale and all plots share the same legend.

Executing the High Precision Correction Loop on the CPU

Two settings exist in the scope of this thesis where the high precision correction loop has to be executed on the CPU and not on the GPU, yielding a hybrid solver. On older GPUs, native double precision is not available. We have already ruled out the possibility to use the emulated high precision format based on the unevaluated sum of two native single precision values in Section 5.6.3, this variant does not meet our accuracy requirements. Therefore, there is no other possibility than to perform the defect correction steps on the CPU, which incurs a potentially significant performance penalty due to expensive data transfers of right hand sides and solution updates over the comparatively narrow PCIe (and on 'ancient' hardware, AGP) buses. Furthermore, in the context of FEAST, the outer loop is not a simple Richardson iteration, but a much more involved multilevel domain decomposition scheme: The subdomain solves discussed in this chapter constitute the

Schwarz preconditioner (smoother) of the two-layer SCARC solution scheme, see Section 2.3.3 on page 37 and Chapter 6.

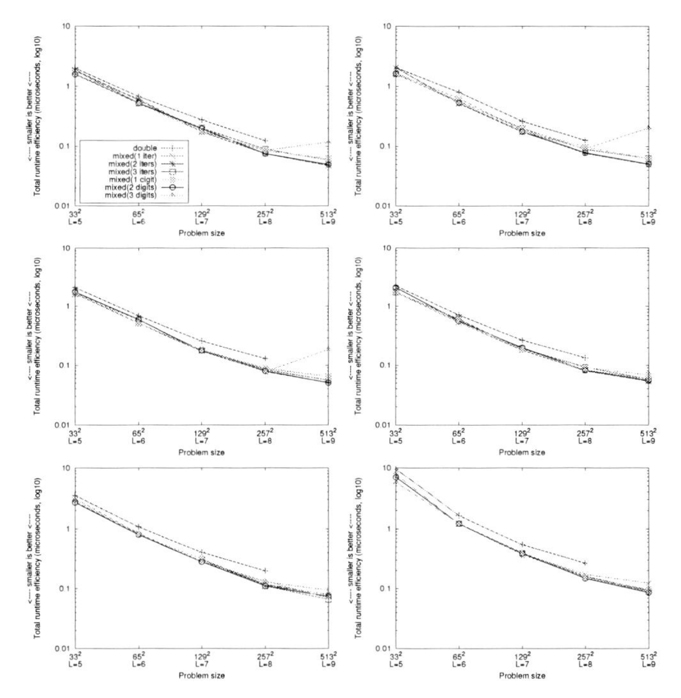

Figure 5.40: Total runtime efficiency of double precision and mixed precision iterative refinement multigrid solvers with varying stopping criteria. Top row: UNI1, ANISOREF1. Middle row: ANISOREF2, ANISOREF3. Bottom row: ANISOREF4, ANISOREF5. The high precision loop is performed on the CPU. The y-axis uses a logarithmic scale, and all plots share the same legend.

The goal of the following tests is thus to gain insight to what extent the bus transfers reduce the total efficiency of our mixed precision iterative refinement solvers. To this end, we repeat the tests depicted in Figure 5.38, but this time execute the refinement loop on the CPU. This setting implies the bus transfer of two vectors per outer iteration. We perform these experiments on the same test system to enable a meaningful comparison with the results depicted in Figures 5.36 and 5.38, in other words, we deliberately ignore that the GPU under consideration natively supports double precision. Double precision results on the device are repeated from Figure 5.38 and included for easier reference.

As expected, the convergence behaviour of these configurations is identical, independent of using double precision on the device or on the CPU. For the first four test problems, the speedup compared to using double precision exclusively reduces to a factor of 1.1 for small problem sizes, and to 1.5–1.6 for the larger ones. For the extremely ill-conditioned test cases, we note a reduction in speedup compared to double precision to factors of 1.3 and 1.7–1.8 for small and large problem sizes, respectively.

More interestingly, the total efficiency of configurations executing hybridly on CPUs and

GPUs compared to iterative refinement solvers performing all work on the device becomes worse with the problem size. For small problem sizes, these variants achieve 90 % of the efficiency, this fraction remains remarkably constant, and then drops rapidly to 80 % for $L = 8$ and 50–80 % for $L = 9$. The concrete loss is dependent on the amount of update steps performed compared to the inner multigrid iterations. The explanation for this behaviour is straightforward: Up to a certain threshold, the high latency of the PCIe bus is the decisive factor, as long as the amount of data being transferred remains sufficiently small, it does not matter how much data is being transferred. The latency of the PCIe bus is of the order of 10–20 µs [132]. As soon as the amount of data is large enough, bandwidth starts to have an impact, and the time spent per transfer is no longer constant, but linear in the data size. Furthermore, for large problem sizes the defect computation and the vector update on the CPU become a bottleneck as well, because not all data fits into cache anymore. In contrast to the variants executing entirely on the device, in this setting it is beneficial to shift more work to the inner loop to reduce the number of outer iterations and thus the amount of expensive bus transfers. For the first four test cases, the setting of performing two iterations or gaining two digits is always optimal. For the last two test cases, gaining three digits or performing three multigrid cycles per outer update step is optimal, in contrast to executing entirely on the device.

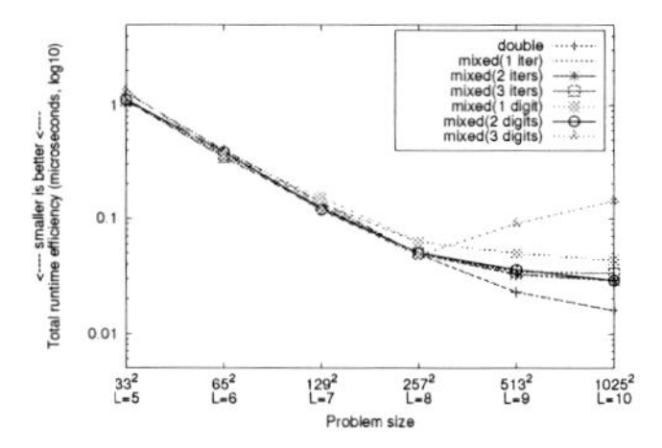

Figure 5.41: Total runtime efficiency of double precision and mixed precision iterative refinement multigrid solvers, UNI1 configuration, JACOBI smoother. The high precision loop is performed on the CPU. The y-axis uses a logarithmic scale.

When using JACOBI as elementary smoother instead of ADITRIDI, the observations and conclusions are qualitatively the same, see Figure 5.41. The faster execution of the JACOBI smoother in single than in double precision on the device however does not cancel out favourably with the penalty of the bus transfers and the slower computations on the CPU, and consequently, executing in double precision alone is always more efficient than employing a *hybrid* mixed precision iterative refinement scheme.

5.7.4. Speedup of GPU-based Solvers

So far, we have discussed the speedup achieved by mixed precision iterative refinement multigrid solvers compared to using native double precision, separately on the CPU and the GPU. In this section we are interested in the speedups obtained by the GPU over CPU-based solvers. We consider all three possibilities on the GPU: Solving entirely in double precision, solving in mixed precision entirely on the device, and solving in mixed precision hybridly on the device (multigrid) and the CPU (iterative refinement loop). We compute speedups against two baseline configurations, either double precision on the CPU, or mixed precision iterative refinement on the CPU. All speedups are computed from the total runtime efficiency of the tests performed in Section 5.7.2 and Section 5.7.3. We do not consider the comparison of double precision on the device against mixed precision iterative refinement on the CPU, and restrict the presentation to the 'optimal' choice of the inner stopping criterion.

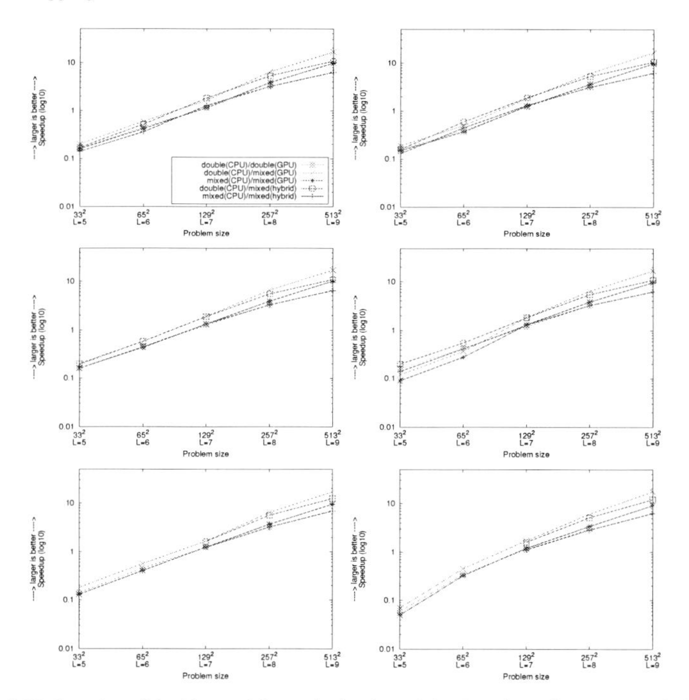

Figure 5.42: Speedup of double precision and mixed precision iterative refinement multigrid GPU solvers against CPU double and mixed precision solvers. Top row: UNI1, ANISOREF1. Middle row: ANISOREF2, ANISOREF3. Bottom row: ANISOREF4, ANISOREF5. The y-axis uses a logarithmic scale, and all plots share the same legend.

Figure 5.42 depicts the speedups we observe. Looking at the comparison between the two double precision solvers (filled squares, cyan) first, the GPU-based solver is for all six configurations more than three times faster than its CPU counterpart on refinement level $L = 8$, the largest possible problem size. This speedup is in line with the measurements presented in Section 5.5.3. The break-even point from which on the GPU outperforms the CPU is refinement level $L = 7$ (16 641 degrees of freedom).

Compared with using double precision on the CPU (the only case available in FEAST, see Chapter 6), the mixed precision iterative refinement solver executing entirely on the GPU is almost twice as fast for refinement level $L = 7$, six times faster for refinement level $L = 8$ and 16–17 times faster for $L = 9$. These factors are achieved for all six test cases, emphasising again that the mixed precision iterative refinement scheme converges identically to the baseline CPU variant.

The comparison with a mixed precision iterative refinement scheme executing entirely on the CPU is more honest, because we have demonstrated in Section 5.7.2 that the CPU benefits from the mixed precision approach to the same extent as the GPU. In this case, the speedup factors reduce to 1.2 ($L = 7$), 3–4 ($L = 8$) and a remarkable factor of 10 ($L = 9$). Again, these factors are achieved independently of the test case and thus of the condition number of the linear system to be solved.

Finally, the hybrid configuration that executes the multigrid solver for the auxiliary problems on the device and the iterative refinement loop on the CPU is still significantly faster than using the CPU alone. The hybrid solver is 10.5–12 times faster than the pure double precision solver on the CPU, for the largest problem size. For smaller levels of refinement, we still see factors of 5.5 ($L = 8$) and 1.8 ($L = 7$). The more honest comparison against mixed precision iterative refinement on the CPU yields factors of 6–7 ($L = 9$), 3 ($L = 8$) and 1.2–1.3 ($L = 7$).

The curves only start to flatten for the largest possible problem size, indicating again that with more shared memory available on future devices, the speedups would be significantly better for larger problem sizes.

JACOBI and ADITRIGS Smoothers

Our implementation of the JACOBI smoother is not limited by the amount of shared memory, and consequently, we can analyse the speedup up to refinement level $L = 10$ for the UNI1 configuration.

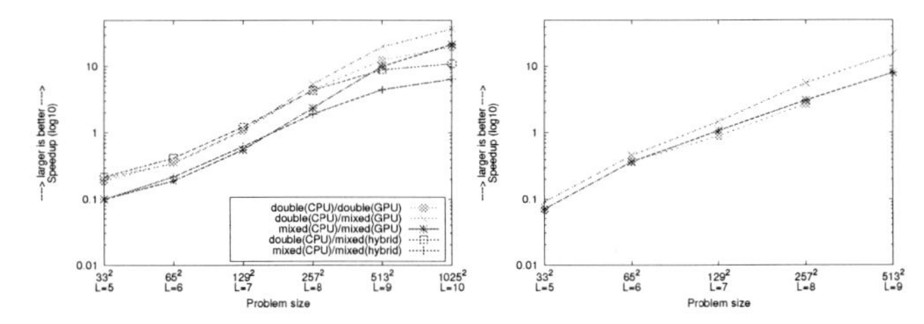

Figure 5.43: Speedup of double precision and mixed precision iterative refinement multigrid GPU solvers against CPU double and mixed precision solvers, UNI1 configuration, JACOBI smoother (left); and ANISOREF5 configuration, bicoloured ADITRIGS smoother (right). The y-axis uses a logarithmic scale and both plots share the same legend.

Looking at the two solvers first that execute entirely in double precision, we observe that the GPU is 1.1, 4, 12.5 and 20 times faster than the CPU for $L = 7, \dots, 10$, respectively. These results are in line with the experiments presented in Section 5.5.3. The mixed precision iterative refinement solver that executes entirely on the device is 20 ($L = 9$) and 37.5 ($L = 10$) times faster than a pure double precision solver on the CPU, these factors reduce to 10 and 22 when comparing against a mixed precision iterative refinement solver on the CPU. Executing the refinement loop on the CPU instead of on the GPU and hence incurring expensive data transfers and slow CPU calculations in each update step reduces the speedup further, and to a greater extent than for multigrid with the ADITRIDI smoother. We observe speedup factors of 9 and 11 against double precision, and of 4.5 (compared to 6–7 for ADITRIDI) and 6.5 against mixed precision, for $L = 9$ and $L = 10$, respectively. For the largest problem size, the speedup curves start to flatten, indicating that the device is sufficiently saturated on refinement level $L = 10$. When employing the ADITRIGS smoother (Figure 5.43 on the right, only the results for the bicoloured variant are shown), we observe speedups of 15.6 (vs. double precision) and 8 (vs. mixed precision) for the largest admissible problem size, in line with the results for the tridiagonal solver.

Older Hardware

We repeat the previous experiment using multigrid with a JACOBI smoother on the best machine used for microbenchmarks of our legacy GPGPU implementation, see Section 5.2.4, and on the 'server' machine with a GeForce 8800 GTX, see Section 5.3.3. Neither of these two GPUs support double precision natively in hardware, so we can only execute the hybrid configurations.

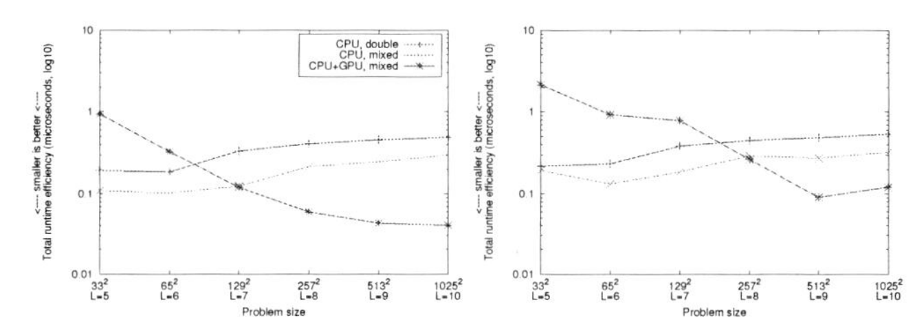

Figure 5.44: Hybrid CPU-GPU mixed precision iterative refinement on the 'server' (left) and 'high-end legacy' (right) machines compared to mixed and double precision on the CPU, UNI1 configuration, JACOBI smoother. The y-axis uses a logarithmic scale.

Figure 5.44 depicts the measured total efficiency of the solvers. On both CPUs, the mixed precision approach is approximately 1.7 times faster than executing the entire computation in double precision. On the 'server' machine, the hybrid CPU-GPU iterative refinement solver is 12.2 times faster (10.7 for $L = 9$) than the solver in double precision on the CPU, and still 7.4 (5.7 for $L = 9$) times faster when comparing to the mixed precision solver that executes only on the CPU. The break-even point from which onwards the GPU achieves speedups is at refinement level $L = 7$. Using our legacy OpenGL implementation on the GeForce 7900 GT based workstation, the break-even point is at refinement level $L = 8$, and the GPU achieves a speedup of 5.3 and 4.4 for the largest two problem sizes compared to executing the computation in double precision on the CPU. The more honest comparison against a mixed precision iterative refinement solver on the CPU alone gives a factor of 3 ($L = 9$) and 2.7 ($L = 10$). These results are in line with the microbenchmarks presented in Section 5.2.4 and Section 5.3.3, and confirm that GPUs are indeed becoming faster at a higher pace than CPUs.

5.7.5. Summary

The most important conclusion that can be drawn from the numerical experiments presented in this section is as follows: *Mixed precision iterative refinement multigrid schemes are always more efficient than using double precision exclusively, both on CPUs and on GPUs.* For sufficiently large problem sizes, these solvers are asymptotically twice as fast while delivering the same accuracy in the computed results. The actual speedup depends on the performance characteristics between single and double precision on the hardware under consideration. For instance, if double precision is not available natively on a co-processor architecture, then the achieved speedup is reduced because of comparatively expensive bus transfers and slow calculations on the CPU.

In the literature, mixed precision iterative refinement has been analysed theoretically mostly for the case of dense matrices and direct solvers based on LU decompositions. In particular, all upper bounds on the number of refinement iterations required until convergence to double precision accuracy are based on the exact solution of the auxiliary systems in single precision. In our setting, we cannot solve the auxiliary problems exactly. The systems are large and *sparse*, and iterative solvers are likely to stall or even fail to converge in single precision. We therefore have to introduce a comparatively weak stopping criterion for the inner solver, which contradicts the idea of constructing black-box solvers and prevents rigorous theoretical analysis. As our numerical experiments indicate, our mixed precision iterative refinement solvers are however very robust with respect to this user-defined parameter, given that it is not chosen too restrictively, i. e., that the auxiliary systems are not solved too accurately. Performing one or two multigrid cycles or reducing the initial residuals of the auxiliary systems by one or two digits turns out to be always a good choice, unless the mixed precision iterative refinement procedure involves expensive bus transfers and less frequent update steps are more beneficial.

In terms of speedups achieved by the GPU over the CPU, mixed precision iterative refinement is very beneficial and even enables the solution of larger problems in case current hardware generations provide too limited resources for double precision alone. Depending on the type of smoother employed by the multigrid solver, we observe significant speedups, reaching up to 37× compared to 20× when using double precision exclusively. *We always observe speedups of at least an order of magnitude for sufficiently large problem sizes* in the honest comparison of mixed precision solvers on the CPU and the GPU. Executing the refinement loop on the CPU rather than on the device has a significant impact on the achievable speedup, but again, the actual reduction depends on how much slower the CPU is compared to the GPU, and of the overhead of expensive bus transfers between host and device. As our proposed *minimally invasive co-processor integration* into FEAST (cf. Section 6.2.2) constitutes a mixed precision scheme as well (with a much more involved outer loop), we return to this topic in the performance evaluations in Section 6.3 (Poisson problem), Section 6.4 (linearised elasticity) and Section 6.5 (stationary laminar flow).

5.8. Influence of the Multigrid Cycle Type and the Coarse Grid Solver

After the discussion of the efficiency of different smoothers and their respective damping factors in GPU-based multigrid solvers in Section 5.5, and of the accuracy and performance of mixed precision solvers in Section 5.6 and Section 5.7, we now analyse the impact of the remaining 'free' algorithmic parameters on performance and efficiency, i. e., the choice of the cycle type and, closely related, the coarse grid solver. The cycle type determines the traversal through the mesh hierarchy and in particular how often the coarse grid problem is visited (cf. Figure 2.5 on page 28). After a brief overview of the test procedure in Section 5.8.1, we demonstrate in Section 5.8.2 that F and W cycles incur a (significant) impact on the achievable speedup by executing multigrid solvers on the GPU rather than on the CPU. Detailed analysis reveals that (1) the loss of parallelism on coarse mesh levels, and (2) the limitation of the current implementation to only solve coarse grid problems with iterative Krylov subspace schemes rather than direct sparse LU solvers, contribute equally to the deficiencies we observe. The remainder of this section is devoted to the analysis of various techniques to alleviate these problems. All numerical studies are performed in double precision, implications for mixed precision iterative refinement solvers are discussed separately in Section 5.8.8. We conclude with a summary in Section 5.8.9.

5.8.1. Overview and Test Procedure

All numerical and runtime efficiency tests in this section are performed for the isotropic and anisotropic test cases introduced in Section 5.5.1 on page 158. We restrict ourselves to using JACOBI and ADITRIDI as smoothers for the multigrid solver or as preconditioners for the coarse grid solver. Our parallelisation of Gauß-Seidel type operators relies on multicolouring and reduces the implicit coupling between the unknowns in the application of the smoothers, potentially introducing deteriorating side effects (cf. Section 5.5) that we want to avoid in our tests in this section. This allows us to focus entirely on the influence of the coarse grid solver and the type of the multigrid cycle. For the same reason, the majority of our experiments is performed with multigrid solvers executing entirely in double precision. In Section 5.8.8 we comment on the implications of our findings for mixed precision iterative refinement multigrid solvers. If not stated otherwise, we use the conjugate gradient solver for the coarse grid problems, its preconditioner is always the same operator as used for smoothing the multigrid algorithm. The multigrid solver is always configured to perform four pre- and postsmoothing steps each (we count one application of an alternating direction implicit smoother as two elementary smoothing steps, i. e., four JACOBI steps count as two TRIDIROW and two TRIDICOL steps). Smoothing is damped with the 'optimal' damping parameter, as determined in Section 5.5.2.

In the comparisons in this section, we are mainly interested in the relative differences between one fixed cycle type either on the CPU or on the GPU. We never explicitly compare, e. g., a V cycle with a W cycle multigrid, simply because a V cycle suffices for the problems previously considered throughout this chapter. However, many problems in practice require F or even W cycles to ensure convergence in competitive time.

To avoid granularity effects, we continue to perform all comparisons in the total (runtime) efficiency which measures the time (in µs) per unknown to gain one digit, see Section 5.5.1 for the exact definition.

5.8.2. Baseline Experiment – Problem Statement

Figure 5.45 depicts the total runtime efficiency of multigrid solvers depending on the choice of the cycle type. The coarse grid solver is always configured to solve 'exactly', i. e., to reduce the initial residuals by eight digits. The mesh hierarchy is always traversed fully, the coarse grid problem

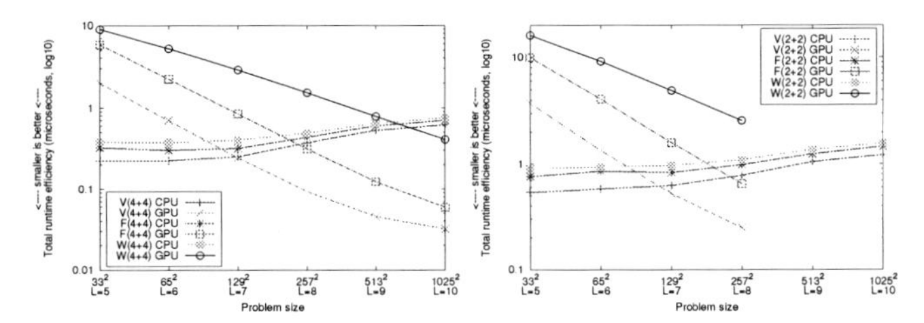

Figure 5.45: Total runtime efficiency depending on the choice of the multigrid cycle. Left: mostly isotropic domain (JACOBI smoother), right: anisotropic domain (ADITRIDI smoother). The y-axis uses a logarithmic scale.

thus consists of nine unknowns; in fact even less, as five unknowns correspond to boundary points with homogeneous Dirichlet boundary conditions and their solution is thus known a priori (the matrix has unit rows corresponding to these unknowns). In case of the JACOBI smoother, we observe a significant reduction in speedup when using more advanced cycle types: The V cycle on the GPU is four, twelve and 20 times faster than the CPU implementation for refinement levels $L = 8, 9, 10$, respectively. With F cycles, the speedup reduces to factors of 1.4, 4.8 and 11.7, and when using W cycles, the GPU only achieves a speedup on the finest level of refinement, the exact factor is 1.8. For multigrid with a stronger smoother on the more distorted, anisotropically refined test domain (cf. Figure 5.24 on the right), the maximum problem size that can be treated in double precision is $L = 8$. While we do see speedups when using a V cycle, the F cycle results in approximately the same efficiency on the CPU and the GPU, and all speedups are lost when using W cycles.

As we have pointed out regularly throughout this chapter, the GPU is not sufficiently saturated for the small problem sizes which do not exhibit enough parallelism. The number of times the coarse grid solver is executed per cycle is one, $L-1$ and 2^{L-1} for the V, F and W cycles, respectively. On the CPU, the additional work does impose some overhead, but all computations on smaller levels of refinement are executed entirely in cache. Consequently, F and W cycles are not much more expensive than V cycles. On the GPU, this is not the case and performing more work on small problem sizes leads to a disproportionate slowdown: For one, the SIMD width of current CUDA-capable GPUs is the warp size (32), and the G200-based GPUs only have one double precision ALU per multiprocessor. Furthermore, not all our kernels exhibit perfectly coalescable memory access patterns for border cases, and on small input sizes these can no longer be 'hidden', as the inner grid points do not outnumber the boundary points by a vast amount. Finally, there is a large amount of synchronisation between the host and the device, many small kernels are launched, their parameters have to be set, and during the iterations of the conjugate gradient solver, scalar values have to be transferred back and forth. It is unclear which of these factors dominates the slowdown, we are however convinced that they all contribute significantly. One could say that for very small input problems, the GPU is not suited, due to its design goal of maximising throughput: In these cases, the GPU can be seen more like a 1.3 GHz SIMD processor with a very high latency to access off-chip memory. We point out that our implementation is purposely designed in a modular way. It would certainly be possible to merge an entire cycle type for one particular smoother and one concrete coarse grid solver into fewer kernel calls, this has for instance been demonstrated by Kazhdan and Hoppe [125] who merged a V cycle multigrid with a fixed number of JACOBI smoothing steps into two kernel calls.

The observed slowdowns in the more distorted test case are of the same order as when using multigrid with JACOBI smoother on the corresponding levels of refinement. We can therefore restrict the discussion in the remainder of this section to the simpler case, which allows us to perform all computations up to refinement level $L = 10$, where the device is fully saturated. In summary, we can say that in terms of speedup, using GPUs for problems that require F cycles is well justified, we still observe more than an order of magnitude faster execution than on the CPU if the device is sufficiently saturated. For W cycle multigrid however, this is not entirely the case, the speedup is less than a factor of two.

Coarse Grid Solver

Detailed analysis of the convergence behaviour of the coarse grid solver indicates a fundamental deficiency of the conjugate gradient algorithm *in floating point* in the context of the test problems we consider.[21] It computes the next search direction $\mathbf{p}$ via

$$\mathbf{p}^{(k)} = \mathbf{z}^{(k-1)} + \frac{\rho_{k-1}}{\rho_{k-2}} \mathbf{p}^{(k-1)},$$

where $\mathbf{z}$ is the result of the preconditioning and ρ denotes the square of the norm of the current incrementally computed residual, see Figure 2.2 on page 25. For this particular test case and for very fine grids in the multigrid cycle, we occasionally observe $\rho_{k-2} \approx 10^{-16}$ and $\rho_{k-1} \approx 10^{-32}$ or even smaller, i. e., the squares of the defects are very small in absolute numbers and are becoming smaller rapidly. The computed steplength parameter, i. e., the quotient of ρ_{k-1} and ρ_{k-2}, becomes very small, and in the scaled addition to compute the next search direction, roundoff and cancellation occurs. Subsequently, the solver stalls or diverges, because the search directions differ only by a tiny amount and are no longer $\mathbf{A}$-orthogonal. In our implementation, we terminate the conjugate gradient solver before this detrimental effect of finite floating point arithmetic occurs, to prevent divergence. An alternative idea is to recompute $\rho = \|\mathbf{b} - \mathbf{Ax}\|^2$, but in our experience, this only shifts the problem on to the next or the subsequent iteration of the conjugate gradient algorithm, as we cannot avoid that the (initial) defect is so small.

As this behaviour often occurs already in the first iteration of the conjugate gradient scheme, in particular when the multigrid solver has performed a number of cycles already, it means that the initial guess (the zero vector) is returned as the solution of the coarse grid solver. After prolongation to the second-coarsest level of refinement, the coarse grid correction (see Figure 2.3 on page 27) thus does not yield a reduction in the defect norm, as required for convergence of the multigrid solver: The prolongation of the zero vector is the zero vector. However, postsmoothing ensures that after prolongating to the next finer level, the corresponding coarse grid correction results in an error reduction. The whole multigrid solver thus converges if the mesh hierarchy is sufficiently deep, i. e., if the error from the coarse grid solve on the coarsest level is 'smoothed away' during the prolongation phase.

Furthermore, we always employ *adaptive coarse grid correction*, see Section 2.3.3 on page 37. While this technique has originally been introduced to FEAST to alleviate the impact of its nonconforming local grids due to FEAST's domain decomposition approach to couple neighbouring subdomains via special boundary conditions [242], it is very beneficial in general. Without adaptive coarse grid correction, the multigrid solver always executes 1–3 (compared to 10–11 total) additional iterations for this particular test case. This is however a side effect that we do not pursue further in this thesis: Adaptive coarse grid correction as well as standard coarse grid correction both add the zero vector to the current solution on the finer grid if the coarse grid solver returns it as the solution.

[21] More correctly speaking, of Krylov subspace schemes in general; we observe the same behaviour when using BiCGStab.

We want to emphasise that the observed problem with the conjugate gradient solver is not specific to this particular problem, we have witnessed it in many situations when investigating the acceleration of FEAST applications on GPU clusters, see Section 6.4 and Section 6.5. At some point during the solution process, the computed quantities like iterates and defects are very small—in absolute numbers—on coarser levels, in particular if several coarse grid solves have been performed already: This is inherent to the multigrid algorithm, which smoothes different error components on different mesh levels, see Section 2.2.3.

In double precision, the deficiency of the conjugate gradient solver never influences the convergence behaviour of the entire scheme, but can very well explain why F or W cycles are not much better than V cycles, in this particular example. The convergence rate of V cycle multigrid for refinement level $L = 10$ is 0.185, and for F and W cycles, we observe 0.148. Initially, both F and Wcycle multigrid converge much faster than the V cycle. After a few cycles have been performed already, this problem appears for the first time, and soon, coarse grid corrections on the finest level are always performed with the zero vector.

We return to this question in Section 5.8.8 where we discuss this topic in the mixed precision iterative refinement case. In the following, we describe and analyse several techniques to improve the speedup when using F or W cycles. Because of this observation, we never compare the efficiency between different cycle types, we restrict all comparisons to the same cycle type on different hardware architectures.

5.8.3. Truncation of Multigrid Cycles

Truncating refers to executing the coarse grid solver not on refinement level $L = 1$, but on finer levels. We have already employed this technique quite early, in our first paper on GPU cluster computing with FEAST [87].[22] This approach has several potential advantages: First, it reduces the depth of the mesh hierarchy and thus the number of coarse grid solves for the F and W cycles. As the problems to be solved are now larger, the computational intensity of the coarse grid solves increases, as there is actual work to do instead of pure overhead.

Figure 5.46 depicts the resulting effects on efficiency, the coarse grid solver is still configured to reduce the initial residual by eight digits. Looking at the CPU results first, we see that truncating decreases the efficiency (increases the runtime) for smaller levels of refinement. The conjugate gradient solver does not converge independent of the mesh width h, its number of iterations is proportional to h and thus approximately doubles on each level of refinement. Thus, the coarse grid solver actually performs a significant amount of arithmetic work. This effect is particularly visible when setting the truncation parameter to 5: The multigrid solver on refinement level $L = 7$ in this case reduces to a two-grid scheme, where each coarse grid problem has 4 225 unknowns. For the largest problem size, truncating does not affect the efficiency at all when using V cycles. The same is true when employing the other cycle types, except that when truncating five levels of refinement, the resulting multigrid solver achieves only approximately 90 % of the efficiency as without truncation. In summary, for F and W cycles and sufficiently large problem sizes, the additional arithmetic work induced by the h-dependency is almost perfectly balanced with the resulting savings in the number of coarse grid solves performed.

On the GPU, the behaviour is different. For the V and F cycles, truncating always decreases the efficiency. This is the opposite to our initial expectation, because the device is more saturated the larger the problem sizes are. Nonetheless, the efficiency reduces the more we truncate, meaning that the h-dependency of the conjugate gradient algorithm has a more severe impact than on the CPU, where all calculations on such small levels of refinement are performed in cache. For

[22] Due to delays in the publication process, it actually was in press later than our second cluster paper [84], but the work was performed first.

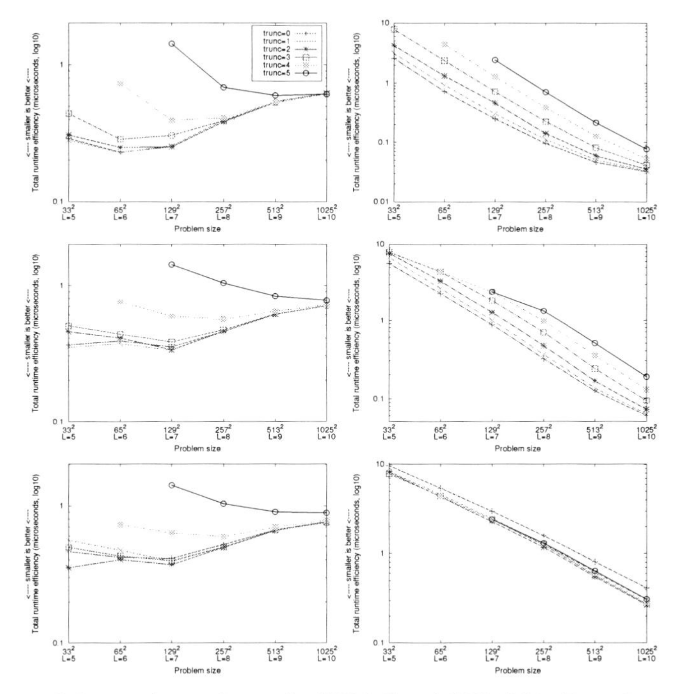

Figure 5.46: Influence of truncating on the CPU (left) and GPU (right). Top to bottom: V, F and W cycle. Notation: `trunc=0` means no truncation, `trunc=1` means coarse grid problems are solved not on level $L=1$ but on $L=2$, etc. All plots share the same legend, and the y-axis uses a logarithmic scale.

instance, with V cycles and truncating so that refinement level $L=5$ (`trunc=4`, 1 089 unknowns) is the coarse grid, is 2.4 times slower than traversing the mesh hierarchy fully to $L=1$. For the W cycle, the reduction of coarse grid solves is exponential in the truncation level, rather than linear or constant as for the other two cycle types. Additionally, the device is severely underutilised on small levels of refinement, levels that are skipped by truncation. Consequently, this technique is always beneficial, we observe an increase in efficiency of at least 30 %. Due to the h-dependency, the increase is maximised when truncating at refinement level $L=4$ (`trunc=3`, 289 unknowns), and in this case, the speedup compared to the best (not truncated) CPU variant increases from a factor of 1.8 to a factor of 2.8.

We finally point out that the truncation of multigrid cycles and thus the reduction of the depth of the multigrid hierarchy completely alleviates our observation of coarse grid corrections by the zero vector on the second-coarsest level of refinement. Truncating the mesh hierarchy by two levels of refinement already removes all these artifacts, even when employing W cycles. This is simply a consequence of the fact that the involved quantities are now much larger in absolute numbers. Of course, our emergency stopping criterion which prevents floating point cancellation and thus divergence due to the loss of **A**-orthogonality still holds occasionally, but only after a sufficient amount of conjugate gradient iterations have been performed. In practice, this means that the coarse grid problems are not necessarily solved 'exactly', i. e., to eight digits accuracy. We return to this topic in Section 5.8.7.

5.8.4. Better Preconditioning of the Coarse Grid Solver

Not only for the Poisson problem, but also for subproblems that occur in the applications we consider from solid mechanics and fluid dynamics (cf. Chapter 6), the condition number of the system matrix $\mathbf{A}$ depends on the mesh width h after discretisation. The idea of preconditioning with a given operator $\mathbf{C}$ is that the condition number of $\mathbf{C}^{-1}\mathbf{A}$ is ideally much smaller than that of $\mathbf{A}$. In our context, this can be exploited, employing a better preconditioner reduces the number of iterations of the coarse grid solver. The iterations of the coarse grid solver are disproportionately expensive on the GPU as demonstrated by the previous experiments. We can therefore hope for an overall efficiency improvement, if the more expensive application of the better preconditioner reduces the number of iterations by a sufficient amount.

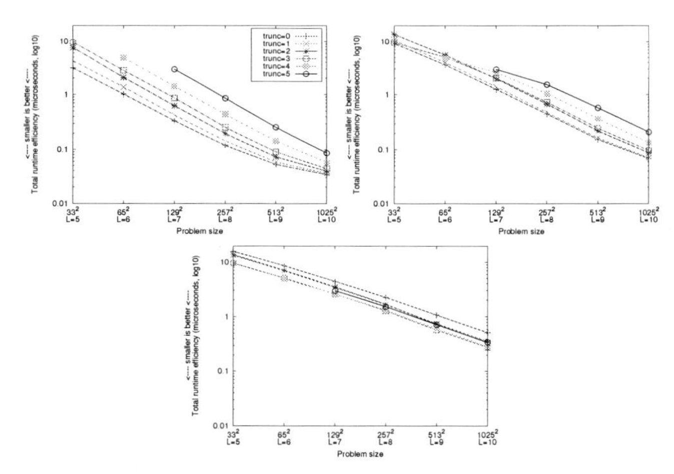

Figure 5.47: Influence of truncating on the GPU when using a better preconditioner for the coarse grid solver. Top: V and F cycle, bottom: W cycle. All plots share the same legend, and the y-axis uses a logarithmic scale.

The conjugate gradient algorithm requires a symmetric and positive definite preconditioner, the former criterion is relevant given the preconditioners we have at our disposal, see Section 5.4. We thus repeat the experiment performed in Section 5.8.3, employing exemplarily the better preconditioner ADITRIDI. Its application involves the solution of one tridiagonal system per mesh row, which is implemented entirely in shared memory on the device and thus comparably fast, shared memory can be interpreted as a cache on the device.

Figure 5.47 presents the total runtime efficiency of the resulting multigrid solvers, we only perform this experiment on the GPU. The results are qualitatively identical to the ones depicted in Figure 5.46. For the V and F cycles, truncation of the multigrid cycle always results in a slowdown. When employing the W cycle, truncation is always beneficial, and best results are obtained when truncating the mesh hierarchy so that the coarse grid solver is executed on refinement level $L = 4$ (`trunc=3`, 289 unknowns).

Compared to the baseline variants on the GPU (cf. Figure 5.46) which precondition the conjugate gradient solver with JACOBI, we however note that for the V and F cycle multigrid, the efficiency is always between 50 and 80 % of the best solver configuration so far for the respective cycle type. The (constant or linear) reduction in the amount of coarse grid solves is not balanced by the (linear) reduction of iterations performed by the coarse grid solver for each solve. The better preconditioner results in a reduction of 20–50 % of conjugate gradient iterations (in the sum of all coarse grid solver iterations until convergence of the entire scheme), but each iteration is more expensive than using the simple JACOBI preconditioner. Our idea is nonetheless justified, as the solvers lose less efficiency by truncating. For W cycles, the best setting of the truncation parameter yields an efficiency that is on par with the best variant on the GPU that does not involve truncation, reaching a speedup against the best CPU variant by a factor of 1.9 compared to 1.8. But this small improvement is outweighed by the improvements we achieve by truncation and the JACOBI preconditioner (a factor of 2.8 versus the CPU).

In summary we can say that employing a better preconditioner for the coarse grid problems does have positive effects. The ADITRIDI preconditioner is however not powerful enough to reduce the number of iterations on the coarse grids by a sufficient amount in balance with its more expensive application. An overall better speedup compared to using the simple JACOBI preconditioner is not achieved, but we expect this technique to be beneficial with a stronger preconditioner.

5.8.5. V Cycles with Variable Smoothing: An Alternative to F and W Cycles

Brenner [39] proves that the contraction property (and thus convergence) of V cycle multigrid can only be guaranteed under the assumption of 'a sufficiently large number of smoothing steps', see also the references therein. In practice, an established technique that exploits this result is to double the amount of smoothing steps in each restriction, and to halve it again during each prolongation, leading to a V cycle multigrid with variable, i. e., level-dependent smoothing. Experience in the multigrid community shows that this modification of the V cycle is applicable in situations that normally require F or even W cycles to ensure convergence, i. e., situations where V cycle multigrid requires many iterations to converge competitively.

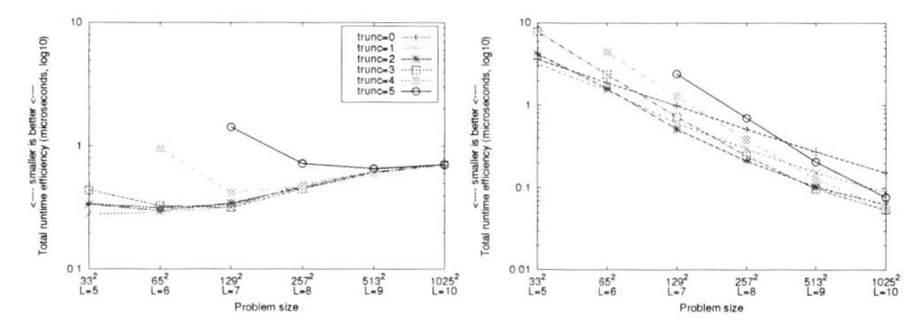

Figure 5.48: Efficiency of V cycle multigrid with variable smoothing on the CPU (left) and GPU (right). Both plots share the same legend, and the y-axis uses a logarithmic scale.

We do not perform numerical tests to quantify the difference in efficiency of this approach compared to using F or W cycles, but instead, we experimentally analyse and quantify the difference in speedup of our GPU-based multigrid solver compared to the CPU executing the same configuration. Figure 5.48 depicts the results of this experiment.

We only discuss the larger problem sizes, as the qualitative impact of truncation on coarser levels of refinement is the same we have observed in our previous experiments. On the CPU, this strategy has no noticeable effect in terms of the truncation parameter, the additional iterations of the (h-dependent) conjugate gradient solver are perfectly balanced by the truncation level. This behaviour is identical to what we have observed previously for V cycle multigrid. On the GPU, truncation always results in a significant increase in efficiency. For the best setting of the truncation parameter (`trunc=3`, coarse grid problems are solved for 289 unknowns), we observe a speedup by a factor of 13 for the largest problem size, and of 6.2 and 1.9 for $L = 9$ and $L = 8$, respectively. The speedups are slightly better than those we observed for the best, i. e., non-truncating configuration of F cycle multigrid, and significantly better than for W cycles. V cycle multigrid with variable smoothing is thus well-suited on GPUs.

5.8.6. Direct Coarse Grid Solvers on the CPU

Given the deficiencies of utilising Krylov subspace schemes as coarse grid solvers, we want to investigate how sparse direct LU decomposition schemes perform in our setting. Such direct solvers are typically based on multifrontal techniques, and the numbering of the unknowns determines the amount of fill-in of nonzero values, see Section 4.5. FEAST employs UMFPACK for these problems, in our experiments in Section 4.5 we use SuperLU because UMFPACK is currently limited to double precision arithmetic only.

To the best of our knowledge, such sparse direct solvers have not been analysed or implemented on the GPU to date, because this is a very challenging task due to the involved irregular memory access patterns and their inherent sequentiality. We will investigate such approaches on GPUs in future work. A simple workaround for the time being is to execute such coarse grid solvers on the CPU. The goal of the following experiment is to evaluate if this workaround is feasible: As the impact of the comparatively narrow PCIe bus between host and device is potentially high, we set up a simple test prior to deciding if implementing such a coupling is worth the effort.

We configure the multigrid solver to perform exactly the same amount of iterations as for our baseline configuration, see Figure 5.46. Instead of executing the coarse grid solver on the device, we copy the restricted right hand side vector to the host, perform a 'dummy operation' on it, and transfer it back to the GPU for prolongation. If we actually solved the coarse grid problems with a sparse direct solver, the convergence rates of the multigrid solvers would be the same as observed for our baseline experiment. We can therefore reuse these original values to postulate the total runtime efficiency. Our results do not take the runtime of the projected sparse direct solver into account, meaning that we establish a relative lower bound for the efficiency of (hybrid) multigrid solvers executing all coarse grid solves on the host instead of the device.

Figure 5.49 depicts the results of this experiment. For small problem sizes and truncation at high levels of refinement, the large improvements are an artifact of our test procedure, as the majority of the work that is normally spent in the coarse grid solver is replaced by a simple dummy operation which takes less time. For sufficiently large problem sizes, the V cycle multigrid achieves only 75 % of the baseline variants depicted in Figure 5.46, in other words, executing the coarse grid solvers on the host is not advantageous. The same is true for F cycle multigrid: While truncating does improve the efficiency up to a factor of two, we only observe between 35 and 70 % of the efficiency of the best configuration that executes the conjugate gradient solver on the device, i. e., of a configuration that actually solves coarse grid problems instead of performing only a dummy operation. For W cycles, where the number of coarse grid solves is exponential in the depth of the mesh hierarchy, truncating the multigrid solver has the biggest impact. Nonetheless, the speedup against the baseline variant on the device is only a factor of one to two, and this estimate does not include the time spent executing the 'virtual' coarse grid solver on the CPU. We can thus conclude that the PCIe bus constitutes too much of a bottleneck in order to make this workaround strategy feasible in practice.

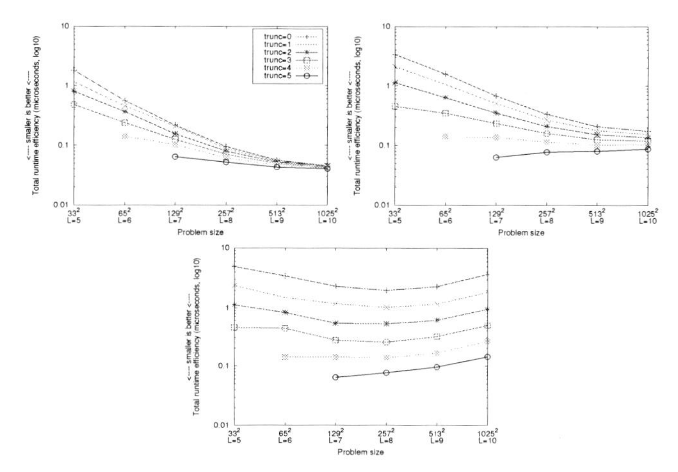

Figure 5.49: Influence of truncating on the GPU when executing the coarse grid solver on the CPU. Top: V and F cycle, bottom: W cycle. All plots share the same legend, and the y-axis uses a logarithmic scale.

5.8.7. Approximate Solution of Coarse Grid Problems

To ensure convergence of the multigrid algorithm, it is not necessary to solve the coarse grid problems exactly, even though this is used in the general convergence proofs [102] that use induction arguments. Looking at the proofs in detail reveals that the coarse grid solves should be as exact as the convergence rate of the entire multigrid solver in order to not degrade the overall convergence. However, this criterion does not lead to a practical parameter setting, because the convergence rate is not known a priori. In all tests presented in Section 5.5, Section 5.7, and in Chapter 6, we have thus heuristically configured the conjugate gradient coarse grid solver to reduce the initial residuals by two digits, this is a compromise between large and small levels of refinement: In our experiments, we usually observe convergence rates between 0.1 and 0.2 for large problems, and below 0.1 for small problems. The setting of two digits is thus slightly too careful for large problem sizes, but we want to be on the safe side and suggest solver configurations with a 'black-box character'.

Alternative choices are basic stationary defect correction schemes, like the Jacobi-, Gauß-Seidel or SOR iterations, but in our experience, the conjugate gradient solver is, for the most part, the best choice, because it typically converges very fast in its first iterations.

Figure 5.50 presents the efficiency of multigrid solvers using this approximate conjugate gradient solver, we compare these results with the measurements obtained when solving the coarse grid problems exactly, see Figure 5.46 on page 195. For the following discussion, it is important to note that in all cases, the amount of multigrid iterations is identical to solving coarse grid problems exactly. Looking at the CPU data for small problem sizes first, we see a significant increase in efficiency: Due to the h-dependency of the conjugate gradient algorithm, the coarse grid solver performs four times less work. For larger problems however, the results are indistinguishable between the two figures. This underlines again that on the CPU and for large problem sizes that do not fit into cache, the computation time is determined by the memory wall problem, the work on the finer grids dominates the runtime. For the GPU, the shape of the curves looks at first sight

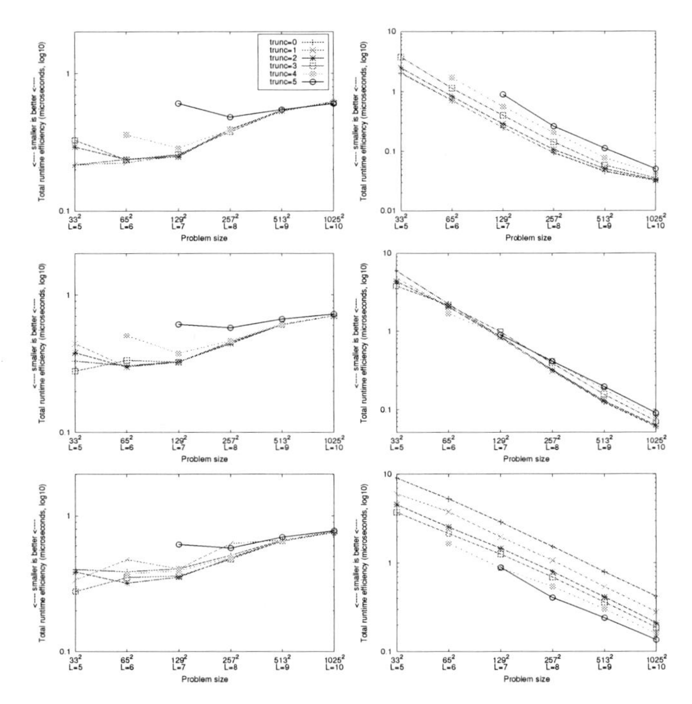

Figure 5.50: Influence of truncating on the CPU (left) and GPU (right) when solving coarse grid problems approximately to two digits accuracy. Top to bottom: V, F and W cycle. All plots share the same legend, and the y-axis uses a logarithmic scale.

reasonably similar to the ones obtained when solving coarse grid problems exactly. To emphasise the subtle yet important differences, we compute a number of different speedup factors, see Figure 5.51.

We look at V cycle multigrid first. The corresponding two lines in the top left plot (red dashes and green crosses) show that we always obtain a speedup compared to the exact coarse grid solution. For a medium truncation setting (`trunc=3`, coarse grid problems have 289 unknowns), we observe a small decrease in speedup with increasing level of refinement. For full truncation (`trunc=5`, coarse grid problems have 4 225 unknowns), the speedup is significantly higher for small levels of refinement, based on the amount of arithmetic work that we save, we expect a factor of four, because the number of iterations performed by the conjugate gradient solver is proportional to the mesh width. The actual factor is smaller, underlining again that the penalty of underutilising the device is disproportionately high. For large problem sizes, this penalty has a smaller impact on overall performance, and the speedup reduces from a factor of 1.5 to 1.2 when truncating the mesh hierarchy less aggressively. These effects translate directly to the speedup comparison against the variant that does not apply truncation at all (top right plot), the slowdown induced by truncation of the mesh hierarchy is much smaller than for solving the coarse grid problems exactly (actual numbers for the latter case are not shown, but can be derived from the data in Figure 5.46): V cycle multigrid achieves 90 % and 42 % efficiency compared to traversing the

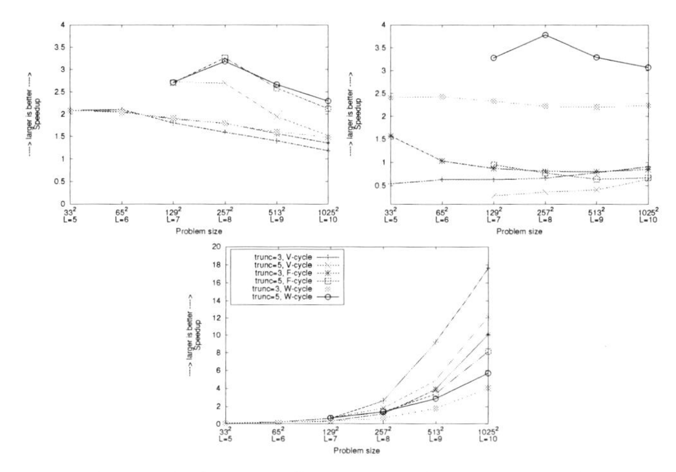

Figure 5.51: Speedup obtained by approximate coarse grid solving: Top left: Approximate vs. exact coarse grid solution on the GPU. Top right: Truncated vs. untruncated multigrid cycles on the GPU. Bottom: Speedup of the GPU over the CPU. The y-axis uses a different linear scaling for all plots to emphasise differences, all plots share the same legend.

mesh hierarchy fully for the two truncation levels under consideration when solving coarse grid problems exactly, and 91 % and 64 % for inexact coarse grid solutions. Consequently, the speedup compared to the CPU (bottom plot), whose efficiency is invariant of truncation, drops less, and we achieve factors of 17.6 and 12.1 compared to 14.9 and 8.0 for exact coarse grid solutions. The best speedup without truncation in both cases is 19.4 and 19.7, respectively; we expect the difference to be in the noise because the coarse grid solver does not dominate the total time to solution at all when fully traversing the mesh hierarchy in a V cycle.

For F cycle multigrid, we observe the same qualitative and quantitative behaviour (blue stars and magenta squares in Figure 5.51), the only noteworthy exception is the speedup compared to not truncating the mesh hierarchy for small problem sizes (top right plot). This effect is however the expected behaviour based on the explanation given above and a direct consequence of the speedup obtained by the faster coarse grid solution. The speedup compared to executing the solvers on the CPU and switching from exact to inexact coarse grid solution is slightly higher than in the case of the V cycle multigrid, because there are more coarse grid problems to be solved. The actual numbers are 11.7 for the exact solution and 12.0 for the inexact solution. Truncation always leads to a reduced speedup.

The effects are much more pronounced for W cycle multigrid (cyan filled squares and black circles in Figure 5.51). The truncation of multigrid cycles has the biggest impact, because the amount of disproportionately expensive work reduces exponentially. The savings compared to solving the coarse grid problems exactly are always maximised in Figure 5.51 on the left, when comparing the corresponding plots for matching truncation levels. The speedups against the baseline variant that does not apply truncation on the GPU reaches a stable factor of 2.3 and 3 for the truncation levels depicted, while for F and V cycles, truncation always leads to a slowdown (see Figure 5.51 (top right)). When the coarse grid solves are exact, this speedup factor is only 1.3, emphasising again that the GPU is disproportionately slow for small problem sizes. Consequently, the speedup of the GPU-based solver against the CPU is maximised when truncating aggressively, reaching a factor of 5.7 compared to a factor of 2.8 when solving exactly.

5.8.8. Implications for Mixed Precision Iterative Refinement Multigrid

In single precision, the exact solution of the coarse grid problems is not possible, as demonstrated in Section 4.5. Therefore, we only have one possibility to execute the mixed precision iterative refinement solver using multigrid for the low precision part, given the limitation of our current implementation to coarse grid solvers of Krylov subspace type: We have to solve the coarse grid problems inexactly. Furthermore, we expect cancellation effects in the computation of the search directions in the conjugate gradient solver to occur earlier, see Section 5.8.2.

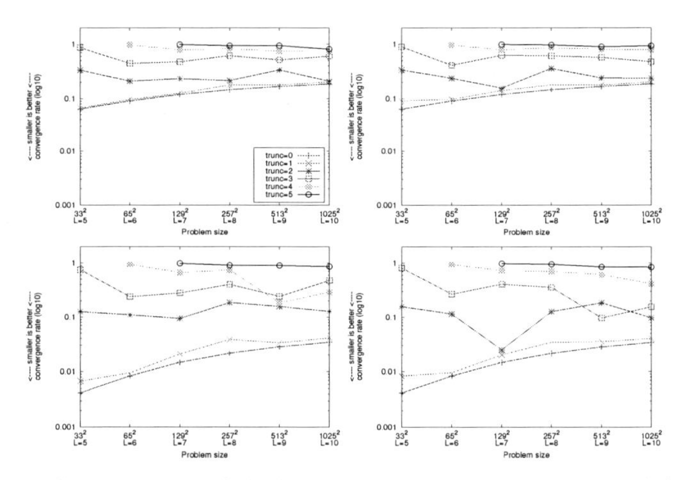

Figure 5.52: Convergence rates of mixed precision iterative refinement V cycle multigrid solvers when applying truncation. Top left: Updates in double precision every multigrid cycle, the coarse grid solver gains one digit (in short, 1:1). Top right: 1:2. Bottom left: 2:1. Bottom right: 2.2. All plots share the same legend, and the y-axis uses a logarithmic scale.

We investigate the latter problem first, by repeating the tests from Section 5.8.7 using exemplarily V cycle multigrid as the low precision inner solver. All presented tests are performed on the CPU only, but we confirm that the GPU exhibits the same behaviour. We vary the update frequency in double precision, i. e., we change the stopping criterion for the inner solver to perform one or two multigrid cycles in between double precision correction steps. Furthermore, we modify the stopping criterion of the inner coarse grid solver to gain either one or two digits. Figure 5.52 presents the results in terms of convergence rates. Obviously, truncation of the mesh hierarchy has severe consequences, in all cases, the mixed precision approach only barely converges when the truncation level is set so that the coarse grid problems are solved on finer levels than $L = 3$ (`trunc=2`, 81 unknowns). By looking at the convergence behaviour of the solvers in detail, we were able to identify the coarse grid solver to cause this problem. With high truncation levels and (depending on the configuration) after 6–8 overall multigrid iterations, the coarse grid solver returns the zero vector because our emergency stopping criterion (appropriately modified by eight orders of magnitude, i. e., the quotient of ρ_{k-1} and ρ_{k-2} evaluates to less than 10^{-8} rather than

10^{-16}). to prevent cancellation and thus divergence holds, see Section 5.8.2. In double precision, we regularly observe this problem as well, but only for F or W cycles and high levels of refinement. Here, the multigrid solver cannot recover from the bad or even nonexisting coarse grid solves, so that the mixed precision iterative refinement scheme converges like a randomly preconditioned Richardson iteration. In double precision, the application of level truncation never results in an increase in the number of multigrid iterations compared to traversing the hierarchy down to the coarsest level of refinement, at least when the coarse grid problems are solved to at least two digits accuracy.

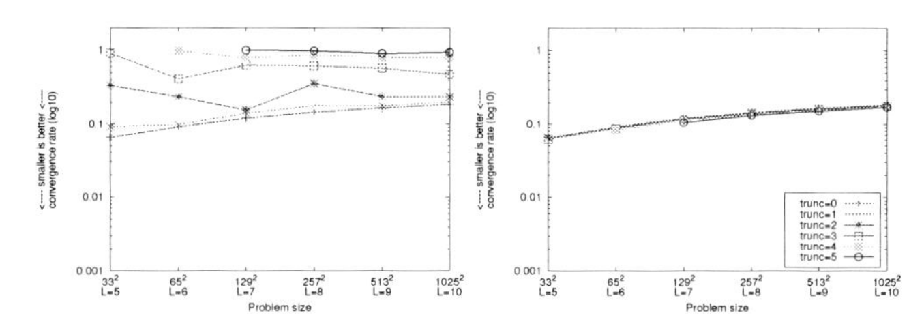

Figure 5.53: Influence of solving coarse grid problems in double precision. Both plots share the same legend, and the y-axis uses a logarithmic scale.

To demonstrate that this problem is only a consequence of the particularities and early termination of the conjugate gradient coarse grid solver, and not inherent to the mixed precision iterative refinement technique, we repeat exemplarily one of these tests, and only switch the coarse grid solver to double precision. Figure 5.53 underlines this impressively. Only by improving the precision of the coarse grid solver, we can recover the robustness of the multigrid solver. Nonetheless, we do not use this setting in the following for two reasons: First, we would potentially lose the comparability to the results in the following and all previous results, in particular the previously published work that is presented in Chapter 6. Second, this trick is only applicable on recent hardware, as older hardware does not support double precision floating point. An alternative idea to avoid the problems of the conjugate gradient coarse grid solver in single precision is to use, e. g., a preconditioned Richardson solver for the coarse grid problems, but since it is also h-dependent, it converges too slowly in our experience. So we do not present any numerical studies along these lines. Looking into using different coarse grid solvers, e. g., sparse direct schemes, is, in summary, mandatory in future work: Sparse direct solvers do not exhibit this problem, and switching to double precision is not necessary.

Influence of Refinement Level Truncation

Figure 5.54 shows the efficiency of various mixed precision iterative refinement solvers, it corresponds to Figure 5.50 on page 200 which depicts the same experiments performed in double precision. The solver is configured to perform two multigrid cycles without convergence control per double precision update step, and the coarse grid solver is configured to reduce the initial residual by two digits. We omit all configurations that suffer from the problem described above for the three largest problem sizes, so that we can use single precision for the entire multigrid solver.

The results are qualitatively identical to the ones in double precision. On the CPU, truncating the mesh hierarchy does not improve efficiency for the interestingly large problem sizes. On the GPU, truncating is always beneficial for F- and W cycles, but leads to the problem described above much earlier so that aggressive truncation cannot be applied: The setting of `trunc=4` (cyan filled squares in the plot) is already impossible for V cycle multigrid, and has severe consequences for

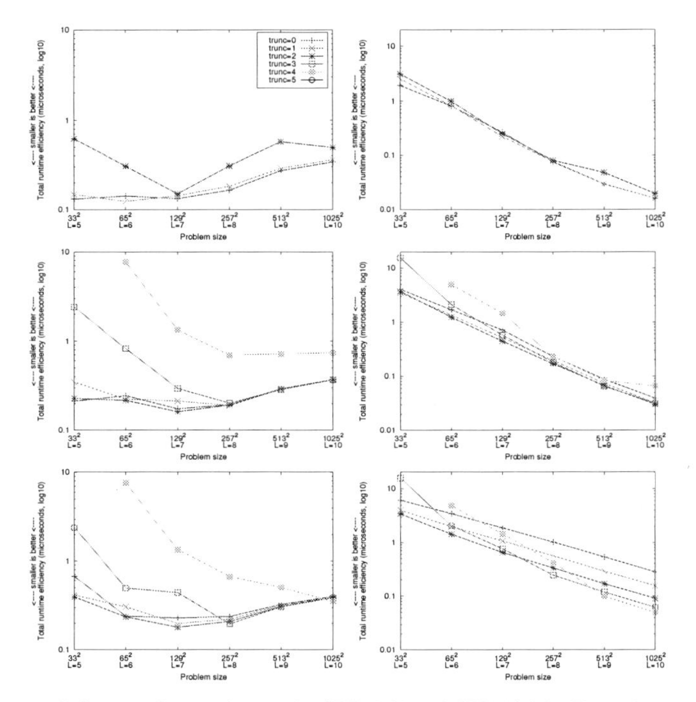

Figure 5.54: Influence of truncating on the CPU (left) and GPU (right). Top to bottom: V, F and W cycle. Notation: `trunc=0` means no truncation, `trunc=1` means coarse grid problems are solved not on level $L = 1$ but on $L = 2$, etc. All plots share the same legend, and the y-axis uses a logarithmic scale.

the other two cycle types and smaller levels of refinement.

Speedup

Figure 5.55 depicts different speedup factors of the best (in terms of the truncation parameter) mixed precision iterative refinement solver for each cycle type. We are only interested in the largest three problem sizes, on smaller ones, the number of iterations might have already started to increase, because of the missing coarse grid solve. Looking first at the top left plot which shows speedups against using double precision exclusively on the device first, we see that when using V cycles, the speedup increases slowly as soon as the device becomes sufficiently saturated, and reaches the expected factor of two. In case of F cycles, the speedup remains remarkably constant at 1.9 for all problem sizes. The W cycle multigrid exhibits a 'superlinear' speedup by a factor of three compared to using double precision, a behaviour that does not occur on the CPU. When truncating the mesh hierarchy at coarser levels, the corresponding factors are 2.1, 1.2 and 0.7 for $L = 10$. All configurations require the same (total) amount of multigrid cycles, independent of the precision. The only 'variable' parameter in our test is the stopping criterion of the coarse grid solver, so this observation can again be explained as an artefact of the conjugate

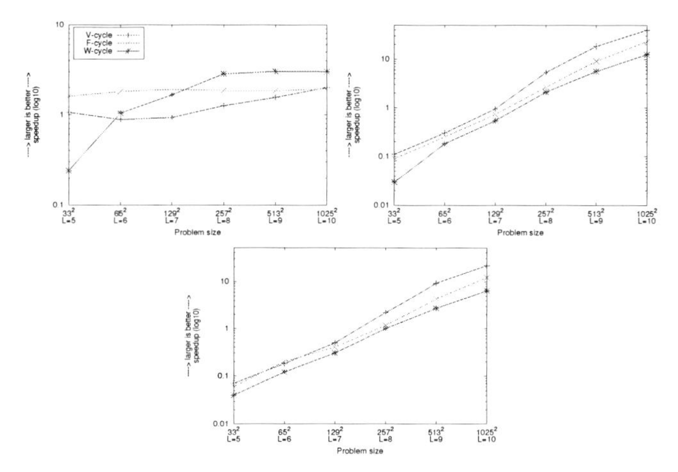

Figure 5.55: Speedup of mixed precision iterative refinement on the GPU over double precision on the GPU (top left), double precision on the CPU (top right) and mixed precision on the CPU (bottom). All plots share the same legend, and the y-axis uses different logarithmic scales.

gradient algorithm: In single precision, the emergency stopping criterion to prevent roundoff and cancellation holds more often, so that overall less work is performed. This underlines again to what extent GPUs are disproportionately inefficient if the current workload is not large enough to exhibit a sufficient amount of data-parallel computations.

Compared to the CPU executing entirely in double precision (top right graph in Figure 5.55), the mixed precision iterative refinement solver on the GPU is 5.3, 18.4 and 39.5 times faster for refinement levels $L = 8, 9, 10$ and V cycles, 2.7, 9.1 and 23.1 times faster for F cycles, and 2.1, 5.6 and 12.5 times faster for W cycles. The break-even point from which on the GPU achieves an overall acceleration is refinement level $L = 7$ for V cycles, and $L = 8$ otherwise. In the more honest comparison against a mixed precision solver that executes entirely on the CPU (bottom graph in Figure 5.55), these factors reduce to 2.2, 9.3 and 21.4 for V cycle multigrid, 1.2, 4.4 and 12.2 for F cycles, and 1.0, 2.8 and 6.5 for W cycles. The break-even point lies at $L = 8$ in all cases.

All these factors are in line with the speedups we measured previously, see Figure 5.43 on page 188 for mixed precision V cycle multigrid on a unitsquare domain, and Figure 5.50 (on the left) for all cycle types in double precision.

Variable Smoothing

In double precision, variable smoothing of V cycle multigrid is very beneficial on the GPU, and we observe speedup factors slightly higher than for F cycle multigrid, see Section 5.8.5. No coarse grid solve is performed on the coarsest grid at all (our emergency stopping criterion always holds already in the first multigrid cycle and the first conjugate gradient iteration) for problems with more than 16 641 unknowns, even in double precision. The exponential increase in smoothing iterations however guarantees in the current configuration that the coarse grid correction after prolongation to the next finer grid results in an advance towards the solution. Consequently, this scheme seems to be very suitable for the mixed precision case as well.

Except for the two largest truncation settings, the curves in Figure 5.56 look qualitatively the same as the ones obtained using double precision, see Figure 5.48 on page 197. V cycle multigrid with variable smoothing is thus also applicable in the mixed precision setting. Both on the CPU and the GPU, and in the pure double and mixed precision setting, the configuration that truncates the mesh hierarchy at refinement level $L = 4$ (`trunc=3`, 289 unknowns) is the most efficient one, and we compute speedup factors for this case only. On the CPU, the mixed precision iterative refinement solver is faster than the multigrid solver that executes entirely in double precision by factors of 1.9–2.3, depending on the level of refinement. The corresponding comparison on the GPU gives approximately the same factors for sufficiently high levels of refinement, only a factor of 1.2 for the problem size with 16 641 unknowns, and is slower otherwise. These factors translate to speedups of 13 and 25 of the GPU over the CPU for the largest problem size, comparing to mixed precision on the CPU, and to double precision on the CPU, respectively. We can conclude that V cycle multigrid with variable smoothing is equally beneficial when used within a mixed precision iterative refinement scheme, and the acceleration of the GPU over the CPU is in line with our findings in Section 5.8.5 and in Figure 5.55.

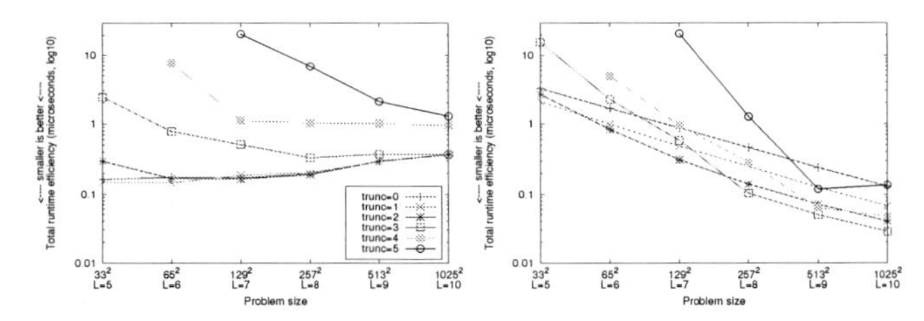

Figure 5.56: Efficiency of mixed precision iterative refinement performing two iterations of V cycle multigrid with variable smoothing per outer loop, on the CPU (left) and GPU (right). Both plots share the same legend, and the y-axis uses a logarithmic scale.

Executing the Coarse Grid Solver on the CPU

As the latency of the PCIe bus between host and device is the dominant factor rather than the bandwidth, and as in the mixed precision setting only half the amount of data is transferred, we expect a larger impact compared to the pure double precision case. Furthermore, computing the correction term in single rather than in double precision is faster, so that the CPU constitutes a more severe bottleneck. Therefore, we do not repeat the experiment from Section 5.8.6 here.

5.8.9. Summary

We conclude this section with a brief summary and discussion of our most important findings.

Main Numerical Result

Our current implementation of GPU-based multigrid solvers is limited to using Krylov subspace schemes as coarse grid solvers. We have identified a problem in these solvers that can easily lead to coarse grid solves by the zero vector in the multigrid scheme. As we always apply adaptive coarse grid correction and typically look at deep refinement hierarchies, this problem does not degrade the convergence of the overall solver in double precision: This effect only occurs after prolongating the coarse grid solution to the next finer level; after postsmoothing and further prolongation or eventually restriction in case of the F or W cycles, the problem is entirely alleviated.

In single precision however, the multigrid solver does not recover easily when this problem occurs, leading to a dramatic loss in efficiency because the mixed precision iterative refinement solver (which is preconditioned by the multigrid iteration) needs much more iterations until convergence than its double precision multigrid counterpart. Here, the mesh hierarchy must comprise more levels in order to recover fast convergence. Furthermore, F or even W cycles are more beneficial to alleviate this problem.

Efficiency and GPU Speedup

The speedup of the GPU over the CPU degrades significantly when using F cycles or even W cycles. While F cycles still yield more than an order of magnitude speedup (and are thus practically relevant in the scope of our acceleration of large-scale fluid dynamics and solid mechanics applications in Chapter 6), for W cycles we initially observe less than a factor of two. These measurements are obtained in double precision. Our main approach to improve the speedup of the GPU over the CPU is to truncate the mesh hierarchy, i. e., to execute the coarse grid solver not on the coarsest level, but already on finer ones. The investigation of different ideas to improve speedups gives a lot of insight into GPU performance characteristics, we discuss our findings separately for the three cycle types.

For V cycles on the CPU, truncation results in a slowdown for small problem sizes, because the number of iterations required by the coarse grid solver is proportional to the mesh width h and thus increases when truncating the mesh hierarchy. For larger problem sizes, the CPU is invariant of truncation, because the time spent on fine levels of the mesh hierarchy dominates the overall time to solution, these computations are not performed in cache. On the GPU, truncation always results in a slowdown, the h-dependency of the coarse grid solver yields a larger loss in efficiency than on the CPU, in other words, *the GPU is disproportionately slow for small problem sizes*. Employing a stronger preconditioner in the conjugate gradient solver in order to reduce the number of iterations until convergence does not result in a net gain in efficiency, it is always slower because the smaller number of iterations is not balanced by the more expensive application of the preconditioner.

These statements are equally valid for F cycle multigrid, both on the CPU and on the GPU.

In contrast to the other cycle types, the number of coarse grid solves performed by W cycle multigrid is exponential in the depth of the mesh hierarchy, and not linear or constant. Consequently, the truncation of the mesh hierarchy has a significant effect. On the CPU, truncation results, similar as for the other cycle types, in a slowdown if too many levels are removed, because the amount of work performed by the h-dependent coarse grid solver is higher than the savings by the reduction in overall coarse grid solves. For small truncation levels, the CPU is again invariant: For sufficiently large problem sizes that are computed out of cache, the fraction of time spent on the finest level of refinement dominates the total runtime. Truncating is however always beneficial on the GPU. The speedup over the CPU depends on the truncation level, and reaches up to a factor of 2.8 for large problem sizes. Stronger preconditioning of the coarse grid solver results in roughly the computation times as for the standard Jacobi preconditioner without truncation, so this strategy does not give returns for W cycles, either.

In practice, a technique called V cycles with variable smoothing is an established alternative to using F or even W cycles. Here, the number of smoothing steps is doubled in each restriction, and halved again during prolongation. For this modified cycle type, the performance characteristic depending on the truncation level is the same as observed for the W cycle multigrid; on the GPU, truncation is always beneficial for sufficiently large problem sizes. The speedups compared to computing on the CPU are in the same range as the ones observed for F cycle multigrid without truncation, reaching a factor of 13.

The best improvements in speedup are however obtained when the coarse grid problems are not solved exactly, but only approximately. In our tests, configuring the conjugate gradient solver to reduce the initial residuals by two digits is a good compromise between the a posteriori bound stemming from convergence theory and practical performance improvements. For V cycles, truncation leads to a loss of efficiency, and therefore, the differences in speedup compared to the exact solution of the very small problems on the coarsest level of the mesh are in the noise. The same holds true for F cycle multigrid. For W cycles however, we observe significantly better speedups, peaking at almost a factor of six compared to two (without truncation) or three (with truncation). This speedup is high enough to justify the use of GPUs in the case of W cycles.

In the mixed precision iterative refinement solver that uses multigrid as its preconditioner, we can solve coarse grid problems only approximately since single precision is not accurate enough for an exact solution. For V cycles on the GPU, these solvers are twice as fast as pure double precision on the GPU, 40 times faster than executing on the CPU in double precision, and 21 times faster than their CPU counterpart. These factors reduce to 2, 23 and 12 when F cycles are employed. For W cycle multigrid, the best speedup against the GPU in double precision is 3, 12 vs. the CPU in double precision, and 6.8 against mixed precision iterative refinement on the CPU. The mixed precision approach benefits equally from V cycle multigrid with variable smoothing, and we achieve speedups of 2, 25 and 13, respectively. In general, we can conclude that the achievable speedups are in line with our previous findings, i. e., the mixed precision approach is always beneficial compared to executing the solvers entirely in double precision. The loss of efficiency on coarse levels of refinement is smaller, which explains the better speedups for the F and W cycles.

'Ideal' coarse grid solvers compute the solution independently of the mesh width. On the CPU, so-called direct sparse solvers are often employed. We are not aware of any efforts in this field on GPUs and will investigate this topic in future work. One idea of a workaround for the time being is to execute the coarse grid solvers on the host and the remaining components of the multigrid solver on the device. Before implementing such a coupling, a small experiment reveals that the low bandwidth and high latency of the PCIe bus between host and device, and the bottleneck due to slow computations on the host, makes this approach impractically expensive in our setting. Even when only performing a dummy operation on the host, the repeated transfers of right hand sides and solution vectors to and from the device always result in worse efficiency compared to using the conjugate gradient algorithm as the coarse grid solver on the device.

Future Work

The main avenue for future work is research into sparse direct methods on GPUs, such as the multifrontal techniques as employed in the UMFPACK and SuperLU packages. Furthermore, we plan to investigate, for problems that do require F or even W cycles to converge sufficiently fast, if the variable smoothing technique can be superior on the device compared to using F or W cycles on the host.

5.9. GPU Performance Guidelines

As mentioned in the introduction already, this chapter has covered a wide range of numerical and implementational topics. Therefore, a complete summary of all of our results is not the aim of this last section, instead, the interested reader is referred to the chapter overview in Section 5.1.1 and the pointers therein to section summaries. Here, we want to provide a comprehensive list of performance guidelines which may be helpful for practitioners in GPU computing, within the limitations to bandwidth-bound operations as they typically occur in the context of finite element discretisations and iterative solvers for the resulting sparse linear equation systems. We aim at keeping this list as general as possible, because *we regard GPUs as current representatives of throughput-oriented many-core processors*. As discussed in Section 3.4.1 on page 77, OpenCL exposes the same architecture model as CUDA [128], and is designed as a unified programming model for multi- and many-core devices. For CUDA-related techniques, we refer to the above-mentioned section summaries, and to NVIDIA's 'Best Practice Guide' [170] for technical details.

5.9.1. Control Flow Between CPU and GPU

At the time of writing, GPUs can only be used as co-processors to the general-purpose CPU, which orchestrates the computation: The CPU triggers (synchronous and asynchronous) memory transfers and (asynchronous) kernel launches, and maintains the control flow of the application in between kernel invocations. A concrete example in the scope of this chapter is: The GPU computes a norm, and the CPU decides if the multigrid solver has converged or if another solver iteration should be executed. Therefore, viewing GPUs as standalone compute devices is not justified (at the moment).

5.9.2. The PCIe Bottleneck

GPUs are included in the host system via the standardised PCIe interface, which constitutes a low bandwidth, high latency bottleneck. Measured numbers are 10–20 µs latency [132], and 3–6 GB/s bandwidth.[23] In comparison, a kernel launch takes approximately half the time of initiating a transfer (see below), and a floating point multiply-add operation of one warp on each of the 30 multiprocessors in this particular GPU, i. e., $32 \cdot 2 \cdot 30 = 1920$ flops, execute in four clock cycles, i. e., in 5.2 ns at a clock frequency of 1.3 GHz.

For many applications (in particular since GPUs have started to support double precision natively in hardware), this is not a problem, the input data is transferred once to the device, and the result is read back only after the computation has finished. This is the configuration we have analysed throughout this chapter, exemplarily for the solution of a Poisson problem on a generalised tensor product domain. In addition, Section 3.2.4 on page 71 provides a bibliography of publications related to successfully mapping applications to GPUs.

In distributed environments such as GPU-enhanced commodity based clusters, the PCIe bottleneck however cannot be ignored. Implementations should therefore avoid transferring data over this bus between host and device too frequently. Due to the high latency, it is advantageous to move data in larger batches, rather than in many small chunks. Finally, modern GPUs allow to overlap PCIe transfers with independent computations, so a suitably adapted and decoupled algorithm can exploit this asynchroneity, almost fully alleviating the PCIe bottleneck. We have discussed this bottleneck in Section 5.7.4 on page 186; and in Section 6.2.3 on page 222 and Section 6.6.4 on page 263 we will present experiments to evaluate the impact.

[23]This number has been measured on the high-end system used for the experimental evaluation throughout this chapter, using the `bandwidthTest` utility that ships with the NVIDIA CUDA SDK.

5.9.3. GPU vs. CPU Performance Characteristics

The most important difference between CPUs and GPUs is that the general performance behaviour is reversed: On the CPU, smaller problem sizes that fit entirely into cache perform best, while larger problem sizes suffer from the relatively low bandwidth to main memory. On the GPU, this is exactly the other way round: Small input sizes yield very poor performance, and only for considerably large input domains, GPUs show superior performance, and significant speedups are achieved. There are two reasons for this behaviour: First, a significant amount of overhead (on the order of 8–10 μs) is associated with launching compute kernels on the device, see for instance Section 5.3.3 on page 133 and the microbenchmarks by Volkov and Demmel [237]. For the kernels we have implemented and analysed, this means that all computations for input sizes below an array length of approximately 1 000–10 000 values (depending on the characteristics of the operation at hand) are dominated by the launch overhead, and not by actual execution speed.

Second, accessing off-chip memory has a high latency because GPUs do not employ hierarchies of caches in contrast to CPUs. Latencies on the order of several hundred clock cycles are typical, and memory performance depends significantly on access patterns, see below. If a given application is memory-bound on the CPU, it will certainly be memory-bound on the GPU, whereas compute-limited applications on CPUs are not necessarily compute-limited on GPUs. At first sight, the main reason for speedup by GPUs for memory-intensive calculations is that GPUs provide at least one order of magnitude more bandwidth to off-chip memory: Commodity based architectures employ standard components, for instance, memory modules from a wide range of vendors can be plugged into the memory sockets of mainboards from many manufacturers. On graphics boards, the connection between the GPU and the off-chip DRAM is hardwired, which enables a higher signal quality, a higher clock of the memory bus, etc. However, raw bandwidth does not tell the whole story: The GPU employs massive hardware-multithreading to keep all ALUs and multiprocessors (cores) equally busy. For small problem sizes, there is simply not enough parallel work available to saturate the device. For operations that are primarily memory-bound, the ability of an architecture to hide stalls due to DRAM latencies is much more important than raw bandwidth: GPUs maximise throughput rather than latency of individual operations, by switching from one 'thread group' (warp in CUDA speak) to the next in case of memory stalls.

If a problem size is too small to saturate the device, then using the GPU instead of the CPU can lead to a slowdown by more than an order of magnitude, even when not taking any bus transfers into account. In the limit case, the GPU behaves more like a 1.3 GHz serial processor with a high latency to access uncached off-chip memory. However, it is often necessary to perform computations on small problem sizes, in particular if the underlying data structure is based on some hierarchy. In this thesis, the most prominent example is the nested multigrid hierarchy; other examples in the scope of our work are reduction-type algorithms (norm and dot product computations, cyclic reduction), for which intermediate sub-tasks are performed on smaller input sizes in the course of a computation.

For sufficiently large problem sizes, the time spent on finer levels (larger input domains) dominates the runtime on the CPU: All work on coarser levels of refinement (small input domains) is performed—effectively in zero time, at least relatively—in cache, and this work is often dominated by the instruction issue rate of the CPU rather than by the compute rate, as CPUs are optimised for low-latency computations. On the GPU in contrast, a large amount of (independent) batches of work per multiprocessor is needed to saturate the device. Since there are many multiprocessors computing concurrently, this amount scales with the number of GPU 'cores'. Achieving good speedups with the GPU means balancing the speedup that is lost on coarse levels of refinement with the large speedups on finer levels. It is thus not always justified to ignore all the work on small levels of refinement, compared to the CPU. Maximising efficiency on the GPU is thus equal to minimising the time spent underutilising the device. This can for instance be achieved by truncating the mesh hierarchy (see Section 5.8 on page 191), or by restructuring the implementation so that the amount of kernel calls for small input problems is minimised (which explains why truncation is not always the best choice). This aspect is discussed in almost all of the section summaries throughout this chapter. In the long run, OpenCL promises efficient CPU-GPU hybridisation: The same code can be compiled for both architectures, and a runtime environment can schedule tasks to the most suitable architecture. We expect such a feature to be provided by middleware developers soon.

5.9.4. Locality, Vectorisation and Memory Access Patterns

On the CPU, spatial locality of memory access patterns is tantamount to achieving good performance, because it immediately determines cache hit rates and thus data reuse. On the GPU, locality is even more important. As described in Section 3.3.2 on page 75, GPUs exhibit SIMD characteristics. On current CUDA-capable GPUs from NVIDIA, the execution on one multiprocessor ('core') is performed on a per-warp basis (the warp is the SIMD unit on these GPUs); the 32 threads per warp execute in lockstep. Branch divergence at the scope of a warp should therefore be avoided. More importantly, accesses to off-chip memory *must* be vectorised as well, because otherwise, the parallel memory subsystem cannot be exploited. In Section 5.3.1 on page 125 we present a simple yet instructive example that highlights this aspect. The same holds true for accesses to the (multi-banked) scratchpad shared memory available per multiprocessor to coordinate threads.

In many applications, it is not immediately possible to arrange memory access patterns for maximum throughput of the memory subsystem (known as 'coalescing' into a minimal number of memory transactions). Rearranging memory accesses on a per-thread basis to enable maximum memory throughput is thus very important, we have applied this technique throughout this thesis, see for instance Section 5.3.1 on page 125.

5.9.5. Resource Allocation

Each 'core' in a GPU only has limited resources (shared memory, register file). By partitioning the available resources per microprocessor specifically for each individual operation, performance can be influenced significantly. This is however not a trivial task, as certain algorithms require careful balancing. For the defect calculation (see Section 5.3.2 on page 129) which uses the on-chip shared memory as a software-managed cache, we could launch one block of threads per multiprocessor and allocate the entire shared memory for it. However, this would mean that the entire multiprocessor stalled at the synchronisation barrier, resulting in a significant degradation of performance. Balancing the amount of communication between threads, the resource usage and the goal to ensure a high number of independent warps (the SIMD granularity) to maximise throughput can be quite an 'art', and carefully tuning the implementation is necessary. In our experience, getting correct results is easy, getting them fast is challenging.

6

Minimally Invasive Co-Processor Integration into FEAST

In Chapter 5 we have demonstrated that graphics hardware has evolved into a very powerful alternative data parallel computing platform for the problems under consideration in the (wider) scope of this thesis. All our experiments have been based on the solution of the prototypical yet fundamental Poisson problem, using finite element discretisations and Krylov subspace and fast multigrid solvers. However, our tests so far have been limited to a single subdomain in FEAST. This chapter deals with the applicability of GPUs in the framework of this large-scale finite element discretisation and solution toolkit. The *minimally invasive* integration suggested here has two important advantages: First, it encapsulates the heterogeneities within the compute node so that MPI sees a globally homogeneous system; and second, applications built on top of FEAST benefit from the GPU acceleration without any code changes. Even though we focus on GPUs and present results on large-scale heterogeneous CPU-GPU clusters, our technique is independent of the accelerator hardware. We consider GPUs the most prominent and the most promising accelerator technology at the time of writing.

6.1. Introduction and Motivation

In Section 6.1.1, we motivate the approach presented and evaluated in this chapter in detail. Section 6.1.2 provides an overview of the remaining chapter.

6.1.1. Motivation

The straightforward approach to port a given application to new hardware architectures such as GPUs is to *rewrite* it, often from scratch. The most important benefit of full application-specific re-implementations obviously is that efficiency (and hence, speedup over conventional, existing CPU implementations) can be maximised by fully exploiting performance tuning techniques specific to the architecture *and* application at hand. This includes the coupling of the co-processor with the orchestrating CPU and, closely related, the network interconnect between cluster nodes.

One could argue that the experiments presented in Chapter 5 fall into this category. Our 'application' has been the accurate solution of the Poisson problem, or more accurately speaking, of linear systems stemming from the finite element discretisation of the Poisson problem on a generalised tensor product domain. The accelerator under consideration has been a single GPU. However, in the scope of this thesis, these experiments serve as building blocks and the Poisson problem is a fundamental model problem in our context, so the analogy is only partly valid.

Accelerate vs. Rewrite

A full rewrite is feasible when one application is to be executed many times on one particular architecture, for instance for parameter studies. Another practically relevant scenario comprises applications for which efficiency is of utmost importance and when real-time constraints exist, for instance tsunami or earthquake predictions, computer-assisted surgery or applications from computational finance. Nonetheless, the programming effort increases drastically with each architecture to be supported. Since the hardware evolves too fast (even in terms of basic feature sets), it is prohibitively expensive and thus practically impossible to re-implement each application for each generation of hardware accelerators. In practice, there is often additional reluctance to re-implement established 'legacy' codes from scratch. Prospective gains in speedup, efficiency and other relevant metrics have to be balanced with the implementational effort, and finally, there is always the risk of having invested into technology that is discontinued early. Section 3.1.5 ('legacy' GPGPU) and Section 3.2.4 ('modern' GPGPU) provide bibliographies of publications that report on successful re-implementations, for a single GPU and for GPU-accelerated commodity based clusters. In addition, we have surveyed the integration of various co-processors in commodity based clusters in detail in a previous publication [84, Section 1].

The alternative to fully rewriting applications is to identify and accelerate crucial portions of the code, i. e., subroutines in which most of the time is spent. This idea is as old as scientific computing itself, the most prominent examples are vendor-tuned BLAS, LAPACK and FFT libraries. Due to the low arithmetic intensity of the operations typically encountered in finite element packages, and due to the narrow bus between host and co-processor, such a library approach on the level of individual operations does not promise performance gains in our case, though. Furthermore, standardisation for sparse codes is poor, even in terms of fundamental data structures.

A weakened variant of this library approach is the concept of tuning and adapting 'application-specific kernels' within a (fully rewritten) general framework that provides all infrastructure required to facilitate the re-implementation of existing applications for a wide variety of hardware platforms. Together with his co-workers, the author of this thesis has pursued this approach and evaluated it for the Poisson problem and an explicit solver for the shallow water equations [236]. Similarly, Cohen and Molemaker [49] present OpenCurrent, a framework that has been designed from scratch to facilitate the simulation of fluid flow on graphics hardware.

Accelerating Established Large Software Packages

In the scope of FEAST, a full re-implementation is not possible for another important reason: FEAST is a toolkit that applications can be built upon. To this end, FEAST provides finite element discretisations, matrix and right hand side assembly routines, linear solvers, and postprocessing routines, e. g. for visualisation. Furthermore, on a lower level, FEAST comprises data structures for distributed matrices and vectors, data storage, numerical linear algebra and finally (MPI) communication routines tailored to the underlying multilevel domain decomposition approach, cf. Section 2.3. These FEAST kernel routines, i. e., all code shared by all applications, comprises more than 100 000 lines of code, which clearly makes a full re-implementation prohibitively expensive.

This approach to build different applications on top of a common framework is not unique to FEAST, and therefore, we believe the methodology we present and evaluate in this chapter can be applied in many other numerical software packages as well. In addition to their already mentioned full re-implementation (cf. Section 3.2.4), Phillips et al. [182] also describe a strategy to integrate GPUs into their Fortran solver MBFLO, using a concept very similar to ours. Erez et al. [66] present a general framework and evaluation scheme for what they term 'irregular scientific applications' (such as finite element computations) on stream processors and architectures like ClearSpeed, Cell BE and Merrimac. Sequoia [69] is a general framework for portable data parallel programming of systems with deep, possibly heterogeneous, memory hierarchies, which can also be applied to a GPU-cluster. Our work distribution is similar in spirit, but more specific to PDE problems and more diverse on the different memory levels.

In summary, our goals are threefold: We want to keep the modifications to the existing codebase of FEAST as small as possible, in particular, we do not want to introduce interface changes to routines that are called by users. The underlying hardware should be encapsulated transparently for the user, i. e., changes to application code should not be necessary, and applications built on top of FEAST should benefit immediately from the faster execution. Finally, we want to retain all functionality, in particular, sacrificing accuracy is not acceptable.

Issues Related to GPU Clusters

A major issue with GPUs in high performance production environments is the lack of error-protected memory, because DRAM errors occur much more frequently in practice than commonly anticipated, cf. the recent large-scale analysis by Schroeder et al. [193]. In all our experiments between 2006 and 2009, we found no indication of such errors. The issue is valid nonetheless, and NVIDIA reacted recently and announced ECC protection on all memory levels of their future HPC products [172].

From a practical perspective, job allocation in large-scale clusters is an issue, in particular under the control of standard multiuser queueing systems such as PBS, because typical GPU-accelerated clusters have more CPU cores than GPUs per node. In our experiments, manual resource scheduling has been problematic initially, especially when MPI implementations were not aware of multicores and process pinning yet. In production environments, data security and privacy are becoming an issue. Kindratenko et al. [132] provide an excellent survey on this topic, and they present a number of tools they developed to address these problems.

6.1.2. Chapter Overview

The remainder of this chapter is structured as follows: In Section 6.2.1, we analyse the two-layer SCARC solvers with respect to accelerating and executing them on co-processors. Section 6.2.2 presents our acceleration approach in a general way, while Section 6.2.3 discusses implementational details and challenges related to an efficient coupling of FEAST with GPUs. Finally, Section 6.2.4 introduces a technique to execute computations hybridly on the CPUs and accelerators in a heterogeneous compute cluster.

In Section 6.3 we analyse and evaluate our approach with detailed numerical experiments. We consider the Poisson problem as an important building block, and discuss the accuracy and efficiency of the accelerated mixed precision two-layer SCARC solvers in Section 6.3.2. As mixed precision techniques on the CPU have not been possible in FEAST prior to this thesis, and as we obtained promising performance improvements on a single subdomain in our tests in Chapter 5, we evaluate the pure CPU and the CPU-GPU case. Section 6.3.3 analyses the obtained speedups, and discusses factors inherent to our approach that limit the speedup. Here, we also develop an analytical performance model to assess and predict how the local speedup achieved in the portions that are offloaded to accelerator hardware translate into global speedup of the entire solver. In Section 6.3.4, we analyse weak scalability and energy efficiency on a large cluster of 160 compute nodes, both for the accelerated and unaccelerated solvers. The hybrid solvers are examined in Section 6.3.5.

Section 6.4 and Section 6.5 assess the applicability and performance of the minimally invasive integration technique, using the two applications FEASTSOLID and FEASTFLOW. These two applications are developed by Hilmar Wobker and Sven Buijssen, respectively, and we only use a subset of their functionality, namely the solution of model problems from linearised elasticity and stationary laminar flow. The two applications have a more diverse CPU/co-processor coupling, a smaller acceleration potential, higher computational demands, and are in general more challenging than the Poisson problem. Both applications build on the extension of the ScaRC solver concept to vector-valued problems as developed by Wobker (see Section 2.3.4 on page 39) and we refer to Section 2.3.5 on page 41 and Section 2.3.6 on page 43 for a brief outline of the theoretical background of these two applications. We emphasise that from the point of view of this thesis, we only *use* these two applications, the author of this thesis has neither implemented them nor contributed substantially to them.

In Section 6.4.2, the accuracy of mixed precision two-layer ScaRC solvers is analysed for various ill-conditioned benchmark problems from the field of solid mechanics, for instance a cantilever beam configuration. Section 6.5.3 describes similar experiments for standard laminar flow benchmark problems. Weak scalability of the accelerated linearised elasticity solver is studied in Section 6.4.3. Section 6.4.4 and Section 6.5.4 are concerned with performance measurements of the accelerated solvers, and a detailed analysis.

This chapter concludes with a summary and a critical discussion including preliminary experiments that examine the feasibility of several avenues for future work, in Section 6.6.

6.2. Minimally Invasive Co-Processor Integration into FEAST

In this section, we present our *minimally invasive* co-processor integration. We start with an analysis of the two-layer ScaRC solvers with respect to accelerability on co-processors in Section 6.2.1. Our approach is detailed in Section 6.2.2, and Section 6.2.3 discusses implementational details and challenges related to an efficient coupling of FEAST with GPUs. Finally, Section 6.2.4 introduces hybrid solvers that execute on the CPUs and accelerators in a heterogeneous compute cluster. The approach has been designed and implemented by the author of this thesis, and has previously been evaluated and published together with a number of co-workers [83, 87].

6.2.1. ScaRC Solver Analysis

The starting point of the integration of co-processors into FEAST is an analysis of its parallel multilevel solver concept. Figure 6.1 depicts a typical scalar two-layer ScaRC solver, see also Section 2.3.3 on page 37. In the following, we use the terms 'outer' and 'global' synonymously for the parts of the computation that involve MPI communication; and 'local' and 'inner' for the multigrid solvers acting on individual subdomains only (highlighted in the figure).

```
global BiCGStab
preconditioned by
  global multilevel (V 1+1)
  additively smoothed by
    for all Ω_i: local multigrid
  coarse grid solver: UMFPACK
```

Figure 6.1: Template two-layer ScaRC solver, communication-free components are highlighted.

The global work is clearly separated from the local work, and furthermore, communication and computation are discrete steps in the implementation. As explained in Section 2.3.2, the domain decomposition is realised by special 'extended Dirichlet' boundary conditions, corresponding to a minimal overlap of one layer of mesh points. Each global operation, for instance a defect calculation, is performed in the same way: First, a local calculation is carried out on each subdomain, which is immediately followed by MPI communication between neighbouring subdomains to ensure consistency of the shared degrees of freedom. This means that only a relatively small amount of work is executed inbetween communication phases, in particular on coarser levels of the multilevel hierarchy.

In commodity based computer systems, co-processors are included via standardised interfaces such as the PCIe or PCI-X buses. These buses provide a relatively narrow bandwidth connection, and furthermore, the latency of data transfers is high. This situation is symmetric to interconnect hardware between cluster nodes, such as (Gigabit) ethernet or Infiniband technology: In order to synchronise a co-processor in one cluster node and a co-processor in another one, data has to travel through three high latency, low bandwidth bottlenecks, first from device memory to host memory, then via the interconnect to the next node, and finally to device memory. Data locality and a favourable ratio between computation and communication is thus even more important in heterogeneous cluster installations than in homogeneous nodes without accelerator hardware. Figure 6.2 illustrates the situation, exemplarily for the bandwidth of data paths in heterogeneous cluster nodes. As a consequence, one design goal of the accelerated ScaRC scheme is to execute a sufficient amount of independent work on the device, without expensive synchronisation or communication with the CPU that orchestrates the computation.

The second, equally important observation we make is that it is impossible to execute the

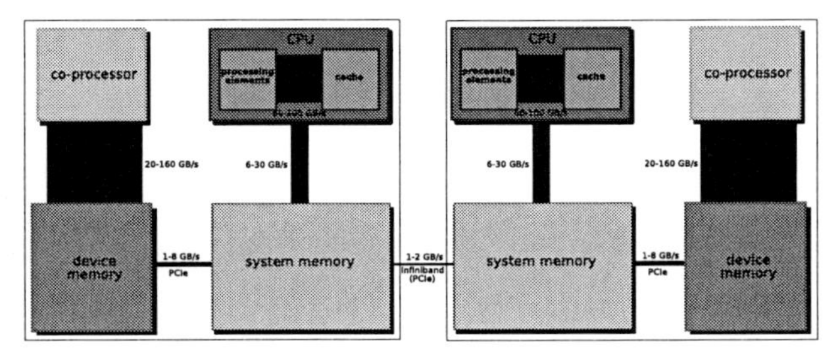

Figure 6.2: Bandwidth between and within heterogeneous cluster nodes equipped with hardware accelerators.

outer solver(s) on hardware that does not support double precision natively: In Section 4.5 on page 95 we have demonstrated that single precision is not accurate enough to compute correct results for the problems in the scope of this thesis. In Section 5.6.3 on page 174 we have shown that the emulated high precision formats using unevaluated sums of native low precision values are not sufficient either. This is of course true independent of the type of co-processor. When the author of this thesis designed the coupling between co-processors and FEAST in early 2006, the consequence of this simple observation has been that GPUs could not be used for any global computation. On newer hardware, it is however possible to execute the outer computations on the device, and the question arises how big the impact of cascaded bottlenecks for device-to-device communication as outlined above is in practice. We return to this topic in Section 6.6.5.

6.2.2. Minimally Invasive Integration

The observations outlined in the previous paragraphs lead to the core idea of the *minimally invasive* co-processor integration into FEAST: *Only the local multigrid solvers are executed on the accelerators, leaving the remaining ScaRC solver unchanged.* The approach is justified, because typically most of the arithmetic work of the entire linear solver is executed by the inner solver. This straightforward idea has many benefits:

- The external interface of the SCARC solvers is not modified. Consequently, all applications that call two-layer SCARC solvers benefit from the co-processor acceleration without any code changes.
- FEAST's MPI infrastructure does not have to be modified, only local work is offloaded to the device.
- The entire SCARC solver can now be interpreted as a mixed precision solver, with a much more involved outer iteration compared to the Richardson loop in the basic solver (cf. Figure 4.4 on page 92).
- Code for new co-processors can be written, tested and tuned on a single workstation, only a multigrid solver for a single subdomain on the device is needed.
- The amount of wrapper code between FEAST and the co-processors is kept small, because all necessary data conversions etc. are already taken care of by our `libcoproc` library which encapsulates different hardware accelerators under one common interface.

We now turn towards implementational aspects on a high level. In FEAST, several subdomains (generalised tensor product patches) can be grouped into one MPI process. The application of the Schwarz preconditioner is then performed in a loop over all subdomains per MPI rank, cf. Figure 6.1. In other words, smoothing on all subdomains by local scalar multigrid solvers is performed independently, prior to communicating entries corresponding to all subdomain boundaries. As explained in Section 2.3.4 on page 39, multivariate problems are brought down to such local scalar subproblems as well.

One additional loop over all subdomains is performed during the initialisation phase of the SCARC solver, immediately after the linear systems have been assembled. In this preprocessing step, auxiliary vectors are allocated and computed, for instance the columnwise representation of the matrix if the local multigrid iteration is smoothed with an alternating direction implicit scheme (cf. Section 5.4.6 on page 154).

To include accelerators, we replace these two loops, and execute either the existing implementation in FEAST, or a replacement implementation, which is specific to each type of accelerator and provided by specialisations within the `libcoproc` library. Via a switch in a configuration file, the user can select which code path is executed. The interface exposed by all implementations is identical, which minimises the code overhead of our approach: In fact, only a branch statement is needed for each supported co-processor; and *we diverge the code paths as late as possible*.

Cross-Language Interoperability

As FEAST is implemented in Fortran 95, and vendor-supplied APIs to access accelerators typically only provide C interfaces, the actual coupling involves technical difficulties. In practice, Fortran-C-interoperability is a solved problem, and most C compilers provide flags to generate symbol tables in a way that C subroutines can be found by Fortran linkers (commonly known as 'Fortran bindings'). Passing data and configuration settings from FEAST to the `libcoproc` library which encapsulates all device-specific code (cf. Section 5.1.2 on page 110) is thus straightforward. In the experience of the author of this thesis, the other way round is more involved, because Fortran 95 is designed around so-called *modules* (compilation units similar to classes without polymorphism and inheritance), and no standardisation exists for the symbol tables and procedure entry points generated by the compiler. The co-processors execute entire subdomain solvers, and we want them to be able to talkback to FEAST, e. g., to terminate execution in case of divergence, or to add convergence information to FEAST's distributed logging mechanism and screen output. Allegedly, the Fortran 2003 standard alleviates most of these issues, but at the time of writing, compiler support for this standard is poor and in parts nonexistent. Our solution to this problem is a straightforward partitioning of the implementation between (partially redundant) Fortran routines which gather the data from FEAST, and the `libcoproc` library which works on the data, accompanied by callback routines for feedback.

Preprocessing and Handling of Matrices

In the preprocessing step, the accelerator is initialised if it has not been used before, and otherwise, all existing data on the device is invalidated. This distinction is important if matrices are re-assembled in the course of a simulation, for instance in the Navier-Stokes solver which assembles new (linearised) systems once per nonlinear defect correction step (cf. Section 6.5). We deliberately do not deallocate all arrays in the latter case, because repeated allocations and deallocations are potentially costly. Instead, we mark these arrays as available and return them to the pool of allocated resources, which is maintained by the memory manager in the `libcoproc` library (cf. Section 5.1.2). The memory manager is able to page out data from the device to the host if the device memory is exceeded; and to page it back in when it is accessed. We assume that the amount of memory on the device is (relatively) small compared to the host, and if there are not enough resources to keep copies of the data in device memory, two fallback solutions exist: If there is only enough memory to keep the data associated with one subdomain on the device, then the data is replaced in each call to the device solver (see below). Otherwise, an error message is generated and the existing CPU solvers in FEAST are executed. Another way to view our memory handling is that we model the device memory as a 'gigantic' level-3 cache, either with manual or automatic prefetching.

The actual implementation of the memory management is device-specific, we discuss the GPU case in Section 6.2.3. In any case, the final preprocessing steps are to compute auxiliary matrix data (e. g., for alternating direction implicit smoothers), to request arrays from the memory manager needed for auxiliary data in the multigrid and Krylov subspace solvers (e. g., depending on the preconditioner), to convert all arrays to single precision and to transfer both matrix and auxiliary data to the device.

Solving and Handling of Right Hand Side and Iteration Vectors

Every time the Schwarz preconditioner is called, we loop over all subdomains in the given MPI process. For each subdomain, the right hand side data and the initial guess (unless it is the zero vector) are converted to single precision and copied to the device. After the preconditioning has been performed, the result is copied back to the host and added to the solution maintained by FEAST. Our system thus requires a device-dependent implementation of a multigrid solver acting on a single subdomain with generalised tensor product property. This solver can be implemented, tested and tuned on a single node or workstation, and the integration of new accelerators into FEAST then reduces to few lines of infrastructure code.

If only the data associated with one subdomain fits into the limited device memory, we also have to transfer the matrix data. Otherwise, we can select the correct data corresponding to the current subdomain.

As each subdomain is treated independently (and FEAST provides all data before the loop over all subdomains is entered), asynchronous data transfers can be exploited if supported by the hardware.

Performance Issues due to the Global Multilevel Solver

The global solver in the two-layer SCARC scheme is a multilevel solver, and we always configure it to execute in a V cycle with exactly one pre- and postsmoothing step (application of the Schwarz smoother). This means that the local multigrid solver is started not only on the finest level of refinement, but also on coarser ones. On the CPU, the ratio between computation and communication thus deteriorates. To alleviate this problem, FEAST applies *truncation* of the mesh hierarchy and starts the coarse grid solver not on the coarsest level of refinement (on the grid comprising only the unrefined subdomains), but already on finer levels, see Section 2.3.2. As the global coarse grid solver is executed serially on a master node while the slave processes are idle, the exact choice of the truncation parameter is problem-dependent, because the coarse grid problems have to be small enough in order not to trade one bottleneck with another one. The same technique can be applied for the local multigrid solves, both on the CPU and on co-processors (cf. Section 5.8.3). However, due to the overhead associated with co-processor configuration, with data transfers over the narrow bus between host and accelerator, and most importantly due to the fact that most fine-grained parallel devices are actually less efficient than the CPU for small problem sizes (see the solver benchmarks in Chapter 5 for the GPU case), level truncation alone is not sufficient.

We therefore apply *level thresholding*, i. e., depending on the currently finest refinement level that the local solver is to be executed on, we reroute the computation from the co-processor to the

CPU, which solves these problems in less time with an identically configured solver. The exact choice of the threshold parameter is again depending on the problem, the solver and the performance characteristics of the CPU, the accelerator and the bus between them. Our implementation currently does not provide an automatic way to select this parameter. Instead, it is supplied by the user based on few calibration runs on a single node, or even heuristically based on microbenchmarks of individual operations.

Summary

Figure 6.3 summarises our minimally invasive integration of co-processors into FEAST. The most important benefits of our approach are:

- User application code does not have to be modified at all in order to benefit from co-processor acceleration.
- The heterogeneity within each compute node is encapsulated in the local layer of the two-layer SCARC scheme, so that MPI sees a globally homogeneous system.
- Support for additional co-processors can be developed on a single node or workstation, the integration into FEAST is straightforward as soon as a multigrid solver acting on a single subdomain is provided.
- The changes applied to FEAST reduce to a minimum.

The currently implemented prototype supports GPUs as co-processors, programmed either through OpenGL or via CUDA. Additionally, a CPU fallback implementation is available. The whole design is independent of the type of accelerator, and in future work we plan to add an interface to OpenCL, a vendor-independent open industry standard to program fine-grained parallel architectures (cf. Section 3.4.1).

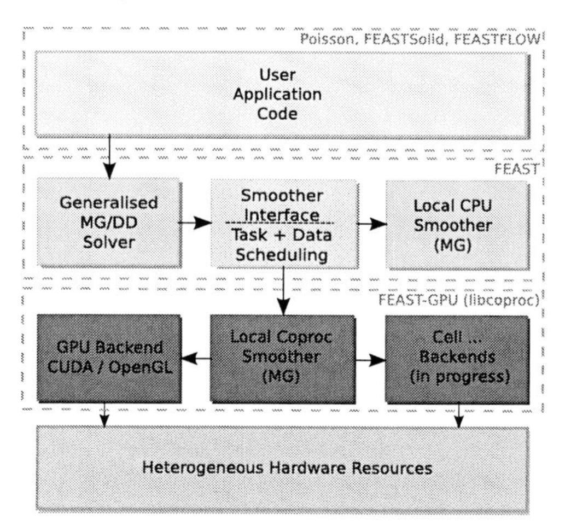

Figure 6.3: Illustration of the minimally invasive co-processor integration into FEAST.

6.2.3. GPU Integration Challenges and Implementational Details

In this section, we discuss challenges we met in an efficient CPU-GPU realisation of our minimally invasive integration, as outlined in Section 6.2.2.

Limited Device Memory

OpenGL virtualises device memory and pages texture data in and out of device memory automatically. To the best of our knowledge, there is no means to query the amount of total or available device memory through the OpenGL API. However, it is possible to assign priorities to the residency of textures in device memory. The paging behaviour and in particular the point in time when paging happens is implementation-dependent and may vary between vendors, or even between driver releases. In our experiments, we found no advantage in using this feature, and we rely on the default behaviour. Our general modelling of device memory as a 'gigantic' cache, and the corresponding manual and automatic replacement policies, are implemented as follows (cf. Section 6.2.2, and a previous publication [87] for more details): The automatic replacement policy is trivially achieved by exploiting that OpenGL virtualises device memory: Textures to hold all matrices per refinement level per subdomain are allocated in a preprocessing step, matrix data is copied to these textures, and control to page data in and out of device memory during the solution process is passed to the GL as implemented in the device driver. The manual replacement policy allocates one set of textures to hold exactly one matrix and copies the data as needed.

CUDA also does not allow to query the amount of dynamically available memory at runtime, while it is possible to query the total amount of device memory via API calls.[1] On all CUDA-capable GPUs we have used so far in our experiments, the amount of device memory has been well-balanced with the host memory: There has always been a sufficient amount of memory to keep all data associated with all subdomains per MPI process on the device. Therefore, we did not implement all features of our memory manager yet, even though it would certainly be possible to realise a bookkeeping procedure for the allocated and used device memory and to page data in and out manually, to realise the automatic prefetching in a general way. The implementation of the manual prefetching policy is a trivial modification of the OpenGL case.

Data Transfers Between CPU and GPU

Fully asynchronous transfers of independent data, i. e., overlapping computation and data copying, are not exposed to the programmer in OpenGL, as the device memory is virtualised. An extension[2] provides support for asynchronous readbacks of data from the device to the host, but these transfers are only asynchronous on the CPU. In our experiments, we found no advantage using these partially asynchronous transfers, when applying an analogous technique to the one described in the following.

CUDA is capable of overlapping independent computation and data copies through its *stream API*, see Section 5.3.1. This technique is only applicable if the host memory is allocated in a page-locked way, i. e., the involved arrays must not be paged out to disk by the operating system. Page-locked memory enables direct memory access by the driver, and thus more efficient transfers. We have not implemented support for this feature yet because it requires significant modifications to the existing code. The obligatory use of page-locked memory does not pose additional difficulties, as we convert the data from double to single precision on the host so that less data is transferred via PCIe, this single precision array can be allocated in a page-locked way. The challenge lies in

[1] This query returns the physical global memory, not all of it may be available for our purposes if for instance there is an X-server running on the device which needs memory, at least for its double-buffered framebuffer. A reasonable estimate is that 90 % of the memory is available to the application.

[2] `ARB_pixel_buffer_object`, see `http://www.opengl.org/registry`

the implementation of the loop over all subdomains in FEAST's implementation of the Schwarz preconditioner of the outer multilevel solver. In the design we envision, one CUDA stream is associated with each subdomain in the MPI rank. Each stream contains the following operations, and the streams are enqueued in a loop over all subdomains:

1. conversion of the right hand side and eventually initial guess from a double precision FEAST array to a single precision page-locked CUDA array on the host
2. asynchronous transfer of these arrays to the device
3. execution of the multigrid solver
4. asynchronous transfer of the result back to the host
5. barrier synchronisation of the stream
6. (scaled) addition of the result to the solution array consumed by FEAST

Each stream executes in-order, so correctness is guaranteed. The hardware can overlap computations from one stream (subdomain) with independent transfers from another one, and thus, we achieve asynchroneity in the entire Schwarz preconditioning step.

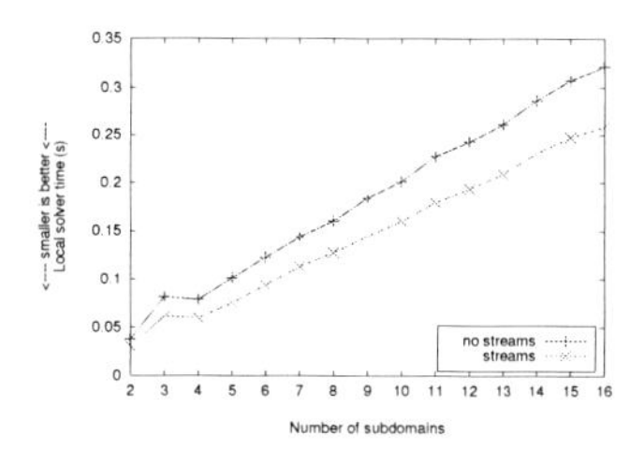

Figure 6.4: Influence of asynchronous transfers and overlapping computation using streams.

Figure 6.4 presents timings obtained when simulating this procedure outside of FEAST in a small test code. We configure the local multigrid solver to perform two V cycles with four pre- and postsmoothing steps each, and measure its execution time for an increasing number of subdomains. The experiment corresponds to the very first Schwarz preconditioning step in the SCARC solver, in this case, exemplarily for a local problem size of refinement level $L = 9$ (263 169 degrees of freedom). We observe that the fully asynchronous application of the Schwarz preconditioner executes between 25 % and 35 % faster than the sequential variant, independent of the number of subdomains as soon as more than four subdomains are present. On this GPU and for this problem size, approximately 64 subdomains are required to fill up the entire device memory. The timings are obtained on a single node of our small GPU test cluster, see Section 6.3.1 for the hardware specifications. The performance improvements are promising, and we will incorporate this technique in future versions of FEAST.

For non-scalar problems, it will be interesting to evaluate a similar software pipelining approach, e. g., to treat the two velocity components in the CFD case in a decoupled way if the block-preconditioner is not sequential (i. e., Block-Jacobi instead of Block-Gauß-Seidel). Such a treatment is not possible in the current implementation of FEAST.

6.2.4. Hybrid Solvers for Heterogeneous Clusters

FEAST has originally been designed for homogeneous compute clusters, and only supports using one type of SCARC solver, at least for scalar problems. On heterogeneous clusters where some nodes include accelerators, or when there are more CPU cores than co-processors in a system, we would like to be able to execute (different) local solvers on CPUs and accelerators. We have extended FEAST by a simple prototypical implementation, which is limited to two different types of local solvers, that are called from the same global multilevel iteration. One of the problems in implementing more generality in this context is the assignment of the solvers to the subdomains. While it is not very difficult to initialise more solvers, the problem is to formulate the rules upon which the assignment takes place.

A similar problem occurs with the assignment of jobs within the MPI framework when faced with heterogeneous architectures. In the absence of explicit directives, the assignment of jobs to nodes is clearly sub-optimal in these cases. For 'pure' co-processor cases (which still require CPU management), and for pure CPU jobs, the assignment is simple since all jobs are equally sized. However, for mixed CPU-accelerator runs, the allocation of jobs across nodes must be managed more carefully, since MPI cannot distinguish between a 'pure' CPU job and a CPU job which is actually managing, e. g., a GPU-based solver.

During the work for our first GPU-cluster publication [87], we often experienced extreme imbalance between nodes, until we identified this problem and inserted explicit assignment directives. As a concrete example, for a combined run with CPU and GPU jobs, instead of the desired partitioning of one CPU subdomain and one GPU subdomain per node, the default scheduling assigned two CPU subdomains each to the first half of the nodes and two GPU subdomains each to the second. Given that each node of this particular cluster had only one GPU, this resulted in both GPU jobs competing for the same co-processor. These clearly incorrect assignments yielded in extremely poor performance, and required significant effort to remedy.

A further complication introduced by the heterogeneous architecture is load balancing between the two different processors. While some of the partitioning is dictated by the solver capabilities or memory considerations, there is still considerable leeway in the assignment of subdomains to either processor. Obviously, the goal is that all subdomains complete simultaneously to minimise wait times of the global solver. As we shall see from the results in Section 6.3.5, the optimal partitioning of subdomains between GPU and CPU is both problem and size dependent, and considerable effort is required to explore the parameter space. At the time of writing, FEAST only supports static load balancing.

6.3. Numerical Building Block: Poisson Problem

In this section, we evaluate the performance and accuracy of double and mixed precision two-layer parallel ScaRC solvers, in particular in view of the *minimally invasive* accelerator integration technique as outlined in Section 6.2. All tests are performed for the Poisson problem, the fundamental model problem in the scope of this thesis. Section 6.3.1 presents an overview of this section, and our findings are discussed and summarised in Section 6.6, for the Poisson problem, the linearised elasticity solver and the stationary laminar flow solver. This section constitutes an extension and update to previously published work by the author of this thesis and his co-workers [83, 84, 87].

6.3.1. Introduction and Test Procedure

In Section 5.6.2 on page 174 we have demonstrated that the mixed precision iterative refinement procedure, i. e., a double precision Richardson solver preconditioned by 'few' single precision multigrid cycles, delivers exactly the same accuracy as a multigrid solver executing entirely in double precision. The experiments in Section 5.7 and Section 5.8.8 have shown that the mixed precision solver is always advantageous in terms of total efficiency, both on conventional CPUs as well as on GPUs from various hardware generations. The two-layer ScaRC solver can be interpreted as a mixed precision scheme as well when the local solvers are executed in single precision (cf. Section 6.2.2). In contrast to the classical iterative refinement procedure, the outer, double precision solver is more involved in this case, and we present a number of numerical experiments in Section 6.3.2 to assess the accuracy and efficiency of *mixed precision two-layer ScaRC solvers* on commodity based clusters comprising CPUs and GPUs.

The core idea of our approach to integrate hardware accelerators into FEAST is to only offload the local multigrid solvers to them, i. e., parts of the Schwarz preconditioner of the global multilevel solver. To alleviate the inefficiency of fine-grained parallel co-processors on small subproblems that are approximately solved in the course of the two-layer ScaRC solver, we have implemented a *thresholding* technique, that schedules local problems either to the CPU or to the accelerator based on the problem size. This heuristic approach is also evaluated in Section 6.3.2.

An important benefit of our minimally invasive integration is that the entire MPI infrastructure and the domain decomposition approach in FEAST do not have to be modified. However, as a consequence the global efficiency improvement due to a prospective reduction in runtime is limited, and potentially significant portions are executed on the CPU. We discuss this question and its implications in much detail in Section 6.3.3, where we also develop an analytical model to assess and predict the acceleration capabilities of our approach.

In Section 6.3.4, we evaluate the weak scalability of both the original ScaRC solver and the GPU-accelerated variant, on a large-scale cluster with 160 nodes. Here, we also discuss energy efficiency and other power considerations, which are increasingly important in terms of 'green computing'.

Finally, Section 6.3.5 is dedicated to the hybrid local solvers distributing the work simultaneously between CPU cores and CPU cores orchestrating co-processors (cf. Section 6.2.4).

Hardware Platform

The majority of the tests in this section are performed on a small four-node test cluster, which allows us to explore and exploit the full functionality of the unified `libcoproc` library (cf. Section 5.1.2). In particular, this includes the CUDA backend, and the ability to execute CPU solvers in single precision, both features have not been available at the time when the original experiments for the underlying publications have been performed [83, 84, 87]. A number of older results are presented nonetheless, results that have been obtained with an OpenGL-based predecessor of the `libcoproc` library on a medium-sized and a large-scale GPU accelerated cluster installed at Los Alamos National Laboratory. Details on these systems are presented as soon as they are needed.

We have already used a single node of this test cluster in several of our microbenchmarks and local solver benchmarks, see for instance Section 5.3.3 and Section 5.7.4.[3] Each node comprises a dual-core AMD Opteron 2214 processor (Santa Rosa, 2×1 MB level-2 cache, 2.2 GHz), DDR2-667 memory (peak shared bandwidth 10.7 GB/s), and an NVIDIA GeForce 8800 GTX GPU (G80 chip, 16 multiprocessors) with 768 MB GDDR3 memory (384 bit interface, 86.4 GB/s bandwidth). Gigabit ethernet is used as interconnect. The cluster is part of the heterogeneous server collection at the Faculty of Mathematics at TU Dortmund, and the network also services a distributed filesystem (NFS) for more than 100 users. In our measurements, we found an average amount of noise of approximately 10 % in the system. The timing results are thus averaged over many runs, and the irregular maximum data points (due to, e. g., some user moving several gigabytes worth of data through the network) are not taken into account. The approach is justified, because dedicated HPC installations typically do not exhibit this problem, and furthermore use lower-latency interconnects such as Infiniband.

Test Problems and Solver Configuration

The function $v(x,y) = 16x(1-x)y(1-y)$ and its Laplacian are used to construct Poisson problems on the unitsquare with analytically known solutions, to be able to measure the L_2 error of the computed solution (cf. Section 2.1.3). We have already used this particular function in the numerical studies in Section 4.5. In Section 4.5.1 we have shown that single precision is insufficient to yield correct solutions of an analogous Poisson problem on a single subdomain.

The computational domain is covered with either four by four, six by four or eight by four congruent subdomains (i. e., the global coarse mesh exhibits mild anisotropies), and each patch is refined uniformly eight to ten times. The corresponding local problem sizes are 66 049, 263 169 and 1 050 625; and the global problem sizes reach up to 32 million degrees of freedom. We refer to these configurations as 4X4, 6X4 and 8X4, respectively.

The 'template' two-layer SCARC solver (see Figure 6.1 on page 217) is configured as follows: The global BiCGStab iteration reduces the initial residuals by eight digits, and is preconditioned with exactly one iteration of the global multilevel scheme, which executes in a V cycle with one pre- and postsmoothing step (damped by $\omega = 0.8$) of the Schwarz preconditioner. The local multigrid solver gains one digit and traverses the hierarchy in a V cycle. The local multigrid solvers employ preconditioned conjugate gradient iterations (set to gain two digits, cf. Section 5.8.7) as coarse grid solver, the global multilevel scheme uses the sparse direct solver UMFPACK, executed serially on a master node. Even though stronger smoothers are not necessary for this simple domain, we employ JACOBI, GSROW, ADITRIDI and ADITRIGS (GSFULLROW instead of GSROW on the GPU, and ADITRIGS in the bicoloured variant, see Section 5.4.3 and Section 5.4.5) as smoothers for the local multigrid solvers for simplicity on these test domains, and explicitly refrain from comparing their efficiency with each other: Different smoothers are employed because they exhibit different compute and memory intensities, and thus allow us to explore the balance between local and global work. The latter remains constant (at least when measured in the total runtime efficiency, see below) because the global solver is configured to perform a fixed amount of arithmetic work per iteration. We count one application of an alternating direction implicit smoother as two smoothing steps, and set the number of elementary smoothing steps to four pre-and postsmoothing steps each for all configurations.

For the three test domains 4X4, 6X4 and 8X4, we statically schedule four, six and eight sub-

[3] In the tests in Chapter 5, this machine has been referred to as the 'server' configuration.

domains per node. The available hardware resources allow us to explore the following parameter space (the outer solver always executes in double precision):

1CORE-DOUBLE one MPI process per node, inner solver double precision

1CORE-MIXED one MPI process per node, inner solver single precision

2CORE-DOUBLE two half-sized MPI processes per node, inner solver double precision

2CORE-MIXED two half-sized MPI processes per node, inner solver single precision

1COREGPU-MIXED one MPI process per node, inner solver single precision on the GPU

HYBRID-MIXED one CPU core and one GPU per node (controlled by the other core, varying load balancing among the two processes), both inner solvers single precision

The GeForce 8800 GTX GPU does not support double precision, so we can only execute mixed precision two-layer SCARC solvers with GPU acceleration. For consistency, the CPU process in the hybrid configuration executes the local solver in single precision as well. As there is only one GPU per node, we cannot test the remaining 2COREGPU-MIXED configuration. It is technically possible to access one device from two processes, the driver then shares the resources via time-slicing. We have performed a few experiments, and found that the runtime increases by a factor of 2.5–3 when scheduling two half-sized tasks simultaneously to one GPU, rather than one full size task. This configuration is therefore not feasible, and we exclude it in the following. Switching from single- to dualcore configurations constitutes a standard *strong scalability* experiment, the problem size stays constant while the amount of resources are doubled. In such a configuration where several MPI processes reside on the same node, OpenMPI (the MPI runtime that we employ) does not route data exchange between processes that reside on the same node through the interconnect, but utilises a shared memory buffer instead. This implementation effectively makes local communication more efficient than interconnect communication. Furthermore, we take advantage of OpenMPI's processor and memory affinity features, i. e., processes are pinned to their respective cores and cannot be migrated by the operating system.

We continue to measure the *total runtime efficiency* rather than the time to solution or other coarse-grained metrics of the various solver configurations, see Section 5.5.1 on page 158 for the exact definition.

6.3.2. Accuracy and Efficiency of Mixed Precision Two-Layer ScaRC Solvers

Accuracy

Table 6.1 lists the L_2 errors of the computed solutions for the three tests problems. Identical results are achieved independent of the partitioning of the problem into MPI processes, of the floating point precision of the inner, local solver, and of the hardware architecture the local solver is executed on.

Efficiency on the CPU

Figure 6.5 depicts the total runtime efficiency of the four CPU-based problem partitionings. For simplicity, we restrict ourselves to the 4x4 test domain, so the global problem sizes are one, four and 16 million unknowns for refinement levels $L = 8, 9, 10$, respectively.

The 1CORE-DOUBLE configuration is always the least efficient one, independent of the choice of the smoother for the local subdomain solver. Partitioning the computation so that both cores

Level	1CORE-DOUBLE	2CORE-DOUBLE	1CORE-MIXED	2CORE-MIXED	1COREGPU-MIXED
4X4					
8	1.0841E-6	1.0841E-6	1.0841E-6	1.0841E-6	1.0841E-6
9	2.7132E-7	2.7132E-7	2.7132E-7	2.7132E-7	2.7132E-7
10	6.8965E-8	6.8994E-8	6.8997E-8	6.8997E-8	6.8997E-8
6X4					
8	8.0989E-7	8.0989E-7	8.0989E-7	8.0989E-7	8.0989E-7
9	2.0248E-7	2.0248E-7	2.0248E-7	2.0248E-7	2.0248E-7
10	5.0168E-8	5.0168E-8	5.0168E-8	5.0168E-8	5.0168E-8
8X4					
8	7.3301E-7	7.3301E-7	7.3301E-7	7.3301E-7	7.3301E-7
9	1.8349E-7	1.8349E-7	1.8349E-7	1.8349E-7	1.8349E-7
10	4.6830E-8	4.6830E-8	4.6830E-8	4.6830E-8	4.6830E-8

Table 6.1: Accuracy of double and mixed precision two-layer SCARC solvers (relative L_2 errors).

of the Opteron processor are responsible for a half-sized problem increases efficiency by approximately 40 % when employing the simple JACOBI smoother, and by 50 % for the other configurations, independent of the problem size. In terms of strong scalability, these results are disappointing, because 50–60 % speedup are lost. The reason for this behaviour is that in the Santa Rosa architecture, both cores share the same memory controller and the bus to off-chip DRAM. The effect is most pronounced when using the simple JACOBI smoother, because it exhibits the highest memory intensity and no potential for data reuse compared to the other smoothers. It is worth noting that we observe the same qualitative and quantitative behaviour on other commodity CPU designs. For instance, the LIDONG cluster operated at TU Dortmund[4] comprises two quad-core Xeon processors (Harpertown) per node, and all eight cores share two Northbridge memory controllers through a QPI bus operating at 1.3 GHz compared to 3 GHz processor clock. In an exhaustive study on parallelised sparse matrix-vector multiply, Williams et al. [241] state that off-chip bandwidth scales rather with the number of sockets (or more precisely, memory controllers) per node than with the number of cores per CPU. In summary we can conclude: *Bus contention prevents multicore CPUs to scale well for codes that are limited in performance by the memory wall problem.*

Switching the local multigrid solvers to single precision is more beneficial in terms of efficiency, as the results in Figure 6.5 demonstrate. The two stronger smoothers in general benefit more from switching to single precision, because they have a higher arithmetic intensity and consequently, more memory bandwidth and cache capacity is saved by halving the data size. For all configurations except when employing the GSROW smoother, the singlecore configuration in single precision for the local solver is at least as efficient as executing a dualcore configuration in double precision, which is remarkable because the singlecore variant executes twice the amount of arithmetic work in the same time (for this test case, all configurations always perform the same amount of global iterations until convergence). The dualcore mixed precision solver is between 1.7 and 1.9 times (refinement levels $L = 8$ and $L = 9$, all smoothers), 1.5 and 1.3 ($L = 10$, JACOBI and GSROW) and 1.9 and 1.7 ($L = 10$, ADITRIDI and ADITRIGS) times faster than the corresponding solvers executing entirely in double precision. These speedup factors underline again the impact of bus contention, because now, the mixed precision variants halve their bandwidth requirements for the local solvers. It is worth noting that these factors are in line with (or only slightly smaller than) previous measurements on a single subdomain, see Section 5.7.2. We discuss this aspect further in Section 6.3.3. Finally, when comparing the baseline double precision singlecore configuration with the mixed precision dualcore configuration, we observe remarkably

[4]ranking at # 404 in the November 2009 Top500 list of supercomputers

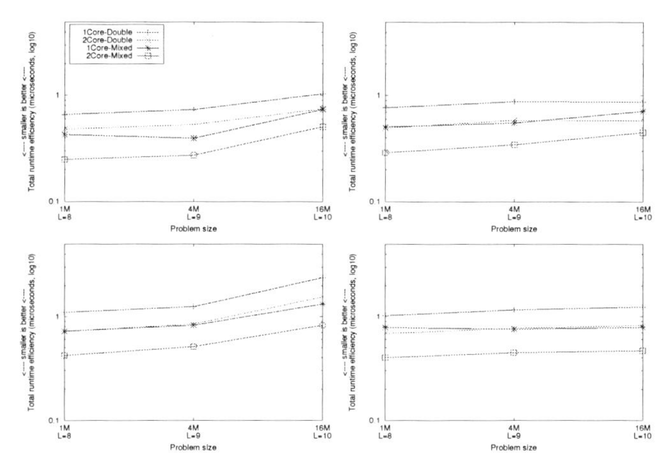

Figure 6.5: Total runtime efficiency of double and mixed precision two-layer SCARC solvers on the CPU, 4X4 test problem, JACOBI (top left), GSROW (top right), ADITRIDI (bottom left) and ADITRIGS (bottom right) smoother for the local multigrid. All plots share the same legend, and the y-axis uses a logarithmic scale.

constant speedups by a factor of 2.5, with two exceptions on refinement level $L = 10$: GSROW only reaches a factor of 2.25, and ADITRIDI peaks at almost 3.0.

Efficiency on the GPU and Influence of Thresholding between Host and Accelerator

Figure 6.6 presents the total runtime efficiency of the 1COREGPU-MIXED accelerated two-layer SCARC solver on the 4X4 configuration. We vary the type of smoother for the inner multigrid, and increase the thresholding parameter to execute small local subproblems on the host and large ones on the GPU. Both stronger smoothers involve the inversion of a tridiagonal system, which limits the maximum admissible problem size to refinement level $L = 9$, cf. Section 5.4.4.

A number of interesting conclusions can be drawn from these results, and we start by looking at the impact of the thresholding parameter alone: For the smaller $L = 8$ test configuration, thresholding between host and device has a noticeable effect, and we gain 20 % and 32 % for the simple, and 10–13 % efficiency for the two stronger smoothers, respectively. The optimal thresholding parameter is `thresh=4225`, i. e., all local problems with a maximum problem size of 4 225 or fewer degrees of freedom are treated on the CPU and not on the GPU (if a problem is scheduled to the GPU, it is solved entirely there, and coarser levels of refinement are not re-scheduled to the CPU). For the largest admissible problem size, the beneficial impact of thresholding diminishes. This can be explained by the fact that the absolute time spent on the two finest levels of refinement is much larger than the time spent treating smaller problems, irrespective of solving them on the CPU or on the GPU (the local solver executes a V cycle, cf. Section 5.8). Finally, setting the thresholding parameter too large results in a significant loss in efficiency, up to a point where the solver is less efficient compared to not applying thresholding at all. Even when taking the comparatively expensive bus transfers into account, the CPU executes too slowly compared to the GPU as soon as the problem sizes are large enough to saturate the device to a sufficient amount. We have already observed this behaviour in the local solver benchmarks, see Figure 5.44 on page 189 for results on the same hardware. Interestingly enough, those measurements indicate a higher break-even point than the one we observe here. The outer solver, which is included in the measurements, is different in the two sets of tests. We return to this topic in Section 6.3.3.

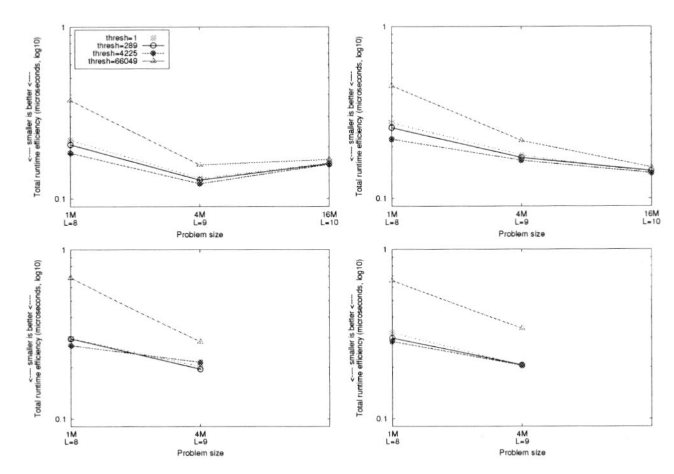

Figure 6.6: Total runtime efficiency of mixed precision two-layer SCARC solvers on the GPU with varying threshold settings, 4X4 test problem, JACOBI (top left), GSFULLROW (top right), ADITRIDI (bottom left) and ADITRIGS (bottom right) smoother for the local multigrid. All plots share the same legend, and the y-axis uses a logarithmic scale.

In the local solver benchmarks of the mixed precision iterative refinement solver accelerated by the GPU, and also in the microbenchmarks of individual operations, we have observed that the GPU always improves efficiency when the problem sizes increases from $L = 9$ to $L = 10$, or at least maintains the same degree of efficiency. Here, this is no longer the case when the local multigrid employs the simple JACOBI smoother. Surprisingly enough, the reason is also the worse efficiency of the CPU, which will become clear in the following section: The local acceleration by the GPU is so high that it results in an unfavourable ratio between computation and communication. We come to this conclusion because we do not observe this behaviour when scheduling more local work per node, see Figure 6.11 on page 239 for a comparison between the 4X4, 6X4 and 8X4 test configurations.

6.3.3. Speedup and Speedup Analysis

Figure 6.7 depicts the speedup factors of the GPU-accelerated 1COREGPU-MIXED configuration over the four CPU variants. The optimal thresholding parameter as determined in the previous experiment is chosen. The GPU-accelerated two-layer SCARC solvers are, for the largest admissible problem sizes, between 5.5 and 6.5 times faster than the 1CORE-DOUBLE configuration. The more honest comparison of the two singlecore mixed precision solvers yields a speedup by a factor of 3.5–4.5, similar to the comparison with the multicore double precision CPU configuration. A speedup of more than a factor of three is maintained over the mixed precision CPU solver that distributes the work across two CPU cores, with the exception of the ADITRIGS case which achieves only a speedup of 2.2.

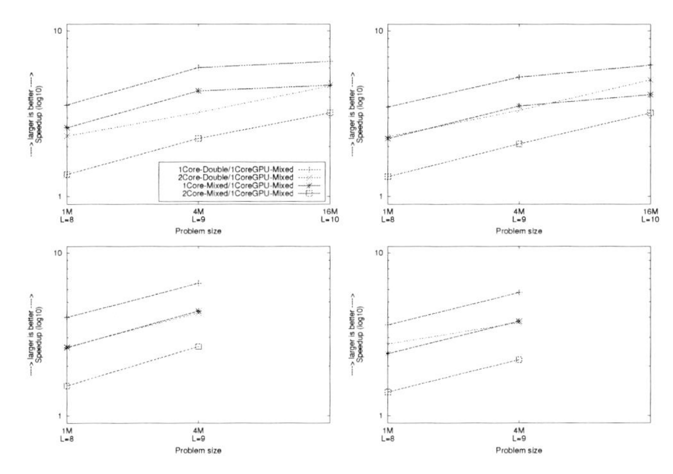

Figure 6.7: Speedup (based on total runtime efficiency) of mixed precision two-layer SCARC solvers on the GPU over the four CPU variants, 4X4 test problem, JACOBI (top left), GSROW vs. GSFULLROW (top right), ADITRIDI (bottom left) and ADITRIGS (bottom right) smoother for the local multigrid. All plots share the same legend, and the y-axis uses a logarithmic scale.

Acceleration Potential

These speedups are impressive and practically relevant. Nonetheless, the question arises why these numbers are more than a factor of two away from the performance we measure with the plain mixed precision iterative refinement solver for a single subdomain on a single node of this cluster: With the same fraction of runtime spent in expensive bus transfers between host and device as for a single subdomain, the mixed precision iterative refinement solver on the GPU is more than twelve times faster than a double precision solver on one CPU core, and still more than seven times faster than its singlecore CPU equivalent, see Figure 5.44 on page 189 for the corresponding local solver benchmarks on the same hardware. Furthermore, the speedup obtained by the mixed precision scheme on the CPU (Figure 6.5) is much more in line with expected numbers than on the GPU.

The answer to this question is surprisingly straightforward, and independent of the GPU, which serves as an exemplary co-processor in our tests. The behaviour we observe is an immediate consequence of our *minimally invasive* approach that does not accelerate the entire solver, but only parts of the Schwarz preconditioner. The correct model to analyse the speedup is *strong scalability* within the node, as the addition of GPUs increases the available compute resources. Consequently, the portions of the application dealing with the coarse-grained parallelism constitute the 'fixed' part of the solution process, i. e., they are not accelerated. In other words, the fraction of the execution time that can be accelerated limits the achievable speedup, this is analogous to *Amdahl's Law* [3]. To improve clarity, the detailed analysis in the following is not based on total runtime efficiency but on absolute time to solution.[5] Due to granularity effects (for the most part, one half-iteration more or less of the outermost BiCGStab solver), the calculated speedups are different to the ones in the previous section. For the sake of brevity, we only perform the analysis for the speedup achieved by the 1COREGPU-MIXED configuration over the four double and mixed

[5]Computing the total runtime efficiency of all inner solves is not feasible, as it would involve an averaging of convergence rates.

precision CPU variants, and furthermore restrict ourselves to the solvers employing the JACOBI and ADITRIDI smoothers locally. We emphasise that the same analysis can be applied, e. g., to assess the speedup obtained by the mixed precision CPU solvers over their double precision counterparts.

To separate the accelerable from the unaccelerable portions of the local solution process and thus to quantify the acceleration potential, we instrumented the code with a special timer T_{local} that measures the local multigrid solvers on each subdomain independently, including all data transfers between the host and the co-processor. T_{total} denotes the total time to solution of the two-layer SCARC solver. This data allows us to calculate the following derived quantities:

- $T_{\text{fixed}} := T_{\text{total}} - T_{\text{local}}$ is the amount of time spent in the outer solver and in MPI communication.
- $R_{\text{acc}} := T_{\text{local}}/T_{\text{total}}$ is the accelerable fraction of a given solver.
- $S_{\text{local}} := T_{\text{local}}^{\text{CPU}}/T_{\text{local}}^{\text{GPU}}$ is the local acceleration achieved by the GPU, i. e., it only takes work into account that can actually be offloaded to the device.
- $S_{\text{total}} := T_{\text{total}}^{\text{CPU}}/T_{\text{total}}^{\text{GPU}}$ is the measured speedup of the entire solver.

Table 6.2 depicts these derived quantities. Looking at the column 'S_{local}' first, we note that the speedup factors we expected to see based on the experiments on a single subdomain are fully recovered. In fact, they are surpassed, because the acceleration we measured in Section 5.7 has been achieved for a mixed precision iterative refinement solver which includes additional arithmetic work amounting to one defect calculation and one vector update in each outer iteration. We expect the T_{fixed} timings to be identical within the single- and dualcore configurations, respectively. This is not the case due to a granularity effect: The mixed precision solvers 1CORE-MIXED, 2CORE-MIXED and 1COREGPU-MIXED are 'lucky' so to speak and need one half-iteration of the global BiCGStab solver less. Such small variations are the normal behaviour of the BiCGStab solver, and justify again the use of the total runtime efficiency to compare solvers on a high level. If we normalise the timings with the number of iterations performed, then all numbers are as expected. Nonetheless, these granularity effects do not impair our argument in the following.

Amdahl's law, suitably paraphrased for our situation, states that if a fraction of $R_{\text{acc}} \in [0,1]$ of a given solver is 'fixed', then the theoretical upper bound for the speedup, $S_{\max}$, is $1/(1-R_{\text{acc}})$, assuming infinite local acceleration. For the $L=8$ test problem, the CPU solvers exhibit an acceleration potential of at least $2/3$ (limiting the speedup to a factor of 3), and for larger problem sizes, 85–90 % of the entire solvers can be accelerated. It is worth noting that these numbers justify again our minimally invasive approach, because the local computations clearly constitute the bulk of the total arithmetic work. Nonetheless, the measured total speedups are still considerably far from the theoretical upper bounds and diminish rapidly for the dualcore solvers. We address this question in a more general framework in the following.

Analytical Performance Model

Given the local speedup factor S_{local} and the accelerable fraction R_{acc}, we can predict the total speedup as:

$$S_{\text{predicted}} = \frac{1}{(1-R_{\text{acc}}) + (R_{\text{acc}}/S_{\text{local}})} \tag{6.3.1}$$

Note the similarity of this formula to Amdahl's law [3]. Let us exemplarily validate this model for the measured acceleration of the 1COREGPU-MIXED configuration compared to the four solvers that execute entirely on the CPU, cf. Table 6.2 for the exact numbers. On refinement level $L=10$ and when employing JACOBI for the local multigrid, the model predicts total speedups of 7.4,

JACOBI	L	T_{total}	T_{local}	T_{fixed}	R_{acc}	S_{local}	S_{total}	S_{max}
1COREGPU-MIXED	8	1.69	0.73	0.96				
	9	4.67	1.64	3.03				
	10	19.38	5.52	13.86				
1CORE-DOUBLE	8	5.94	5.00	0.94	0.84	6.87	3.52	6.34
	9	27.74	24.68	3.06	0.89	15.03	5.94	9.06
	10	157.93	142.53	15.39	0.90	25.84	8.15	10.26
1CORE-MIXED	8	3.92	2.64	1.28	0.67	3.62	2.32	3.06
	9	15.10	11.75	3.34	0.78	7.16	3.23	4.51
	10	90.63	76.65	13.98	0.85	13.90	4.68	6.48
2CORE-DOUBLE	8	4.32	3.51	0.81	0.81	4.82	2.56	5.33
	9	20.09	17.59	2.50	0.88	10.71	4.30	8.03
	10	113.98	102.78	11.20	0.90	18.63	5.88	10.18
2CORE-MIXED	8	2.29	1.58	0.71	0.69	2.17	1.36	3.24
	9	10.47	8.20	2.27	0.78	5.00	2.24	4.62
	10	62.25	52.75	9.50	0.85	9.56	3.21	6.55
ADITRIDI	L	T_{total}	T_{local}	T_{fixed}	R_{acc}	S_{local}	S_{total}	S_{max}
1COREGPU-MIXED	8	2.56	1.39	1.17				
	9	7.04	3.83	3.21				
1CORE-DOUBLE	8	10.18	8.96	1.22	0.88	6.46	3.98	8.36
	9	46.17	42.97	3.21	0.93	11.22	6.56	14.40
1CORE-MIXED	8	6.69	5.47	1.22	0.82	3.95	2.61	5.50
	9	30.06	26.88	3.18	0.89	7.02	4.27	9.46
2CORE-DOUBLE	8	6.60	5.79	0.81	0.88	4.17	2.58	8.17
	9	30.99	28.69	2.30	0.93	7.49	4.40	13.50
2CORE-MIXED	8	3.89	3.11	0.78	0.80	2.24	1.52	4.97
	9	18.60	16.33	2.27	0.88	4.27	2.64	8.19

Table 6.2: Detailed analysis of the speedup achieved by the 1COREGPU-MIXED configuration over the four CPU variants: Breakdown of the speedup into acceleration potential, local and global acceleration. The disproportionately high T_{fixed} values for the 1CORE-DOUBLE and 2CORE-DOUBLE configurations compared to their mixed precision counterparts are a granularity effect.

4.7, 6.7 and 4.2 over 1CORE-DOUBLE, 1CORE-MIXED, 2CORE-DOUBLE and 2CORE-MIXED, respectively. The measured values are 8.1, 4.7, 5.9 and 3.2. When factoring out the differences in BiCGStab iterations for the five solver configurations, i. e., when normalising the timings by this number (which is a first order approximation to the total runtime efficiency), the model predicts speedups of 7.2, 4.7, 6.5 and 4.0, respectively; so the 'error' introduced by not evaluating performance based on the total runtime efficiency is negligible and the model is accurate, at least when only comparing singlecore configurations.

The differences between the measured and predicted values for the two dualcore configurations are due to an unfair disadvantage of the singlecore GPU solver in our setting, as we only have one GPU per node: When scheduling two half-sized MPI processes to the two CPU cores in each node, we benefit from strong scalability effects. The comparison of the T_{local} timings measured for one GPU compared to two CPU cores is of course fair (one chip vs. one chip). When using two cores, the local speedup reduces from 25.8 to 18.6 (double precision CPU) and from 13.9 to 9.6 (single precision CPU), i. e., the GPU loses approximately 30 % in speedup. However, the 'fixed' portion of the solver benefits to the same extent from these strong scalability effects in the dualcore configuration, while the GPU solver does not. This can be seen in the differences between the T_{fixed} timings, and amounts to a fixed penalty for the GPU solver of three to four seconds in this test case. As this number is a large fraction of the total runtime of the GPU solver (approximately 20 %), the local speedups do not translate to global speedups by the same degree as in the comparison of singlecore configurations alone. Unfortunately, FEAST does not allow to consolidate MPI processes within each node (that execute the 'fixed' part of the solvers on several

CPU cores, two in our case) at each local solution step to sequentially treat the local computations on a single GPU per node. Partitioning the computation in a different way for disjunct stages of the solution process would require fundamental restructuring of FEAST, modifications that essentially mean a full re-implementation of FEAST's domain decomposition approach, because such a requirement has never been considered in the initial design. We conclude that an ideal GPU-accelerated cluster for FEAST needs one GPU per CPU core. If the number of CPU cores and GPUs in a given node is not equal, we could schedule several smaller MPI processes per node (so that the 'fixed' part benefits from strong scalability effects), and partition the local work between the GPU(s) and the remaining CPU cores. We investigate this approach in the scope of our hybrid solvers, see Section 6.3.5.

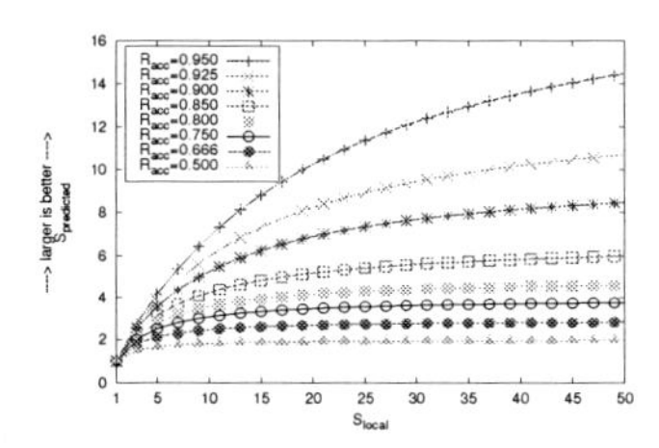

Figure 6.8: Illustration of the total speedup $S_{\text{predicted}}$ as a function of the local acceleration S_{local} and the accelerable fraction R_{acc}.

Figure 6.8 illustrates the predicted total speedup as a function of the local speedup and the accelerable fraction of the entire solver. If we match the (measured) data points for the three quantities S_{total}, S_{local} and R_{acc} from Table 6.2 with the predicted values in the figure, it becomes clear again that our measured total speedups are remarkably close to the predicted limit. The accelerable fraction, which is a property of the CPU-based solver, determines the speedup once the local speedup is large enough. For sufficiently large local problem sizes, our implementation of the GPU-based local solvers is fast enough for this argument to hold.

Implementing, e. g., the asynchronous transfers of data, which in a simple experiment promised an increase in efficiency of 25–30 % (cf. Figure 6.4 on page 223), is expected to yield only diminishing returns, because the local speedup by the GPU is already substantially high. Analogously, Figure 6.8 also underlines why the mixed precision two-layer SCARC scheme on the CPU is inferior to the GPU variant: The theoretical limit of S_{local} is always a factor of two for the former case, while the GPU, based on microbenchmarks or local solver benchmarks, is always more than an order of magnitude faster. We discuss consequences of these observations in more detail in Section 6.6.

6.3.4. Weak Scalability on a Large-Scale Cluster

The small four-node test cluster we have employed throughout this section in order to benefit from the new features of the unified `libcoproc` library is clearly not large enough to perform any meaningful weak scaling experiments. In the following, we therefore summarise previously published results on solving large Poisson problems on an older cluster [84].

Hardware and Solver Configuration

For the test cases presented in the following, 160 compute nodes and one master node of the DQ visualisation cluster at Los Alamos National Laboratory's Advanced Computing Lab have been used. Each node comprises two singlecore Intel Xeon EM64T processors (3.4 GHz, 1 MB level-2 cache), 6–8 GB DDR-667 memory (6.4 GB/s shared bandwidth), and one NVIDIA Quadro FX1400 graphics board (NV40 chip, 128 MB DRAM, 19.2 GB/s bandwidth). The graphics cards in this cluster are rather old, in fact, they are one generation behind the CPUs and equivalent to the GeForce 6 boards which we have evaluated in Section 5.2.4. The nodes are connected via a full bandwidth DDR-1 Infiniband switch.

At the time when these tests were performed, the problem of nonconforming local grids in the SCARC scheme had not been treated by the adaptive coarse grid correction as described in Section 2.3.3. Instead, all local multigrid solvers have been executed in an F cycle with four pre- and postsmoothing steps to alleviate this deficiency, even for isotropic meshes and operators (see also Wobker [242]). Apart from this, the 'template' SCARC solver (cf. Section 6.3.1) has been used without modifications.

Bandwidth Reduction

The GPUs in the DQ cluster are unfortunately one generation of technology older than the rest of the system, so we do not expect substantial speedups. Comparing the peak bandwidths alone, we expect $19.2/6.4 = 3.0$ as an upper bound for the accelerated parts of the solution procedure minus the time to transfer data via the PCIe bus and other related overhead. Despite being restricted to these old GPUs, we want to gain insight into and predict the behaviour of more modern hardware. The observed peak bandwidth roughly doubles with each generation of GPUs (Quadro FX1000 9.6 GB/s, Quadro FX1400 19.2 GB/s, Quadro FX4500 33.6 GB/s, Quadro FX5600 76.8 GB/s). Thus, we want to lower the bandwidth requirements of the computations on the old GPUs. Additionally, these GPUs have only very limited on-board memory and run into problems when storing the full matrix data for large problems; the data required for a multigrid solver on one subdomain and refinement level $L = 10$ just barely fits into the video memory. On better GPUs with more local memory, we can store the matrix data for more than one subdomain, and we still have enough memory available to software-pipeline the code to reduce delays due to data transfer (see Section 6.2.2, and also a previous publication for tests on 16 more powerful GPUs [87]). Here, we have to manually page matrix data in and out of the device memory via comparatively expensive PCIe transfers, every time a new subdomain is treated.

We therefore implemented two versions of the matrix-vector multiplication on the GPU. The first one, labeled *var*, reads all nine bands of the matrix and hence requires a lot of (PCIe and memory) bandwidth and storage, this is the standard variant. The second one, labeled *const*, assumes that the bands are almost constant and only stores a stencil and the 81 values deviating from it (values of the subdomain edges, vertices and the interior for each matrix band). The actual matrix entries are reconstructed during each matrix-vector multiplication, substantially reducing bandwidth and storage requirements while slightly increasing the computational overhead. This second implementation relies on an equidistant mesh discretisation and is not applicable for subdomains that are not parallelogram-shaped, use anisotropic refinement and when the differential operators depend on spatial locations. We should point out that we use a similar bandwidth-friendly implementation on the CPU [15], so no artificial advantage is given to the GPU.

Weak Scalability

To test weak scalability, we statically schedule eight subdomains per node, and increase the number of nodes from 8 to 160. The stencil-based matrix-vector multiplication is used, and we only execute the largest refinement level $L = 10$, i. e., the problem sizes increase from 64 to 1280 million unknowns. We either use one GPU, one CPU or both CPUs per node. Figure 6.9 presents the results of this experiment. All three test cases scale very well, and the outdated GPU outperforms the two (newer) CPUs by 66 %. As the chipset in DQ's nodes shares the memory bus between the two processors, the dual-CPU configuration is only 25 % faster than the single-CPU variant, even though the latter treats eight subdomains per CPU and the former only four. This underlines again, similar to the experiments on more modern multicore CPUs in Section 6.3.2, that the bandwidth to off-chip memory is the decisive performance factor for the type of applications under consideration in this thesis, and performance degrades due to shared buses and resulting bus contention.

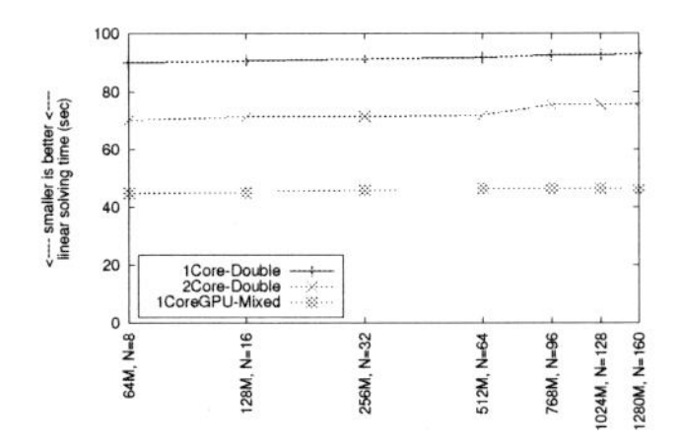

Figure 6.9: Weak scalability, constant workload of eight $L = 10$ subdomains per node. The x-axis uses a logarithmic scale.

Energy Efficiency and Power Considerations

Under full load, each base node consumes approximately 315 W and the GPU consumes 75 W, so the addition of the GPUs increases the power requirements of the cluster by 20 %, which is substantially smaller than the increase in performance. Newer generation GPUs generally not only improve raw performance but also performance per Watt (see Section 5.2.4 and Section 5.3.3), so adding GPUs to an existing cluster is clearly a worthwhile alternative to a more traditional cluster extension in this metric. The improvements in performance per Euro are also remarkable. Due to the limitations of the older GPUs in the DQ cluster, we are not able to perform any meaningful tests for strong scalability as well. FEAST has been shown to scale well [15, 233], and we have previously performed limited strong scalability tests on few, but more powerful GPUs [87]. Thus, we can assume that the strong scalability is maintained on more GPU nodes as well. Under this assumption, we can estimate how an accelerated GPU cluster will scale in the metric 'energy to solution'. As we do not have exact numbers, we calculate an energy efficiency ratio for strong scalability, which we define as the ratio of modified and base system speed divided by the same ratio for power consumption (higher values are better). For the outdated GPUs in DQ, we achieve an energy efficiency ratio of 1.67 (doubling of speed, 20% increase in power consumption). The ratio for doubling the number of unaccelerated nodes is 1.0 (doubling execution speed, two times the power consumption), slightly higher in fact as each node requires only half the amount of memory and hence slightly less power. We estimate 5 W/GB power consumption for the memory modules in DQ. The higher power consumption of newer GPUs is balanced by their speed: NVIDIA's Quadro FX 4500 GPUs consume 95 W each and execute the test problems (without the stencil-based matrix-vector multiplication) three times faster than the CPUs [87], achieving an efficiency ratio of roughly 3.0/1.3 = 2.3. Consequently, GPUs are also favourable in view of the costs related to operating a cluster, not only in terms of acquisition cost.

Impact of Refinement Level and Type of Matrix-Vector Product

Figure 6.10 (left) compares the scheduling of eight $L = 10$ subdomains against 32 $L = 9$ subdomains per node, shared between the two CPUs, and the difference between the stencil-based and the full matrix-vector product. Figure 6.10 (right) makes the same comparisons for the GPU.

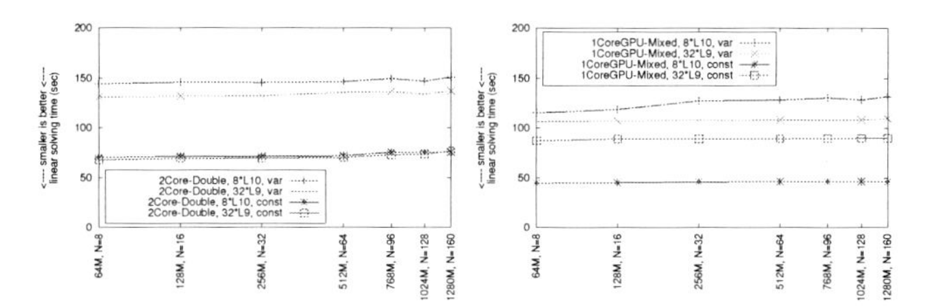

Figure 6.10: Impact of refinement level and type of matrix-vector product on the CPU (left) and the GPU (right), constant workload of 8 million degrees of freedom per node. The x-axis uses a logarithmic scale.

For the CPU and the bandwidth-friendly stencil-based matrix-vector product (denoted with 'const' in the figure), we see that it makes almost no difference whether the problem is partitioned with subdomains of one million unknowns or with four times the amount of subdomains with 0.25 million unknowns each. The overhead of having four times as many calls into the inner multigrid solver is equalled by the reduced bandwidth requirement for the smaller subdomains, as a higher fraction of the operations is performed in cache. For the full matrix-vector product ('var' in the figure), the increased bandwidth requirements for the larger subdomains have a higher impact, and consequently, the computation takes 10 % longer for the higher refinement level. The stencil-based computations are almost twice as fast due to the bandwidth reduction.

The numbers on the GPU tell a different story. The configuration with 32 $L = 9$ subdomains is two times slower for the stencil-based matrix-vector product. The high number of subdomains implies a configuration overhead for the GPU, less parallel work is available in each step and the data is transferred via the PCIe bottleneck more frequently, but in smaller batches. In case of the full matrix-vector product, more local work must be done on the small subdomains ($L = 9$, *var*) and the communication overhead is a smaller fraction of the overall execution time. Therefore, the difference between *const* and *var* is smaller than expected from pure bandwidth considerations and the same comparison on the CPU.

The plot of the full matrix-vector multiplication ($L = 10$, var) has several flaws. First, there is more parallel work per subdomain, so this configuration should be faster than the $L = 9$ case as the total problem size per node is the same, while the two runs are switched in the graph. Second, in contrast to the other plots, there is a higher variance for different numbers of nodes. The limited memory of the graphics cards[6] is most probably responsible for this strikingly different behaviour. If data cannot be represented locally it must be moved back and forth through the narrow connection between the host and device memory. Theoretically, there is (just) enough video memory for a full $L = 10$ subdomain, but we do not have sufficient control over its allocation and usage on the GPU to verify this. We attribute this problem to the memory capacity nonetheless, because graphics cards with more local memory perform very well for this configuration, as could be tested on up to 16 nodes [87].

[6] Even though these boards have 128 MB physical memory, it is unclear how much memory is actually available and not consumed by, e. g., the X-server that is running on them. Another aspect is that OpenGL is free to pad textures, to rearrange them etc., so the memory footprint on a CPU is not necessarily the same as on a GPU.

Despite these anomalous timings, the graphs clearly demonstrate that the heterogeneous cluster enhancement with GPUs scales very well. Given more local memory on the graphics card these results are transferred to the full matrix-vector product case, which is required for less regular grids and anisotropic operators.

Scale-out to Larger Clusters

FEAST executes the coarse grid solver of the global multilevel scheme sequentially on a master node, while the compute nodes are idle. Consequently, we expect to lose scalability at some point, because the coarse grid solves become the bottleneck. However, the experiments in this section demonstrate that this is not the case for large-scale computations on up to 160 nodes, and coarse grid problems consisting of 1280 scalar subdomains. UMFPACK, the sparse direct solver employed by FEAST, is efficient enough for this problem size. As the plots are almost perfectly flat, we do not expect to lose scalability for moderately larger runs. For large-scale runs on thousands of cores however, FEAST will have to be modified accordingly. Bergen et al. [23] come to the same conclusions for a finite difference code on a unitcube domain, their results have been updated in subsequent work by Gradl and Rüde [96] on close to 10 000 cores.

6.3.5. Hybrid CPU-GPU Solvers

Our implementation supports executing two different local solvers within one global two-layer SCARC scheme, see Section 6.2.4. We have used this feature successfully in earlier published work, at a time stronger smoothers were not available on the GPU: The CPU executed a stronger smoother on few, anisotropic subdomains, and the GPU treated many isotropic subdomains with the simple JACOBI smoother [87].

In the following, we perform a similar experiment. For the three test configurations 4X4, 6X4 and 8X4, we statically schedule one subdomain to the CPU, and the remaining three, five and seven to the co-processor. This is achieved by scheduling two different MPI processes per node, one for the CPU and one for the GPU, so that the total load of each node is four, six and eight subdomains, respectively. Both local solvers use the simple JACOBI smoother, which enables us to also run the $L = 10$ case. Figure 6.11 depicts the total runtime efficiency results we obtain, we compare the 1COREGPU-MIXED and the HYBRID-MIXED variants with the 2CORE-MIXED solver configuration, which is the most efficient variant that executes on the CPU alone.

The comparison between the two non-hybrid configurations confirms our findings from Section 6.3.3: The speedup achieved by the GPU-accelerated solver reaches a factor of 2.8, 2.7 and 3.1 for the three test cases and the largest problem sizes. This is a direct consequence of the fact that the local acceleration factor, S_{local}, is—for practical purposes—identical in all three test cases, because it is independent of the number of subdomains. We attribute the small differences we observe nonetheless to the small amount of noise in our test cluster that is unavoidable. The GPU-based solver does not lose efficiency when going from refinement level $L = 9$ to $L = 10$ when more than four subdomains are scheduled per node. In this case, there is still a favourable ratio between computation and communication (cf. Section 6.3.2).

The hybrid configuration is, as expected, faster than the CPU-based solver, but less efficient than executing on the GPU alone. The measured speedup factors are 2.2, 2.3 and 2.5 for the three test domains, respectively. Compared to the GPU-based solver, the hybrid variant loses 46 %, 18 % and 22 % efficiency. This can be explained by the fact that the local speedup on the GPU alone is already very high, and that not enough subdomains are available, even for the 8X4 case, to balance the local execution time on the GPU and the CPU within each node so that one does not have to wait for the other and introduce stalls. To quantify this, we make use of the local timer that we have already used in the analysis of the speedups (cf. Section 6.3.3). This timer gathers results

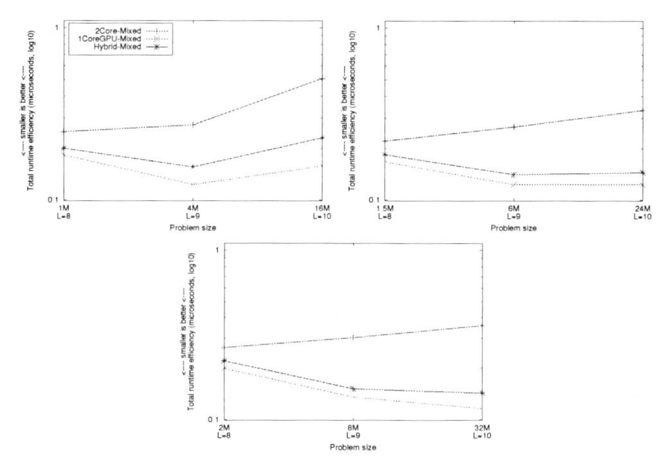

Figure 6.11: Total runtime efficiency of mixed precision two-layer SCARC solvers on the CPU, the GPU and hybridly on both, for the configurations 4x4 (top left), 6x4 (top right) and 8x4 (bottom). All plots share the same legend, and the y-axis uses a logarithmic scale.

for each MPI process separately, and thus, we can collect the shortest and longest local execution time for all eight MPI processes (not taking into account the master process which executes the global coarse grid solver). This approach has been beneficial throughout all our experiments already to determine the amount of noise in the results induced by the fact that the test cluster is not a dedicated HPC installation. Table 6.3 presents a breakdown of these derived quantities, additionally normalised by the number of subdomains treated by the corresponding MPI processes. As the hybrid solver always schedules one subdomain to the CPU and the remaining ones to the GPU, there is no need to normalise the former local measurements. For brevity, we only discuss timings for the largest problem size.

	2CORE-MIXED		1COREGPU-MIXED		HYBRID-MIXED		
	T_{local}	$T_{\text{local}}^{\text{normalised}}$	T_{local}	$T_{\text{local}}^{\text{normalised}}$	$T_{\text{local}}^{\text{GPU}}$	$T_{\text{local}}^{\text{GPU, normalised}}$	$T_{\text{local}}^{\text{CPU}}$
4x4	50.62	12.66	5.53	1.38	4.92	1.64	15.93
6x4	57.32	9.55	6.94	1.16	6.89	1.38	13.46
8x4	80.74	10.09	9.67	1.21	9.96	1.42	16.84

Table 6.3: Detailed breakdown of the local computation time for the three solver configurations on the three test domains.

The measurements for the 4x4 configuration are slightly disturbed, because the 2CORE-MIXED solver executes one additional half-iteration of the outermost BiCGStab solver. For all other cases, the solvers converge identically. The normalised timings for the 2CORE-MIXED and 1COREGPU-MIXED solvers are consistent, and the normalised timings of the 1COREGPU-MIXED run are almost recovered by the GPU portion of the HYBRID-MIXED solver. However, in the hybrid case, the local timings are imbalanced between CPU and GPU. The imbalance decreases with increasing number of subdomains that are scheduled to the GPU, which explains why the 6x4 and 8x4 configurations improve efficiency when going from refinement level $L = 9$ to

$L = 10$ (cf. Figure 6.11) while the others do not: For the three test problems and the hybrid solver, the CPU portion is slower than the GPU portion (comparing $T_{\text{local}}^{\text{GPU}}$ and $T_{\text{local}}^{\text{CPU}}$) by factors of 3.2, 2.0 and 1.7, respectively. A second reason why the hybrid solver is less efficient than the pure GPU solver can be found in the difference between the normalised CPU timings of the hybrid and CPU-only solver. The latter is 25–67 % faster, even though the timings correspond to the exact same amount of arithmetic work. We suspect that the shared memory bus between the CPU cores is the primary reason. The configuration that executes the local solvers on the GPU puts a significant amount of pressure on the host's memory bus: All matrix data is transferred during the preprocessing stage, but for each subdomain, iteration and right hand side vectors have to be consumed from FEAST, have to be converted to and from single precision, have to be copied to and from device memory (it is not possible to execute these transfers via DMA), and finally, the result of the local solve has to be accumulated to FEAST's global solution. On the other hand, the CPU-only solvers treat each subdomain sequentially, so the memory access patterns are more coherent, even when using both CPU cores. In summary, all these operations, that are performed for several subdomains, induce a significant amount of memory bus contention and can explain the slowdown of the CPU solvers that we observe.

Summary

The possibility to execute different local solvers on different architectures is very beneficial, as it supports heterogeneous configurations such as, for instance, cluster nodes with several co-processors with varying capabilities. Nonetheless, if the speedup by the GPU is too high, adding additional CPU subtasks does not always pay off and can be slower than executing on the GPUs alone. In any case, load balancing between the different MPI processes is increasingly challenging, and crucial for optimally exploiting the available hardware with different performance characteristics. We discuss these aspects further in Section 6.6.

6.4. Application Acceleration: Linearised Elasticity

For the Poisson problem, we are able to achieve substantial speedups by virtue of our minimally invasive acceleration approach, see Section 6.3 and previously published work [84, 87]. Here and in Section 6.5 we demonstrate that even fully developed non-scalar applications can be significantly accelerated, without any changes to either the application or the accelerator code. These application specific solvers have a more complex data flow and more diverse CPU/co-processor interaction than the Poisson problem. This allows us to perform a detailed, realistic assessment of the accuracy and performance of co-processor acceleration of unmodified code within our concept. In this section, we consider the application FEASTSOLID, which is built on top of FEAST and enables the large-scale solution of solid mechanics problems (*computational solid mechanics, CSM*). It has been written by Hilmar Wobker and constitutes a large part of his thesis [242]. In Section 2.3.5 on page 41 we have summarised the theoretical background of FEASTSOLID. Section 6.5 is then dedicated to a fluid dynamics application.

The remainder of this section is structured as follows: In Section 6.4.1 we provide a brief introduction, details of the hardware and the solver we employ, and define the test configurations. Section 6.4.2 is dedicated to accuracy studies of our mixed precision approach for large-scale calculations on 64 compute nodes. We present three tests, targeting the comparison with analytical reference solutions, global anisotropies that rapidly increase the condition numbers of the systems, and local anisotropies. Weak scalability is evaluated in Section 6.4.3. Finally, Section 6.4.4 discusses performance and speedup of the GPU-accelerated solver, using the same analysis we already applied to the accelerated solution of the Poisson problem in Section 6.3.3.

Most of the results that are presented in this section have been published previously by the author of this thesis, Hilmar Wobker and their co-workers [86, 90].

6.4.1. Introduction, Test Configurations and Solution Strategy

We restrict ourselves to the fundamental model problem of linearised elasticity for compressible material, i. e., the small deformation of solid bodies under external loads. We use a *displacement formulation* in which the only unknowns are the displacements in x- and y-direction, see Section 2.3.5 on page 41 for details on the mathematical background and the high-level solution strategy. After discretisation and the application of a suitable renumbering strategy, the resulting non-scalar (vector-valued) problem has the following block structure:

$$\begin{pmatrix} \mathbf{K}_{11} & \mathbf{K}_{12} \\ \mathbf{K}_{21} & \mathbf{K}_{22} \end{pmatrix} \begin{pmatrix} \mathbf{u}_1 \\ \mathbf{u}_2 \end{pmatrix} = \begin{pmatrix} \mathbf{f}_1 \\ \mathbf{f}_2 \end{pmatrix}, \tag{6.4.1}$$

where $\mathbf{f} = (\mathbf{f}_1, \mathbf{f}_2)^\mathsf{T}$ is the vector of external loads and $\mathbf{u} = (\mathbf{u}_1, \mathbf{u}_2)^\mathsf{T}$ the (unknown) coefficient vector of the finite element solution. The matrices $\mathbf{K}_{11}$ and $\mathbf{K}_{22}$ of this block-structured system correspond to scalar elliptic operators (cf. Equation (2.3.3) in Section 2.3.5), and thus, the generalised SCARC solvers (cf. Section 2.3.4 on page 39) are applicable.

Test Configurations

We evaluate four configurations that are prototypical for practical applications. Figure 6.12 shows the coarse grids, the prescribed boundary conditions and the partitioning for the parallel solution of each configuration. The BLOCK configuration is a standard test case in CSM, a rectangular block is vertically compressed by a surface load [185]. The PIPE configuration represents a circular cross-section of a pipe clamped in a bench vise. It is realised by loading two opposite parts of the outer boundary by surface forces. With the CRACK configuration we simulate an industrial test environment for assessing material properties. A workpiece with a slit is torn apart by some device attached to the two holes. In this configuration the deformation is induced by prescribed horizontal displacements at the inner boundary of the holes, while the holes are fixed in the vertical direction. For the latter two configurations we exploit symmetries and consider only sections of the real geometries. Finally, the STEELFRAME configuration models a section of a steel frame, which is fixed at both ends and asymmetrically loaded from above. The latter two configurations employ unstructured coarse grids.

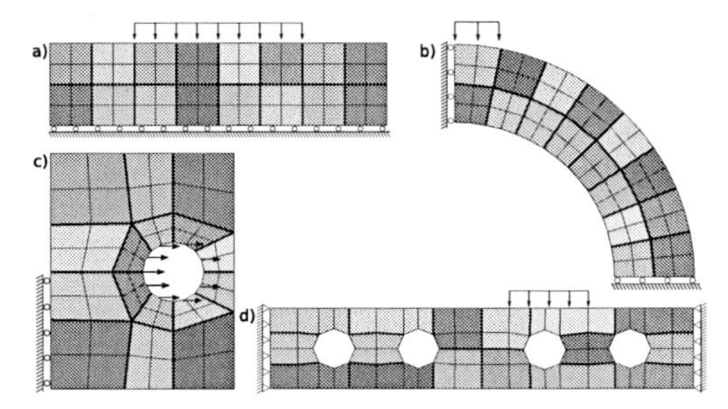

Figure 6.12: Coarse grids, boundary conditions and static partition into subdomains for the configurations (a) BLOCK, (b) PIPE, (c) CRACK and (d) STEELFRAME.

Figure 6.13 shows the computed deformations of the four geometries and the von Mises stresses, which are an important measure for predicting material failure in an object under load. Both the accelerated and the unaccelerated solver compute identical results, according to a comparison of the displacement of selected reference points and the volume of the deformed bodies.

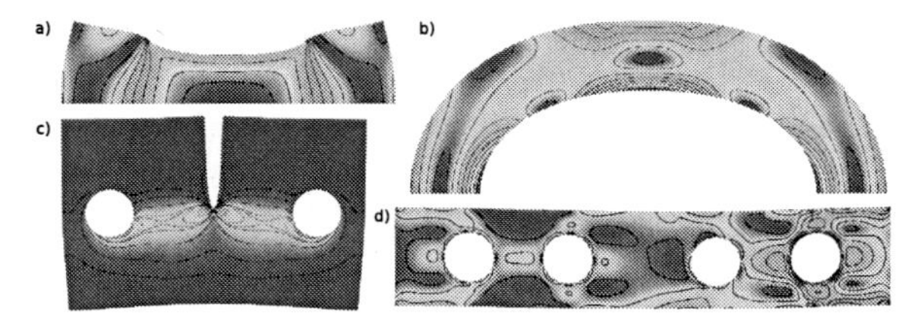

Figure 6.13: Computed displacements and von Mises stress for the four main test configurations.

Our fifth test case, a cantilever beam (BEAM), is a standard benchmark configuration in CSM, and is known to be difficult to solve numerically [34, 173, 242]. A long, thin beam is horizontally attached at one end and the gravity force pulls it uniformly in the y-direction (see Figure 6.14).

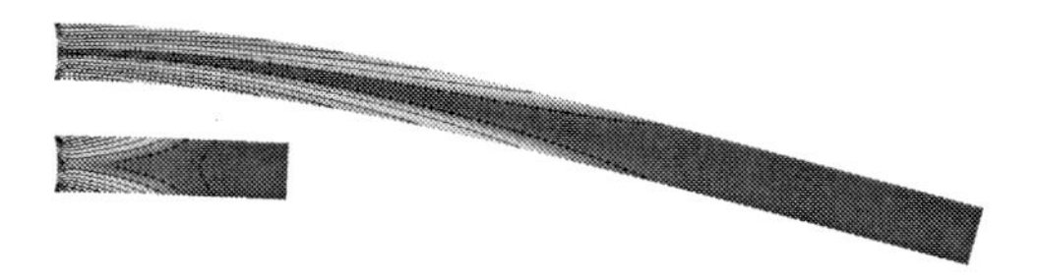

Figure 6.14: Computed displacements and von Mises stresses for the BEAM configuration with anisotropy 1:16 (top) and 1:4 (bottom).

We partition this geometry in such a way that the number of subdomains (and thus compute nodes) is proportional to the length of the beam, and test two configurations: one consisting of 8×2 square subdomains (distributed to 4 nodes), the other of 32×2 square subdomains (16 nodes), resulting in a global domain anisotropy of 4:1 and 16:1, and a maximum problem size of 32 and 128 million degrees of freedom, respectively.

Solver Setup

In all tests we configure the solver scheme (cf. Figure 6.15) to reduce the initial residuals by six digits, the global multilevel solver performs one pre- and postsmoothing step each in a V cycle, and the local multigrid uses a V cycle with four smoothing steps each and executes exactly one multigrid cycle. This solver is a straightforward generalisation of the scalar 'template' two-layer SCARC solver depicted in Figure 6.1 on page 217, suitably modified to boil the solution of block structured systems down to scalar solves as explained in Section 2.3.4.

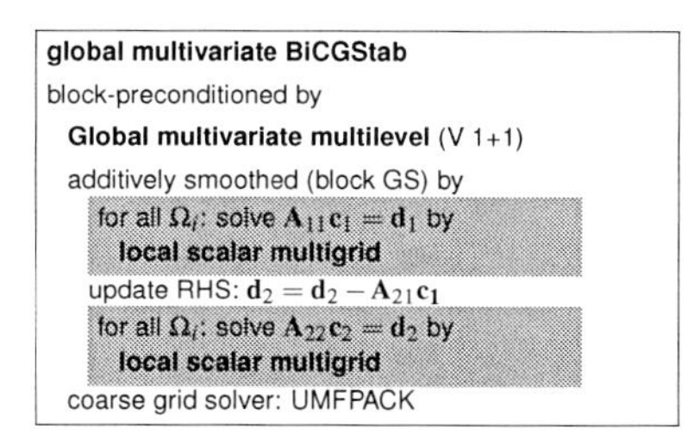

Figure 6.15: Multivariate two-layer SCARC solver for the linearised elasticity problem, accelerable components are highlighted.

The subdomains per node are either collected in one MPI process, leaving the second core of the processors in our test clusters (see below) idle, or distributed to two MPI processes per node. These scheduling schemes are referred to as 1CORE-DOUBLE and 2CORE-DOUBLE, respectively. As we only have one GPU per node, we perform GPU tests in the 1COREGPU-MIXED configuration. We could also run one large GPU process and one small CPU process per node in parallel, but previous experience (cf. Section 6.3.5) has shown that this is only feasible if there is enough memory to allow for many subdomains per node. As explained in Section 6.2.2, we accelerate the plain CPU solver with GPUs by replacing the scalar, local multigrid solvers with their GPU counterparts, as highlighted in Figure 6.15. Otherwise, the parallel solution scheme remains unchanged. All computations on the CPUs are executed in double precision, i. e., we do not test the 1CORE-MIXED and 2CORE-MIXED cases, because at the time these experiments have been performed, the unified `libcoproc` library has not been available yet. The GPU-based multigrid is based on a legacy implementation and employs OpenGL and the Cg shading language to access the device.

Hardware Platforms

The majority of the performance tests in this section are carried out on the 16 nodes of an advanced cluster called USC. This machine is installed at Los Alamos National Laboratory. Each node comprises a dualcore AMD Opteron 2210 processor (Santa Rosa, 2×1 MB level-2 cache, 1.8 GHz), DDR2-667 memory (peak shared bandwidth 10.7 GB/s), and an NVIDIA Quadro FX5600 GPU (G80 chip, 16 multiprocessors) with 1.5 GB GDDR3 memory (384 bit interface, 76.8 GB/s bandwidth). DDR-4 Infiniband is used as interconnect. The scalability experiments are carried out on the older DQ cluster, see Section 6.3.4. Updated results using the unified `libcoproc` library are obtained on our small four-node test cluster, see Section 6.3.1 for hardware details.

6.4.2. Accuracy Studies

On the CPU, we use double precision exclusively. For the Poisson problem, we have already demonstrated an important benefit of our solver concept and the corresponding integration of hardware acceleration: The restriction of the GPU (and local solvers in general) to single precision has no effect on the final accuracy and the convergence of the global solver. To verify this claim for the more challenging linearised elasticity problems, we perform three different numerical experiments and increase the condition number of the global system and the local systems. As we want to run the experiments for the largest problem sizes possible, we use the 'outdated' cluster DQ, of which 64 nodes were still in operation when these experiments were performed in early 2007. As newer hardware generations provide better compliance with the IEEE 754 single precision standard, the accuracy analysis remains valid for better GPUs and for mixed precision solvers on the CPU.

Analytical Reference Solution

This numerical test uses the BLOCK configuration, covered by 16, 64 or 256 subdomains which are refined seven to ten times. We define a parabola and a sinusoidal function for the x and y displacements, respectively, and use these functions and the linearised elasticity equation (2.3.2a) (cf. Section 2.3.5) to prescribe the right hand side for the global system, so that the exact analytical solution is known. This allows us to compare the L_2 errors (Section 2.1.3), which according to finite element theory are reduced by a factor of four (proportionally to the square of the mesh width h) in each refinement step.

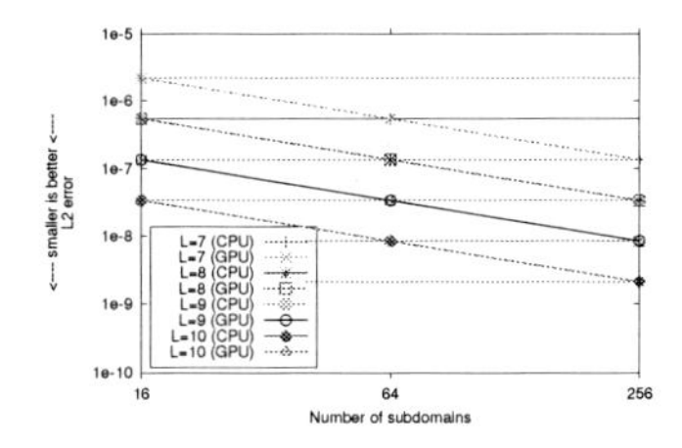

Figure 6.16: Error reduction of the linearised elasticity solver in the L_2 norm, BLOCK configuration. Both the x- and y-axis use a logarithmic scale.

Figure 6.16 illustrates the main results. We first note that all configurations require exactly four iterations of the global solver. Most importantly, the differences between CPU and GPU runs are in the noise, independent of the level of refinement or the number of subdomains. In fact, in the figure the corresponding CPU and GPU results are plotted directly on top of each other. We nevertheless prefer illustrating these results with a figure instead of a table, because it greatly simplifies the presentation: Looking at the graphs vertically, we see the global error reduction by a factor of four with increasing level of refinement L. The independence of the subdomain distribution and the level of refinement can be seen horizontally, for instance 16 subdomains on level $L = 8$ are equivalent to 64 level-7 subdomains.

Global Ill-conditioning: Cantilever Beam Configuration

The high degree of domain anisotropy in conjunction with the large ratio between free Neumann boundary and fixed Dirichlet boundary (only the narrow side at one end is fixed) and the high level of refinement found in the BEAM configuration results in a very ill-conditioned global system [9, 173, 242]. To illustrate this, we first use a simple unpreconditioned conjugate gradient method whose iteration count grows with the condition number of the system and is thus a suitable indicator.

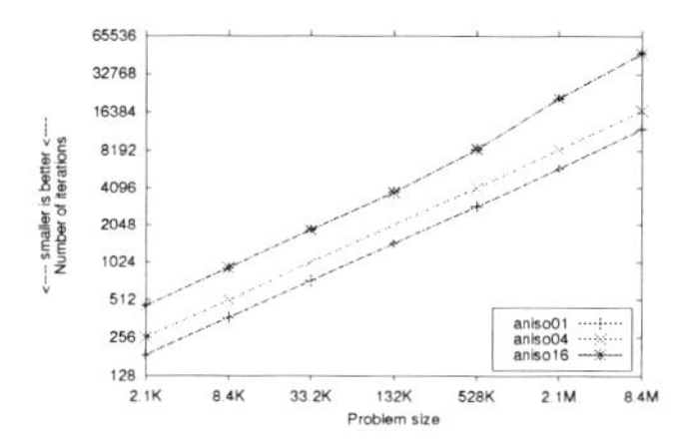

Figure 6.17: Illustration of the ill-conditioning of the BEAM configuration with a simple CG solver. The y-axis uses a logarithmic scale.

For small problem sizes we solve the two beam configurations described above, as well as a third variant – a very short 'beam' with global 'anisotropy' of 1:1. The latter configuration is used exclusively in this test to emphasise the dependency on the global anisotropy. Figure 6.17 shows the iteration numbers of the conjugate gradient solver for increasing problem size. Reading the graphs vertically (for a fixed number of degrees of freedom), shows a significant rise of iteration numbers due to the increasing degree of anisotropy (note that the y-axis uses a logarithmic scale). For the isotropic and mildly anisotropic beam we can clearly observe that each grid refinement doubles the number of iterations, which is the expected behaviour of the conjugate gradient solver. For the strongly anisotropic beam, however, this is no longer true on the highest refinement levels, where the precise iteration counts are 8 389, 21 104 and 47 522, showing a factor which is clearly greater than two.

Table 6.4 contains the results we obtain with the generalised two-layer SCARC solver for the cantilever beam configuration. The fractional iteration count is a consequence of the global BiCGStab solver permitting an 'early exit' after the first of the two applications of the preconditioner (an entire parallel multilevel iteration), if the scheme has already converged. As we have no analytical reference solution in this case, we use the displacement (in y-direction) of the midpoint of the free side of the deformed beam and its volume as features of the solutions to compare the computed CPU and GPU results, for increasing levels of refinement L and corresponding problem sizes.

We see no accuracy difference except for floating point noise between the CPU and the GPU configurations, although the CPU configuration uses double precision everywhere and the GPU configuration executes the local multigrid in single precision. This is a very important advantage of our minimally invasive co-processor acceleration and corresponding solver concept. The global anisotropies and ill-conditioning are entirely hidden from the local multigrids, which see locally isotropic problems on their respective subdomains.

Overall, we see the expected rise of iteration numbers when elongating the beam and thus increasing the system's condition number. But apart from some granularity effects that are inevitable for the BiCGStab solver, we see good level independence and, in particular, identical convergence of the CPU and the GPU variants.

aniso04	Iterations		Volume		y-Displacement	
L	CPU	GPU	CPU	GPU	CPU	GPU
8	4	4	1.6087641E-3	1.6087641E-3	-2.8083499E-3	-2.8083499E-3
9	4	4	1.6087641E-3	1.6087641E-3	-2.8083628E-3	-2.8083628E-3
10	4.5	4.5	1.6087641E-3	1.6087641E-3	-2.8083667E-3	-2.8083667E-3
aniso16						
8	6	6	6.7176398E-3	6.7176398E-3	-6.6216232E-2	-6.6216232E-2
9	**6**	**5.5**	6.7176427E-3	6.7176427E-3	-6.6216551E-2	-6.6216552E-2
10	5.5	5.5	6.7176516E-3	6.7176516E-3	-6.6217501E-2	-6.6217502E-2

Table 6.4: Iterations and computed results for the BEAM configuration with anisotropy of 1:4 (top) and 1:16 (bottom). Differences are highlighted in bold face.

Local Anisotropies: Towards Incompressible Material

While the previous experiment examined how our solver concept performs for ill-conditioned system matrices due to global domain anisotropies, we now analyse the impact of local anisotropies. These tests and the ones for the cantilever beam configuration are thus symmetric to the uniform refinement of anisotropic domains, and of the anisotropic refinement of isotropic and anisotropic domains in Section 5.6 on page 171.

We use the BLOCK configuration for the following tests. To induce local anisotropies, we increase the Poisson ratio ν of the material, which according to Equation (2.3.5) (page 42) changes the anisotropy of the elliptic operators in Equation (2.3.3). Physically, this means that the material becomes less compressible. Since we use a pure displacement finite element formulation, the value of ν must be bounded away from 0.5, otherwise the effect of *volume locking* would hinder convergence to the correct solution [34]. Equation (2.3.1) shows that the critical parameter λ tends to infinity for $\nu \to 0.5$. As the legacy code used for these experiments only provides a simple JACOBI smoother on the GPU, we have to increase the number of JACOBI smoothing steps of the inner solver (the outer solver's configuration remains unchanged) to improve convergence. All experiments are again performed on 64 nodes of the DQ cluster, yielding a maximum problem size of 512 million degrees of freedom for the highest refinement level ($L = 10$). As this cluster is outdated and its GPUs come from an even older technology generation than its CPUs, only changes in relative timings (CPU/GPU) are relevant, and we do not examine speedups.

For clarity, Table 6.5 does not contain accuracy results, but we confirm that in this series of tests we get identical results (up to numerical noise) for the CPU and the GPU just as in the previous two experiments.

Several important observations can be made from the data listed in Table 6.5. First, as the increasing value of ν affects the complete system (2.3.3), the number of solver iterations rises accordingly. Second, the number of iterations required to solve the system is reduced by increased local smoothing (the anisotropy of the elliptic operators is hidden better from the outer solver), and occasional granularity effects disappear. Finally, the accelerated solver behaves identically to the unaccelerated one, in other words, even if there are effects due to the GPU's reduced precision, they are completely encapsulated from the outer solver and do not influence its convergence. Looking at the (normalised) timings in detail, we see that on the CPU, performing twice as many smoothing steps in the inner multigrid results in a 40–50 % increase in runtime for the highest level of refinement, while on the GPU, only 20–25 % more time is required. There are two reasons for this behaviour: On the CPU, the operations for high levels of refinement are performed completely out of cache, while on the GPU, full bandwidth is only available to the application for large input sizes, as the overhead costs associated with launching compute kernels is less dominant. Second, as the amount of video memory on these outdated GPUs is barely large enough to hold all ma-

L	V-4+4 JACOBI				V-8+8 JACOBI				V-1+1 ADITRIGS			
nu=0.40	Iters.		Time/Iter.		Iters.		Time/Iter.		Iters.		Time/Iter.	
$a_{op} = 6$	CPU	GPU	CPU	GPU	CPU	GPU	CPU	GPU	CPU	GPU	CPU	GPU
8	4	4	3.3	4.9	3.5	3.5	4.5	5.7	3.5		3.6	
9	4	4	11.1	11.2	3.5	3.5	15.9	13.4	3.5		12.0	
10	4	4	48.2	41.2	3.5	3.5	69.7	52.2	3.5		49.3	
nu=0.45	Iters.		Time/Iter.		Iters.		Time/Iter.		Iters.		Time/Iter.	
$a_{op} = 11$	CPU	GPU	CPU	GPU	CPU	GPU	CPU	GPU	CPU	GPU	CPU	GPU
8	4.5	4.5	3.2	4.9	4.5	4.5	4.3	5.5	4.5		3.4	
9	5	5	11.0	11.1	4.5	4.5	15.6	13.2	4		11.8	
10	5	5	48.1	41.1	4.5	4.5	69.3	52.1	4		49.4	
nu=0.48	Iters.		Time/Iter.		Iters.		Time/Iter.		Iters.		Time/Iter.	
$a_{op} = 26$	CPU	GPU	CPU	GPU	CPU	GPU	CPU	GPU	CPU	GPU	CPU	GPU
8	7.5	7.5	3.0	4.6	6.5	6.5	4.1	5.5	6.5		3.1	
9	7.5	7.5	10.8	11.0	6.5	6.5	15.4	13.5	6.5		11.5	
10	7	7	47.9	42.1	6.5	6.5	69.2	52.4	6.5		49.0	

Table 6.5: Convergence behaviour of the CPU and the GPU solver with increasing degree of operator anisotropy, BLOCK configuration. To concentrate on general tendencies, the timings are normalised by the number of iterations.

trix data associated with one subdomain, the high cost of paging data in and out of memory is amortised much better when more operations are performed on the same data.

To verify that our test results are not influenced by the weakness of the JACOBI smoother, we perform all tests again with a very powerful alternating direction implicit tridiagonal Gauß-Seidel smoother (ADITRIGS, see Section 2.3.1), for which two smoothing steps (one in each direction) suffice. Table 6.5 shows the expected results: The type of the inner smoother has no effect on the outer convergence behaviour, as long as the inner smoother is strong enough to resolve the local operator anisotropy. This justifies—at least for this particular test case—our 'emulation' of a stronger smoother on the GPU by performing eight JACOBI smoothing steps.

We finally note that a similar type of anisotropy occurs in fluid dynamics simulations, where it is often necessary to resolve boundary layers more accurately than the inside of the flow domain. The resulting high element aspect ratios typically influence the condition number of the system matrix comparably to anisotropies in the elliptic operators, see also Section 5.6 and Section 6.5.

6.4.3. Weak Scalability

In this section, we analyse weak scalability of our GPU-enhanced solver in comparison to the unaccelerated case. The configurations used in these tests comprise two variations of the standard BLOCK benchmark test case, modified so that each subdomain remains square when doubling the number of subdomains. We increase the number of nodes (and hence, unknowns) from 4 to 64 (32 to 512 million, $L = 10$). As we do not have access to enough nodes with modern GPUs, the runs are executed on the older DQ cluster (cf. Section 6.3.4).

Figure 6.18 demonstrates good weak scalability of our approach for both the accelerated and the unaccelerated solver. The relatively poor performance gain of the GPUs is attributed to the outdated GPUs in the DQ cluster, in particular, their small amount of local memory can only hold the data associated with a single subdomain, and consequently, the entire matrix data is paged in and out from video memory for each subdomain (cf. Section 6.3.4).

As described in Section 2.3.3, the parallel computation is completely decoupled, data is exchanged only between neighbouring subdomains. The only exception is the solution of the global

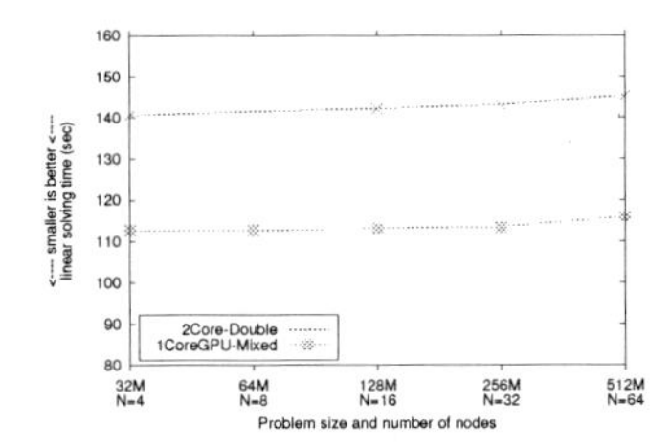

Figure 6.18: Weak scalability on up to 64 DQ nodes, BLOCK configuration. The x-axis uses a logarithmic scale.

coarse grid problem of the data-parallel multigrid solver, which we perform on the master node while the compute nodes are idle. The size of this coarse grid problem depends only on the number of subdomains (to be more precise, on the number of degrees of freedom implied by the coarse grid formed by the collection of subdomains) and is in particular independent of the level of refinement and the number of parallel processes. Due to the robustness of our solver, comparatively few global iterations suffice. In accordance with similar approaches (see, for example, Bergen et al. [23] and Gradl and Rüde [96]), we can safely conclude that the global coarse grid solver, which is the only sequential part of the otherwise parallel execution, is not the bottleneck in terms of weak scalability. As a concrete example, the solution of the global coarse grid problems in the scalability tests in Figure 6.18 contributes at most 3 % to the total runtime.

As previous experiments with the prototypical scalar Poisson equation on up to 160 nodes resulted in equally good scalability (Section 6.3.4), combining these results we may argue that our heterogeneous solution approach for the more demanding application discussed in this section would also scale well beyond 100 GPU nodes.

6.4.4. Performance and Speedup

To evaluate absolute performance and the speedup of the accelerated solver, we consider only the refinement level $L = 10$ (128 million degrees of freedom) of the four test configurations BLOCK, PIPE, CRACK and STEELFRAME, and statically assign four subdomains per node, such as to balance the amount of video memory on the USC cluster with the total amount of memory available per node. We measure the absolute time to solution for the three scheduling schemes 1CORE-DOUBLE, 2CORE-DOUBLE and 1COREGPU-MIXED, see Figure 6.19. Consistently for all four configurations, the accelerated solver is roughly 2.6 times faster than the unaccelerated solver using only one MPI process per node, and the speedup reaches a factor of 1.6 if we schedule two half-sized MPI processes per node. Finally, the 2CORE-DOUBLE configuration runs 1.6 times faster than the 1CORE-DOUBLE configuration. For the Poisson problem on similar CPUs from the same design generation, we have measured only 40 % gain, cf. Section 6.3.2. But these absolute timings do not tell the whole story, and favour the CPU in the 2CORE-DOUBLE vs. 1COREGPU-MIXED comparison. We investigate this effect further in the following, based on updated results that incorporate the full functionality of the new `libcoproc` library.

Updated Results with the Unified libcoproc Library and CUDA

Figure 6.20 depicts updated timing measurements for the BLOCK configuration, suitably downscaled for the four nodes of our test cluster (see Section 6.3.1): Two multivariate subdomains are scheduled to each node, so that the workload per node corresponds to the 4x4 configuration in our

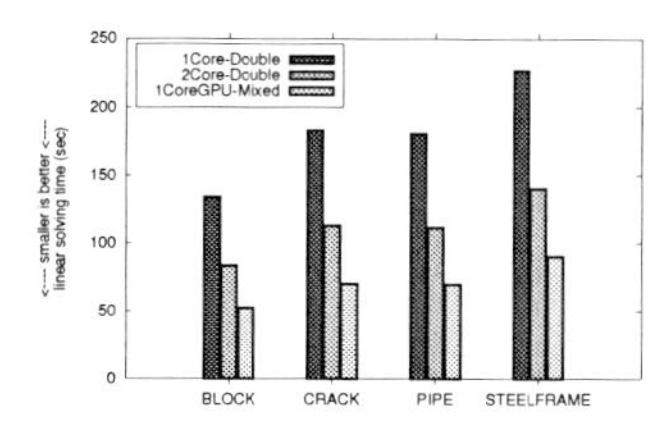

Figure 6.19: Execution times for the linearised elasticity solver with and without GPU acceleration on the USC cluster.

tests for the Poisson problem, cf. Section 6.3.2.

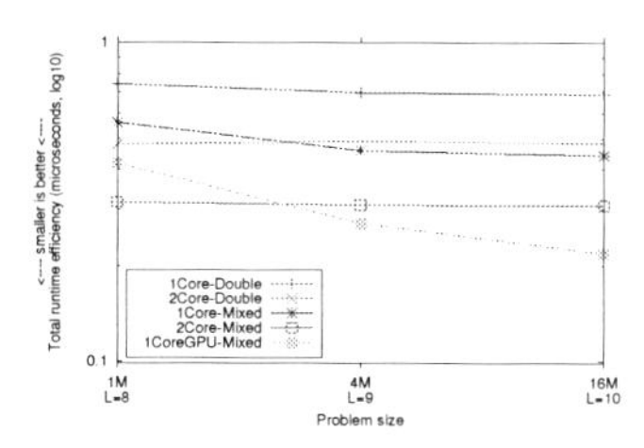

Figure 6.20: Total runtime efficiency for the BLOCK configuration on the four-node test cluster. The y-axis uses a logarithmic scale.

The GPU is always more efficient than the CPU, with the exception of the 2CORE-MIXED solver and the smallest refinement level. The dualcore configuration in double precision is approximately 40 % faster than its singlecore counterpart. The mixed precision singlecore solver is between 1.33 and 1.55 times more efficient than the corresponding double precision variant, for $L = 8$ to $L = 10$. Finally, the dualcore mixed precision solver achieves a speedup of 2.3 over the baseline singlecore double precision configuration. These measurements are slightly below the factors we observe for the Poisson problem, which is the expected behaviour because the 'fixed' portions of the solver are more involved than for the Poisson problem.

GPU Performance Analysis

Table 6.6 elucidates the dependencies between the local and the global solver in more detail for the speedup achieved by the GPU over the four CPU configurations. The measurements correspond to the computation of the BLOCK configuration on the four-node test cluster, see Figure 6.20.

The local speedup numbers are slightly below those we measured for the Poisson problem (Table 6.2), which is surprising at first because the scalar subproblems have the same structure and size. It can be explained by the different stopping criteria of the inner solvers, for the Poisson tests the inner solver gained one digit, and here, the solver performs a fixed number of iterations. In contrast to the Poisson problem, the linearised elasticity solver has a slightly smaller acceleration potential. We only discuss the largest problem size: The accelerable fraction of the double precision configurations is approximately 70 %, compared to 60 % for the mixed precision variants. The local speedups of 18 and 9 against the double and single precision singlecore CPU solvers thus translate to total speedups of 3.2 and 2.2, in line with our analytical performance model and

	L	T_{total}	T_{local}	T_{fixed}	R_{acc}	S_{local}	S_{total}	$S_{predicted}$	S_{max}
1COREGPU-MIXED	8	4.10	0.70	3.40					
	9	9.30	1.40	7.90					
	10	30.50	3.90	26.60					
1CORE-DOUBLE	8	7.30	4.10	3.20	0.56	5.86	1.78	1.87	2.28
	9	23.90	16.40	7.50	0.69	11.71	2.57	2.69	3.19
	10	96.10	69.40	26.70	0.72	17.79	3.15	3.14	3.60
1CORE-MIXED	8	5.50	1.80	3.70	0.33	2.57	1.34	1.25	1.49
	9	15.70	8.00	7.70	0.51	5.71	1.69	1.73	2.04
	10	61.90	35.40	26.50	0.57	9.08	2.03	2.04	2.34
2CORE-DOUBLE	8	4.70	2.90	1.80	0.62	4.14	1.15	1.88	2.61
	9	16.80	12.00	4.80	0.71	8.57	1.81	2.71	3.50
	10	67.60	50.30	17.30	0.74	12.9	2.22	3.19	3.91
2CORE-MIXED	8	3.10	1.30	1.80	0.42	1.86	0.76	1.24	1.72
	9	10.60	5.80	4.80	0.55	4.14	1.14	1.71	2.21
	10	43.10	25.70	17.40	0.60	6.59	1.41	2.02	2.48

Table 6.6: Detailed analysis of the speedup achieved by the 1COREGPU-MIXED configuration over the four CPU variants, BLOCK configuration: Breakdown of the speedup into acceleration potential, local and global acceleration.

sufficiently close to the respective theoretical maxima S_{max}. The difference between the (predicted and theoretical) acceleration and the measured one is much larger for the dualcore configuration, a direct consequence of the unfair disadvantage of the singlecore GPU configuration: Due to strong scaling effects when going from one CPU core to two, the T_{fixed} dualcore timings are only 2/3 of the singlecore timings, which leads to a constant reduction in total runtime by nine seconds (30 % of the total time to solution of the GPU solver) that the GPU solver cannot benefit from (cf. Section 6.3.3). Nonetheless, the GPU-accelerated solver is still 2.2 and 1.4 times faster than the 2CORE-DOUBLE and 2CORE-MIXED CPU solvers, the latter one has not been possible in FEAST prior to this thesis.

A similar analysis has been carried out for the experiments on the USC cluster [90]: For the 16 node runs and the highest level of refinement, the accelerable fraction is approximately 66 %, and local speedups by a factor of 9 and 6.5 (obtained by our legacy OpenGL implementation) are achieved against one and two cores, respectively.

We finally note that the measured acceleration potential is still very good in view of our minimally invasive integration technique, because substantial speedups are achieved without changing a single line of application code.

6.5. Application Acceleration: Stationary Laminar Flow

In this section we explore the limitations of our approach by accelerating the Navier-Stokes solver FEASTFLOW, which is built on top of FEAST. It is being developed by Sven Buijssen and constitutes a large part of his thesis [44], see also Section 2.3.6 on page 43 for a brief summary of the theoretical background. This nonlinear saddle point problem is much more involved than our previous tests, and does not exhibit an equally favourable acceleration potential: Not all computational work is concentrated inside the linear solver, as linearised systems need to be assembled in each nonlinear step. The general discretisation and solution scheme has been presented in Section 2.3.6 on page 43. Section 6.5.1 provides an overview of this section and introduces the test problems and concrete solver configurations. In Section 6.5.2 we examine the linear solver alone by solving the Stokes problem, a simplification of the general Navier-Stokes equations. In all other experiments in this section, we solve the Navier-Stokes equations. The accuracy of different double and mixed precision solvers is examined in Section 6.5.3. In Section 6.5.4 we analyse performance and the speedup of mixed precision solvers on the CPU and the GPU over configurations that execute entirely in double precision. A critical review of our approach is presented in Section 6.6, where we also discuss conclusions and future work.

The presented results are significantly updated from previously published work by the author of this thesis, Sven Buijssen and their co-workers [88].

6.5.1. Introduction, Test Configurations and Solution Strategy

Many PDE problems of practical interest are nonlinear. After linearisation, focussing entirely on accelerating (portions of) linear solvers is justified, because the approximate solution of linear subproblems typically constitutes the most time-consuming task in the solution process, especially for saddle point problems arising, e. g., in CFD. However, this focus naturally limits the achievable speedups, as the system matrices for the linearised problems have to be assembled in each nonlinear sweep. In this section, we explore the limitations of our minimally-invasive integration approach by applying it to the acceleration of a Stokes and Navier-Stokes solver for small Reynolds numbers, computing a stationary laminar flow. We refrain from investigating non-stationary flows for two reasons: First, the result in terms of speedups would be qualitatively comparable and the time-stepping loop only increases the runtime of our experiments. Second, the stationary approach significantly worsens the condition number of the matrices, which can be critical in view of our mixed precision techniques. Hence, by confining ourselves to the stationary case, we actually examine the more difficult configuration.

Test Configurations

Our first test problem is a lid-driven cavity simulation on a unitsquare domain—a standard benchmark for CFD codes. Figure 6.21 depicts the computed results for two different Reynolds numbers, $Re = 100$ and $Re = 250$. Note that the problem is harder to solve the higher the Reynolds number gets since the pressure mass matrix becomes a less feasible preconditioner for the chosen Schur complement approach (cf. Section 2.3.6). The outer BiCGStab method requires more iterations, leading in turn to an increase in the total number of global multilevel iterations. We refer to these tests as DRIVENCAVITY100 and DRIVENCACITY250, respectively. The domain is partitioned into eight subdomains, each of which is refined regularly L times, leading to global problem sizes of 1.5, 6.3 and 25.1 million degrees of freedom for $L = 8, 9, 10$, respectively. The corresponding local, scalar subproblems comprise 66 049, 263 169 and 1 050 625 unknowns. The global problem sizes are chosen to fill the available host memory of our test cluster, and we statically schedule two subdomains per node.

Figure 6.21: Stream tracer plots of the velocity field for lid-driven cavity simulations at Reynolds number $Re = 100$ (left) and $Re = 250$ (right).

Our second test is based on the 1996 DFG benchmark configuration *Benchmark computations of laminar flow around a cylinder* [226]. Figure 6.22 shows the domain, the unstructured coarse grid and the computed pressure and velocity. As the Schur complement solver approach implemented in the FEASTFLOW application is designed for non-stationary flows and is not the optimal choice for the computation of steady flows [225], we reduce the Reynolds number Re from 1000 to 100, but otherwise, we solve exactly the same configuration as described in the benchmark. The coarse grid comprises 24 subdomains, so we cannot perform the simulation for refinement level $L = 10$ due to memory limitations. The largest problem size for this CHANNELFLOW test is 18.8 million degrees of freedom.

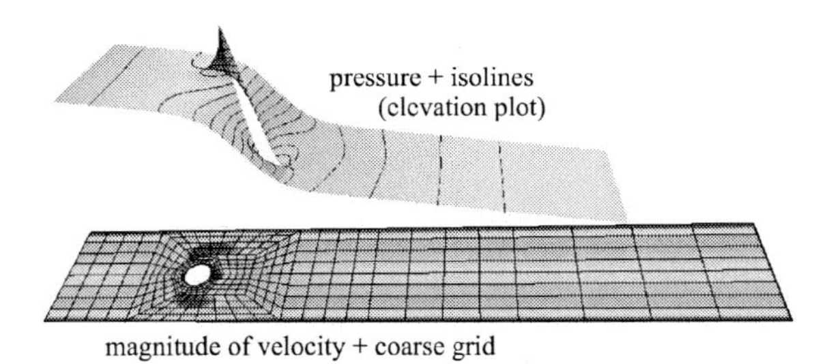

Figure 6.22: Channel flow around a cylinder.

Solver Setup

The general solution process implemented by Buijssen [44] is described in Section 2.3.6 on page 43. The diagonal block matrices $\mathbf{A}_{ii}$ (for the velocities) correspond to scalar operators that stem from the momentum equations, so FEAST's tuned scalar solvers can be applied. Our solution scheme (Figure 6.23) is a two-layer SCARC solver, generalised to treat the vector-valued block system (2.3.9) (see page 45). The outer layer comprises a global multilevel solver configured to perform a V cycle with one pre- and no postsmoothing step (more smoothing only results in longer total runtimes). The global multilevel solver is additively smoothed by local multigrid solvers issued on every subdomain that run either on the CPU or GPU. This inner layer performs a V cycle with four pre- and postsmoothing steps each. Its coarse grid solver is a conjugate gradient method, as a sparse direct solver like UMFPACK has not been implemented yet on the GPU. The same linear solver is used for the Stokes problem, which is linear. For the Stokes and driven cavity simulations, it suffices to equip the local multigrid with the simple JACOBI smoother, we found no performance improvement when switching to stronger smoothers. The CHANNELFLOW benchmark requires a stronger smoother; the solver converges too slowly otherwise due to a substantial increase in calls to the local solvers. We employ the ADITRIDI smoother, as ADITRIGS gives no substantial improvement, and count one application of this alternating direction implicit scheme as two elementary smoothing steps.

To speed up the computations in the Navier-Stokes case, we prolongate the solution computed on refinement level $L = 6$ to the finest grid, and use it as initial guess for $L = 7, \dots, 10$.

fixed point iteration
solving linearised subproblems with
global BiCGStab (reduce initial residual by 1 digit)
Block-Schur complement preconditioner
1) approximately solve for velocities with
global multivariate multilevel (V 1+0), additively smoothed by
for all Ω_i: solve for $\mathbf{u}_1$ with **local scalar multigrid**
for all Ω_i: solve for $\mathbf{u}_2$ with **local scalar multigrid**
2) update RHS: $\mathbf{d}_3 = -\mathbf{d}_3 + \mathbf{B}^\mathsf{T}(\mathbf{c}_1, \mathbf{c}_2)^\mathsf{T}$
3) scale $\mathbf{c}_3 = (\mathbf{M}_\mathbf{p}^\mathsf{L})^{-1}\mathbf{d}_3$

Figure 6.23: Stationary laminar flow solver, the accelerable components are highlighted.

Hardware Platforms

All tests in this section are carried out on the small four node test cluster that is also used for other tests in this chapter, see Section 6.3.1 on page 225 for the exact specifications.

6.5.2. Analysis of the Linear Solver

Our first experiment focuses on the linear Schur complement solver alone, so we consider the (linear) Stokes problem instead of performing a full Navier-Stokes simulation. This test is thus comparable with the studies of the Poisson and linearised elasticity solvers in Section 6.3 and 6.4.

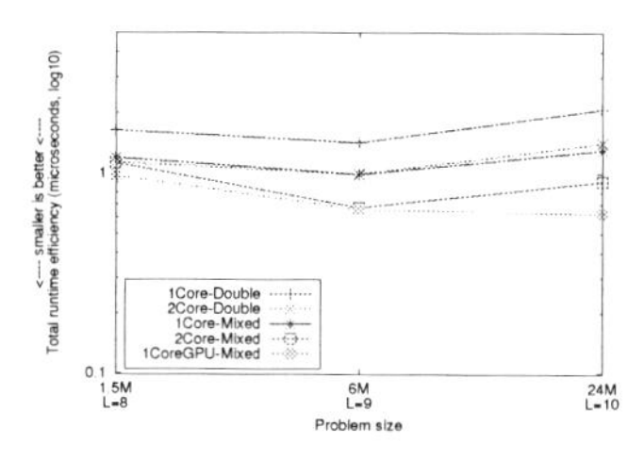

Figure 6.24: Total runtime efficiency of the Stokes solver in five different hardware configurations. The y-axis uses a logarithmic scale.

Figure 6.24 depicts the total runtime efficiency of the five different scheduling schemes that we test throughout this chapter, see Section 6.3.1 for their definition. All five configurations show the exact same number of iterations and consequently, the same convergence rates for each level of refinement. Therefore, we can combine the discussion of the total speedup (which is computed from the total runtime efficiency) and the local speedup (which is measured in seconds) in the following without the deteriorating influence of too large granularity effects.

The CPU variants all lose efficiency when going from refinement level $L = 9$ to $L = 10$, while the GPU-accelerated solver is most efficient for the largest problem size. We have observed the latter behaviour throughout all tests in this thesis, whereas the loss of efficiency by the CPU has not been observed for the linearised elasticity solver (Figure 6.20 on page 249), but to an even greater extent for the Poisson problem (cf. Figure 6.5 on page 229).

We look at the two CPU variants that execute entirely in double precision first: Using two cores instead of one improves efficiency by 43–47 % for the three refinement levels under consideration, in line with the measurements for the linearised elasticity (Figure 6.20) and Poisson (Figure 6.5) solvers. The same holds true when switching the local solvers in these two configurations to single precision. When both subdomains are scheduled to the same core, the mixed precision CPU solver achieves speedups of 1.37, 1.44 and 1.59 for $L = 8, 9, 10$, these factors are consistent with the linearised elasticity and Poisson solvers. The mixed precision dualcore configuration is 1.01, 1.47 and 1.55 times faster than its double precision counterpart. Finally, the best 2CORE-MIXED configuration is faster by factors of 1.45, 2.11 and 2.28 than the baseline 1CORE-DOUBLE variant. The speedups obtained by the mixed precision solvers are remarkable, and they have not been available in FEAST prior to this thesis. It is worth noting that the dualcore configuration in double precision is as fast as the mixed precision solver on a single core, which executes twice the amount of arithmetic work. This behaviour has been observed for the solid mechanics and Poisson test problems already.

	L	T_{total}	T_{local}	T_{fixed}	R_{acc}	S_{local}	S_{total}	$S_{\max}$
1COREGPU-MIXED	8	12.60	1.90	10.70				
	9	33.70	4.20	29.50				
	10	127.70	12.40	115.30				
1CORE-DOUBLE	8	21.20	10.20	11.00	0.48	5.37	1.68	1.93
	9	72.60	43.00	29.60	0.59	10.24	2.15	2.45
	10	423.70	297.50	126.20	0.70	23.99	3.32	3.36
1CORE-MIXED	8	15.50	4.70	10.80	0.30	2.47	1.23	1.44
	9	50.50	21.30	29.20	0.42	5.07	1.5	1.73
	10	265.70	149.10	116.60	0.56	12.02	2.08	2.28
2CORE-DOUBLE	8	14.70	7.30	7.40	0.50	3.84	1.17	1.99
	9	50.70	31.10	19.60	0.61	7.4	1.5	2.59
	10	287.80	208.70	79.10	0.73	16.83	2.25	3.64
2CORE-MIXED	8	14.60	3.20	11.40	0.22	1.68	1.16	1.28
	9	34.40	15.40	19.00	0.45	3.67	1.02	1.81
	10	185.70	106.40	79.30	0.57	8.58	1.45	2.34

Table 6.7: Detailed analysis of the speedup achieved by the 1COREGPU-MIXED configuration over the four CPU variants, Stokes problem: Breakdown of the speedup into acceleration potential, local and global acceleration.

Table 6.7 presents a breakdown of the speedup achieved by the 1COREGPU-MIXED configuration over the four solvers that execute entirely on the CPU. We only discuss the largest problem size: The accelerable fraction of the double precision configurations is approximately 70 %, compared to 56 % of the mixed precision variants. The local speedups of 24 and 12 against the two singlecore variants thus translate to total speedups of 3.3 and 2.1, in line with our analytical performance model and sufficiently close to the theoretical maximum $S_{\max}$, see also Figure 6.8 on page 234. For the two dualcore variants, the difference between the (predicted and theoretical) acceleration and the measured one is much larger; and the performance improvement of the GPU is thus smaller. This is again a direct consequence of the strong scalability benefits in the 'fixed' part when switching from one to two CPU cores. The reduction by approximately 1/3 in the S_{local} numbers is honest, a single GPU in single precision is still 17 and 8.5 times faster than two CPU cores in double and single precision, respectively. However, the T_{fixed} dualcore timings are only 2/3 of the singlecore timings in absolute numbers, which leads to a constant reduction in total runtime that only the CPU solvers can benefit from (cf. Section 6.3.3). Nonetheless, the GPU-accelerated solver is still 2.25 and 1.45 times faster than the 2CORE-DOUBLE and 2CORE-MIXED CPU solvers, respectively.

As we can only use two subdomains per node due to memory limitations, we cannot test the hybrid solver configuration in a meaningful way, see Section 6.3.5.

6.5.3. Accuracy of the Mixed Precision Stationary Laminar Flow Solver

We measure the accuracy of the computed results by means of the kinematic energy $\frac{1}{2}\int_\Omega ||\mathbf{u}||^2 d\mathbf{x}$, a grid independent quantity often used to compare driven cavity tests [232].

		JACOBI		ADITRIDI
	L	DRIVENCAVITY100	DRIVENCACITY250	CHANNELFLOW
1CORE-DOUBLE	8	0.03425	0.03805	0.02468
	9	0.03437	0.03829	0.02467
	10	0.03442	0.03857	
1CORE-MIXED	8	0.03425	0.03807	0.02468
	9	0.03437	0.03823	0.02467
	10	0.03435	0.03858	
1COREGPU-MIXED	8	0.03425	0.03809	0.02468
	9	0.03437	0.03825	0.02467
	10	0.03439	0.03918	
2CORE-DOUBLE	8	0.03425	0.03805	0.02468
	9	0.03437	0.03829	0.02467
	10	0.03442	0.03857	
2CORE-MIXED	8	0.03425	0.03805	0.02468
	9	0.03437	0.03829	0.02467
	10	0.03438	0.03858	

Table 6.8: Accuracy (measured in kinematic energy) of the five solver configurations for the three test problems.

Table 6.8 shows the computed energy values for the two driven cavity and the channel flow benchmark. We observe small differences between the solvers executing in different precisions on refinement level $L = 10$, and for the DRIVENCACITY250 configuration even for all considered problem sizes. The results computed entirely in double precision are always identical. The deviations are small, and we attribute them to floating point noise that builds up due to the variations in the convergence behaviour of the different solver configurations. This is underlined by the fact that we obtain different results, even in double precision, when switching the scalar multigrid solver to the ADITRIDI smoother. For instance, the computed kinematic energies for the DRIVENCACITY250 problem and the 2CORE-DOUBLE solver are 0.03806, 0.03823 and 0.03805 for $L = 8,9,10$ and the ADITRIDI smoother, and 0.03802, 0.03824 and 0.03822 for the 2CORE-MIXED solver with this smoother, different from the results depicted in the table. It is unclear which set of results is the 'best' one. For all practical purposes however, all computed results are 'identical'.

The results endorse our previous findings, that even for very ill-conditioned problems, the restriction of the innermost linear solver to single precision does not have any negative side effect.

6.5.4. Performance Analysis of the Stationary Laminar Flow Solver

The solvers do not converge in exactly the same way, occasionally the linear solves are slightly cheaper for one configuration, and on a different refinement level and for a different test problem, the local multigrid solvers in another configuration converge slightly better. This is actually the expected behaviour, due to the huge amount of floating point operations and the ill-conditioning of the problem. In the course of the nonlinear loop, these differences build up. Table 6.9 shows the number of nonlinear iterations, the number of Schur complement solves (which are in turn executed twice per outer BiCGStab iteration, i. e., twice per linear iteration), and the number of calls to the local solvers in the course of the global multilevel scheme.

		DRIVENCAVITY100			DRIVENCACITY250			CHANNELFLOW		
	L	#NL	#SC	#LOC	#NL	#SC	#LOC	#NL	#SC	#LOC
1CORE-DOUBLE	8	5	89	12384	4	230	35952	2	20	13248
	9	5	76	16896	4	226	51776	2	17	13920
	10	5	91	25760	4	204	66320			
1CORE-MIXED	8	5	89	12384	4	230	35712	2	20	13248
	9	5	76	16896	4	228	52736	2	17	13920
	10	5	89	25440	4	204	65520			
1COREGPU-MIXED	8	5	89	12384	4	216	34416	2	20	13248
	9	5	76	16896	4	226	53440	2	17	13920
	10	5	91	26000	4	204	64400			
2CORE-DOUBLE	8	5	89	12384	4	230	35904	2	20	13248
	9	5	76	16896	4	226	51776	2	17	13920
	10	5	91	25760	4	204	66320			
2CORE-MIXED	8	5	89	12384	4	218	33792	2	20	13248
	9	5	76	16960	4	225	51904	2	17	13920
	10	5	92	25600	4	204	65440			

Table 6.9: Number of iterations (nonlinear (NL), Schur complement (SC) and number of local (LOC) multigrid calls) until convergence of the five solver configurations for the three test problems.

The double precision solvers all perform the same amount of work, independent of a partitioning for a single or both cores per node. The single precision solvers sometimes execute more, and sometimes less work in the inner solver, and occasionally one or two Schur complement steps more or less. In the following, we ignore these small differences and base our discussion on the total time to solution, measured in seconds.

Runtime of the Driven Cavity Simulations

Figure 6.25 depicts the total time to solution required by the five different solvers. For brevity, we only discuss the largest problem size in detail. Looking at the DRIVENCAVITY100 test case first, we observe the following: The dualcore variant in double precision is 70 % faster than its singlecore counterpart, this value is surprisingly high. Switching the local solver to single precision improves performance by 35 % and 23 % compared to the single- and dualcore solvers, respectively. Finally, the 1COREGPU-MIXED accelerated solver is twice as fast as the 1CORE-DOUBLE solver, but the speedup reduces quickly for the remaining CPU configurations, and the mixed precision dualcore solver is even faster than the singlecore GPU-accelerated solver. The reason for this disappointing result, an unfair disadvantage due to only one GPU being available in the test machine, is explained in the discussion of the channel flow benchmark.

For the DRIVENCACITY250 problem, the dualcore double precision solver is 57 % faster than using only one core. The mixed precision solvers increase performance by 56 % and 51 % over

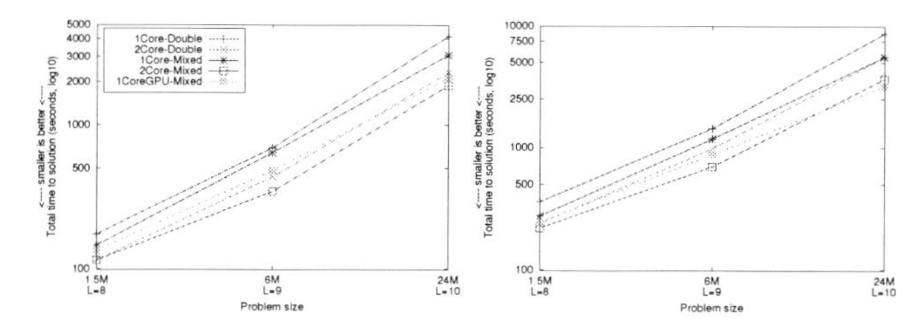

Figure 6.25: Total time to solution of the two driven cavity simulations DRIVENCAVITY100 (left) and DRIVENCACITY250 (right) with different solvers. Both plots share the same legend, and the y-axes use different logarithmic scales.

their double precision counterparts. The speedup factors we measure for the GPU-based solver over the four CPU configurations are 2.7 (1CORE-DOUBLE), 1.7 (1CORE-MIXED and 2CORE-DOUBLE) and 1.15 (2CORE-MIXED).

Runtime of the Channel Flow Simulation

Figure 6.26 shows the measured total time to solution of the CHANNELFLOW benchmark for the different solver configurations. For the largest problem size, the 2CORE-DOUBLE configuration is 70 % faster than the 1CORE-DOUBLE one. Switching these two solvers to single precision for the local work results in a speedup of 35 %. The GPU-accelerated solver is 76 %, 31 % and 5 % faster than the 1CORE-DOUBLE, 1CORE-MIXED and 2CORE-DOUBLE configurations, while the 2CORE-MIXED solver outperforms it by 27 %.

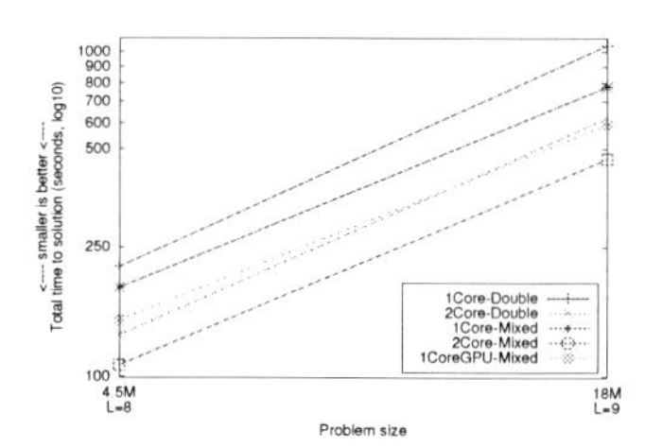

Figure 6.26: Total time to solution of the CHANNELFLOW simulations with different solvers. The y-axis uses a logarithmic scale.

Speedup Analysis

In the course of the nonlinear defect correction loop (which is included in the total time to solution), the linearised problems have to be assembled in each iteration. To facilitate a detailed analysis, a number of additional derived quantities are required compared to the linear case:

- T_{total}: time to solution of the nonlinear loop
- T_{ass}: time spent in matrix assembly
- T_{lin}: time spent in linear solver

- T_{local}: time spent in local scalar subproblems
- $T_{\text{fixed,lin}} = T_{\text{lin}} - T_{\text{local}}$: time spent in unaccelerable parts of the linear solver
- $T_{\text{fixed,total}} = T_{\text{total}} - T_{\text{fixed,local}}$: time spent in unaccelerable parts of the entire solver
- $R_{\text{ass}} = T_{\text{ass}}/T_{\text{total}}$: fraction of the total time spent in matrix assembly
- $R_{\text{acc,lin}} = T_{\text{local}}/T_{\text{lin}}$: accelerable fraction of the linear solver
- $R_{\text{acc,total}} = T_{\text{local}}/T_{\text{total}}$: accelerable fraction of the entire solver

Table 6.10 summarises these quantities for the three stationary laminar flow benchmark problems, for the largest admissible level of refinement. Small deviations of timings that should normally be identical (e. g., $T_{\text{fixed,lin}}$, singlecore) are due to small variations in the convergence behaviour, see Table 6.9. We first look at the two driven cavity test cases, because they both employ the simple JACOBI smoother and are solved for refinement level $L = 10$. Looking at the ratio between linear solver and matrix assembly, the linear solver dominates the runtime in the CPU case. On the GPU, this is only true for the DRIVENCACITY250 test problem. These fractions do not necessarily add up to 100 because we do not explicitly list the few percent needed by the nonlinear defect correction loop and the initialisation phase that, e. g., reads in the solution of a coarser mesh from disk and prolongates it as the initial guess. The acceleration potential of the linear solver alone is in line with the potential we measure for the linear Stokes problem and the same solver configuration. Since the assembly contributes significantly to the overall runtime, the acceleration potential of the entire nonlinear solver is substantially smaller. The values are higher for the higher Reynolds number (DRIVENCACITY250), due to the linear problem being harder to solve (see above). For the CHANNELFLOW benchmark, in essence the same arguments and conclusions hold, taking into account the different smoother and the fact that this simulation is carried out on refinement level $L = 9$.

	T_{total}	T_{ass}	T_{lin}	T_{local}	$T_{\text{fixed,lin}}$	$T_{\text{fixed,total}}$	R_{ass}	R_{lin}	$R_{\text{acc,lin}}$	$R_{\text{acc,total}}$
DRIVENCAVITY100										
1CORE-DOUBLE	4165	1224	2888	1992	895	2173	0.29	0.69	0.69	0.48
1CORE-MIXED	3078	1249	1768	1036	732	2042	0.41	0.57	0.59	0.34
2CORE-DOUBLE	2321	594	1697	1185	513	1136	0.26	0.73	0.70	0.51
2CORE-MIXED	1883	595	1258	745	513	1138	0.32	0.67	0.59	0.40
1COREGPU-MIXED	2128	1193	880	90	791	2038	0.56	0.41		
DRIVENCACITY250										
1CORE-DOUBLE	8660	1018	7597	5357	2240	3303	0.12	0.88	0.71	0.62
1CORE-MIXED	5563	1045	4474	2658	1816	2906	0.19	0.80	0.59	0.48
2CORE-DOUBLE	5529	499	5004	3764	1240	1765	0.09	0.91	0.75	0.68
2CORE-MIXED	3652	498	3129	1890	1239	1762	0.14	0.86	0.60	0.52
1COREGPU-MIXED	3190	992	2152	218	1934	2972	0.31	0.67		
CHANNELFLOW										
1CORE-DOUBLE	1047	401	621	502	120	546	0.38	0.59	0.81	0.48
1CORE-MIXED	778	394	360	266	94	513	0.51	0.46	0.74	0.34
2CORE-DOUBLE	624	215	394	329	65	294	0.35	0.63	0.83	0.53
2CORE-MIXED	467	215	238	173	65	294	0.46	0.51	0.73	0.37
1COREGPU-MIXED	594	401	168	40	127	554	0.68	0.28		

Table 6.10: Breakdown of the solver components of the five solver configurations for the three test problems (largest admissible problem size).

A number of interesting speedup factors can be computed from this data. We first look at strong scalability effects that occur when moving from a singlecore to a dualcore CPU solver. In previous experiments for the Poisson problem and test cases from linearised elasticity, we have measured approximately 40 % when employing the JACOBI smoother, and 50 % for ADITRIDI. These values are recovered when computing the quotient between the corresponding T_{local} timings, independent of the precision (taking for instance the DRIVENCACITY250 problem in double precision, we have $5357/3764 = 0.42$). The corresponding factors for the entire linear solver are slightly larger, because the speedup of the $T_{\text{fixed,lin}}$ timings is higher. The operations included in this timer are less data-intensive than the local solves and thus scale better to two cores, because computations in one MPI process can be overlapped with communication in the other. The matrix assembly scales perfectly to two cores, and we observe the ideal factor of two for refinement level $L = 10$. Table 6.11 presents these derived speedup factors for the two dualcore CPU solvers, computed against their singlecore counterparts.

	S_{ass}	S_{lin}	S_{local}	$S_{\text{fixed, lin}}$	$S_{\text{fixed,total}}$
DRIVENCAVITY100					
2CORE-DOUBLE	2.06	1.70	1.68	1.75	1.91
2CORE-MIXED	2.10	1.40	1.39	1.43	1.8
DRIVENCACITY250					
2CORE-DOUBLE	2.04	1.52	1.42	1.81	1.87
2CORE-MIXED	2.10	1.43	1.41	1.47	1.65
CHANNELFLOW					
2CORE-DOUBLE	1.86	1.58	1.52	1.84	1.85
2CORE-MIXED	1.83	1.51	1.54	1.44	1.74

Table 6.11: Speedups of the dualcore CPU solver configurations over their singlecore counterparts for the three test problems (largest admissible problem size).

We finally compute local and global factors of the GPU-accelerated solver over the four CPU configurations, see Table 6.12 for the exact numbers. The local and singlecore linear solver speedups are in line with previous measurements, cf. Table 6.2 on page 233 and Table 6.6 on page 250. For instance, the local acceleration of the multigrid with JACOBI smoother is more than a factor of 20 compared to double precision, and approximately 12 for the more honest single vs. single precision comparison. In absolute numbers, the reduction is even more impressive: The local solver time reduces from 5357 seconds (2658 in single precision) to slightly more than 200 seconds, i. e., from the order of hours to the order of minutes. These speedups translate to more than a factor of three for the entire linear solver in double precision, and still a factor of two for the mixed precision linear solver using only a single CPU core per node. Due to the large fraction of the total time to solution that is spent in the matrix assembly (see Table 6.10 for the exact numbers), these local speedups diminish rapidly. The obvious conclusion that can be drawn from these observations is that future work needs to address the question how to speed up the matrix assembly using GPUs as well. This can only be done on GPUs that support double precision, and we return to this topic in Section 6.6.

The speedups obtained over the dualcore configurations are flawed in the sense that they suffer disproportionately by the excellent strong scalability of the unaccelerable parts of the linear solver and in particular of the matrix assembly. From the $T_{\text{fixed,total}}$ measurements in Table 6.10 and the speedups in Table 6.11, we can identify a fixed penalty of approximately 500 seconds for the GPU-accelerated solver, exemplarily for the DRIVENCACITY250 test case and the matrix assembly alone. Another approximately 400 seconds are added from the $T_{\text{fixed,lin}}$ timings. Together, these 900 seconds are almost half of the total time to solution of the accelerated solvers, clearly demonstrating that the dualcore CPU solvers have an unfair advantage. On page 233 we explain why these flaws are unavoidable in the current implementation of FEAST, given the limitation of

	S_{loc}	S_{lin}	S_{total}	S_{loc}	S_{lin}	S_{total}	S_{loc}	S_{lin}	S_{total}
	DRIVENCAVITY100			DRIVENCACITY250			CHANNELFLOW		
1CORE-DOUBLE	22.23	3.28	1.96	24.63	3.53	2.71	12.51	3.71	1.76
1CORE-MIXED	11.56	2.01	1.45	12.22	2.08	1.74	6.63	2.15	1.31
2CORE-DOUBLE	13.22	1.93	1.09	17.31	2.33	1.73	8.21	2.35	1.05
2CORE-MIXED	8.32	1.43	0.89	8.69	1.45	1.15	4.31	1.42	0.79

Table 6.12: Speedups of the GPU-accelerated solver over the four CPU configurations for the three test problems ($L = 10$ for the driven cavity tests, $L = 9$ for the channel flow benchmark).

our test cluster of only providing one GPU per node. Based on these numbers, it is safe to assume that the speedup factors over the singlecore solvers will be recovered on a cluster with dual-GPU nodes, or more generally, with a ratio of 1:1 CPU cores to GPUs.

6.6. Summary, Critical Discussion, Conclusions and Future Work

We have presented a *minimally invasive co-processor integration* into FEAST. Our approach is prototypical for the general strategy to *accelerate* the most time-consuming portions of a given large-scale, established code of more than 100 000 lines of Fortran code instead of *re-implementing* it over and over again for each new emerging and promising architecture. We have presented the approach and evaluated it for the Poisson problem and several benchmark problems from linearised elasticity and (stationary) laminar flow. In the following, we summarise our approach in Section 6.6.1. Section 6.6.2 discusses the accuracy of mixed precision two-layer SCARC solvers and Section 6.6.3 summarises our weak scalability studies. In Section 6.6.4 we review our approach based on a summary of our performance studies, and discuss various avenues (backed up by preliminary experiments) for future work, which require a fundamental restructuring of FEAST and might be useful for the design of FEAST II, the successor package currently in its early design specifications in Stefan Turek's group at TU Dortmund. We conclude in Section 6.6.5 with a critical discussion of future scalability of numerical software in general, in particular in view of the trend towards many-core and heterogeneity.

6.6.1. Summary of the Minimally Invasive Integration Approach

The core idea of our minimally invasive acceleration of two-layer SCARC solvers is to execute the local layer on co-processors and leave the outer multilevel (MPI-) layer completely unmodified. If the local work is performed in single precision, then the entire two-layer SCARC solver can be interpreted as a mixed precision solver, with a multilevel outer solver that is additively preconditioned (smoothed) in single precision. This simple idea has many benefits, we list them in no particular order because these aspects have different priorities for the user, the application programmer and the kernel developer:

- Applications built on top of FEAST benefit from the acceleration without changing a single line of code. This is a direct consequence of not modifying the outer layer of the solver, because we do not introduce interface changes that affect user application code.
- Hardware acceleration is transparent to the applications and can be enabled by a simple switch in a configuration file.
- Special 'tricks' associated with a specific co-processor can be exploited, but the design in itself is invariant to the type of hardware accelerator.
- Accelerator-specific code reduces to a set of kernels needed to construct multigrid and Krylov subspace solvers, specialised to treat one subdomain exhibiting the generalised tensor product property. The unified `libcoproc` library handles the kernel calls and the control flow. Furthermore, the memory manager needs to be extended by conversion and data transfer operations between host and each device. The current prototypical implementation provides backend implementation for singlecore CPUs (double and single precision), in CUDA (double and single) and in OpenGL (emulated double-single and single precision, only a subset of the preconditioners).
- As a direct consequence, the addition of new hardware backends, e. g., for the Cell processor by treating the SPEs as on-chip co-processors, or in OpenCL, is straightforward once the kernels for a multigrid solver exist.
- Changes to FEAST reduce to a minimum, because the heterogeneities within the node and hardware-specific details are fully encapsulated in the `libcoproc` library.

- The CPU backend of the library enables mixed precision solvers also on the CPU, which would require a vast amount of nontrivial modifications of FEAST otherwise.

- Admittedly, the porting of a scalar multigrid solver to new hardware architectures is challenging, each co-processor has its own associated programming model, fine-grained parallelisation requirements, etc. In Chapter 5 we have presented the GPU case. The important advantage of our minimally invasive approach (and of the modular, hierarchical solver structure in SCARC) is that as soon as such a scalar multigrid solver is available, it takes almost no effort to transfer the acceleration to *all* applications built on top of FEAST.

We have evaluated our approach in terms of accuracy, performance, speedup and scalability on various GPU-accelerated clusters, using the Poisson problem as an important building block, and the two FEAST applications FEASTSOLID and FEASTFLOW, using only a subset of the supported functionality of these applications (linearised elasticity and stationary laminar flow). We emphasise that from the point of view of this thesis, we only *use* these two applications, the author of this thesis has neither implemented them nor contributed substantially to them.

Furthermore, special attention has been placed on mixed precision solvers on the CPU and the GPU, because we know from our benchmark computations on a single subdomain in Chapter 5 that the mixed precision technique is always beneficial in terms of performance.

6.6.2. Accuracy

We have evaluated the accuracy of mixed precision two-layer SCARC solvers on the CPU and the GPU. For all our tests, we achieve the same accuracy as if executing the solver entirely in double precision, without sacrificing numerical performance. In other words, the mixed precision schemes are always more efficient than their double precision counterparts.

In detail, we have performed the following tests: For the Poisson problem, we have used right hand side data that leads to wrong results in single precision for a single subdomain already; these tests are summarised in Section 6.3.2. The majority of our accuracy tests are carried out for problems from linearised elasticity, cf. Section 6.4.2: We have prescribed analytical displacements for a large-scale simulation on 64 compute nodes, for an overall problem size of half a billion unknowns, and compared the computed results with an analytically known reference solution. The cantilever beam configuration is a standard benchmark problem in linearised elasticity, its high degree of domain anisotropy in conjunction with the large ratio between free Neumann boundary and fixed Dirichlet boundary and the high level of refinement results in a very ill-conditioned global system. Finally, we have examined local anisotropies of the underlying differential operator by moving towards the incompressible limit. For the nonlinear Navier-Stokes problem, we have measured the kinematic energy of the computed solutions for two different lid driven cavity and a 'flow around a channel' configuration, see Section 6.5.3.

6.6.3. Weak Scalability

In Section 6.3.4 and Section 6.4.3 we have demonstrated excellent weak scalability of the Poisson and elasticity solvers, respectively. For the Poisson tests, we have used up to 160 nodes of an older cluster installed at Los Alamos National Laboratory; a year later when the tests for the linearised elasticity solver have been performed, 64 of these nodes were still available. The largest problem sizes we have solved were more than 1.25 billion unknowns for Poisson, and over half a billion for elasticity, corresponding to using as much memory per node as possible.

6.6.4. Performance of Mixed Precision and Accelerated Two-Layer ScaRC Solvers

The tests corresponding to this summary are presented in Section 6.3.2, 6.3.3, 6.4.4 and Section 6.5.4.

Mixed Precision on the CPU

Due to bandwidth considerations, the theoretical upper bound of efficiency improvements is a factor of two. For the Poisson problem, we observe factors between 1.7 and 1.9, these factors are slightly lower for the linearised elasticity and the laminar flow solver, because their acceleration potential is lower: A smaller fraction of the total time to solution is spent in the scalar local solves, we discuss this aspect below.

Due to the memory wall problem, the CPU solvers also do not scale well in the strong sense when switching from a singlecore to a dualcore configuration. Multicore CPUs share the memory bus between the cores, and we only observe a 40–50 % increase in efficiency due to bus contention, instead of the factor of two if the solvers were compute bound rather than memory bound.

GPU-Accelerated Solvers

The GPU-accelerated solver is more than eight times faster than an equivalent CPU solver executing entirely in double precision, and still 4.7 times faster than its mixed precision counterpart on the CPU, for the Poisson problem. The speedup factors are smaller for the two other applications, but still reach 3 and 2 for the elasticity solver, and between 1.3 and 2.7 for the laminar flow solver, depending on the configuration.

The comparison against dualcore CPU variants is less favourable, and 'unfair' in the sense that in our experimental setup, the CPU solvers benefit from strong scalability effects in the unaccelerated portions of the code. This gives an unfair disadvantage to the GPU solvers: as we only have one GPU per compute node in the clusters we used to evaluate our approach, we are restricted to running the GPU experiments with only one MPI process per node, i. e., in a singlecore configuration. In the current implementation of FEAST, it is impossible to benefit from the strong scalability of the parts of the code that are not executed on the GPU. An ideal GPU cluster for our applications includes one GPU per CPU core to maximise performance improvements.

Analysis

The question arises why these speedups are substantially lower than those we have measured on a single subdomain, even when taking the bus transfers of right hand side data and iteration vectors into account. Since we only accelerate portions of the entire solution scheme, the achievable speedups are naturally limited, and the fraction of the time spent in these parts induces an upper bound for the overall acceleration: Even when assuming infinite local acceleration in the accelerable fraction $R_{\text{acc}} \in [0,1]$ of the solver, the total speedup is at most $1/(1-R_{\text{acc}})$, For instance, if 3/4 of a given application can be accelerated, the speedup is bound by four.

The Poisson solver exhibits an acceleration potential close to 90 %, which is already very good. For the elasticity solver, still approximately 2/3 of the entire solution process can be accelerated, this value is also comparatively high when taking into account that no changes to the application code are necessary. For the nonlinear laminar flow problems, the fraction is 50–60 %, because the repeated assembly of linearised problems is not accelerated.

The main goal for future work is thus the design of solver schemes that exhibit an acceleration potential well over 90 %, on the application level and not only for model problems. In the following, we discuss three different ideas how this goal can be achieved: We could implement the entire (linear) two-layer SCARC solver on accelerator hardware, the two main current issues in this approach are the PCIe bottleneck and the fact that (at least) double precision is required for the outer layer. The second idea is to experiment with alternative domain decomposition and substructuring techniques that possibly concentrate more arithmetic work locally, it is however unclear at this stage how such schemes would scale (computational aspect) and if they converge grid-independently (numerical aspect). Finally, we could offload more components of the entire solution scheme than only parts of the linear solves to accelerator hardware, for instance (parts of) the assembly of system matrices. This also needs double precision, which is not an issue anymore on newer hardware. The first two ideas require fundamental modifications to the existing FEAST implementation, the last one is closest in spirit to our minimally invasive approach.

Future Work (I): Entire Linear Two-Layer ScaRC Solver

The outer layer of the SCARC solvers mandatorily requires native double precision support. On GPUs, this is a relatively recent feature, see Section 3.2.5. We perform a small experiment to assess if executing outer solvers on the device is feasible, despite the PCIe bottleneck (see Section 6.2.1). To this end, we repeat the CUDA microbenchmarks from Section 5.3.3 on page 133 (high-end workstation, GeForce GTX 280), but this time include the transfer of $4M$ values to and from the device into our timings. This amount corresponds to the data that is exchanged between neighbouring subdomains of size $M \times M$ via MPI. The experiment is optimistic in the sense that it does not include the efficient extraction of the boundary data from the one-dimensional data layout, which we expect would slightly degrade performance further: A possible implementation that we have used successfully in the seismic wave propagation application [136] is to use device-to-device memory copies with stride one or stride M prior to transferring the data via PCIe.

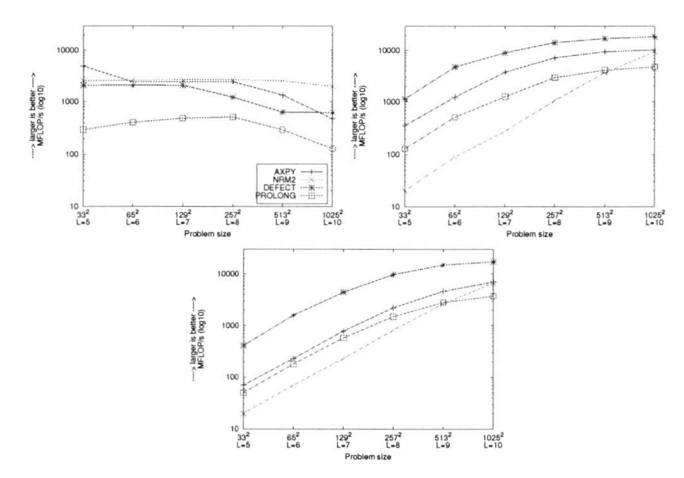

Figure 6.27: Microbenchmark results in double precision on the high-end test platform, on the CPU (top left), on the GPU without PCIe transfers (top right) and on the GPU with PCIe transfers of boundary data (bottom). All plots share the same legend, and the y-axis uses a logarithmic scale.

Figure 6.27 depicts the MFLOP/s rates we measure for the CPU and for the GPU, with and without the additional transfers of boundary data via PCIe (the two plots in the top row are repeated from Figure 5.9 for easier reference). Figure 6.28 presents the same measurements in terms of speedup over the singlecore CPU variant.

The penalty induced by the bus transfers is constant for all four operations under consideration because the amount of data is constant, so the impact in terms of slowdown varies. The `defect` kernel is the most computationally intense one and consequently, the PCIe penalty is smallest, the speedup over the CPU reduces from a factor of 30 to a factor of 27.5 for the largest problem size under consideration. For the other kernels, the break-even point from which on the GPU achieves a speedup over the CPU increases by at least one level of refinement when introducing PCIe transfers corresponding to the boundary data. For the largest problem size, we measure 50 %, 41 %, 8 % and 26 % less speedup for the `axpy`, `nrm2`, `defect` and `prolongate` kernels, respectively. In absolute numbers, the performance gains by the GPU are still large enough to justify the execution of the outer solver(s) on the device, at least for refinement levels $L = 9$ and $L = 10$. For medium and small problem sizes, the bottleneck introduced by the comparatively narrow bus between host and device is too big. Since the outer solver in the SCARC scheme always encompasses multiple levels, a thresholding technique similar to the one we introduced in Section 6.2.3 will be necessary, resulting in a hybrid CPU-accelerator solver.

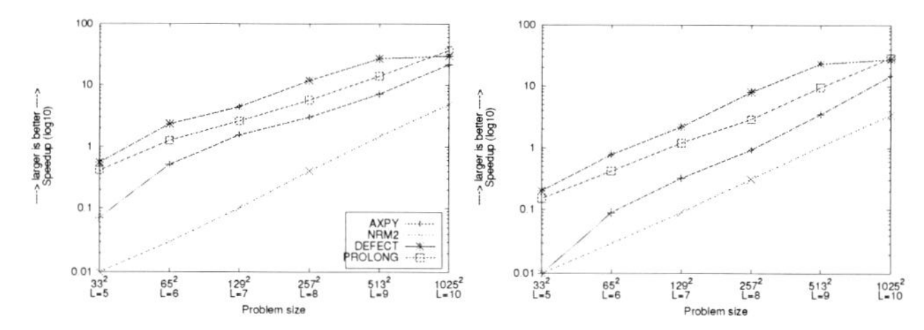

Figure 6.28: Speedup of the GPU over the CPU without (left) and with PCIe transfers of boundary data (right). Both plots share the same legend, and the *y*-axis uses a logarithmic scale.

One additional idea, which is however not possible in FEAST without a major restructuring of its communication routines, is to overlap communication and computation as much as possible. In the seismic wave propagation application that the author of this thesis contributed to [136], we have been able to completely hide the additional latency of the PCIe bus by mesh colouring and the use of non-blocking MPI, and this idea could be employed in FEAST as well. Along these lines, we could also software-pipeline the different scalar subproblems by exploiting FEAST's approach to bring the treatment of nonscalar systems down to their scalar blocks, e. g., transfer the data associated with the velocities in one direction while computing on the data associated with the velocities in another direction, see Section 6.2.3.

Future Work (II): Alternative Domain Decomposition Techniques

In the domain decomposition approach realised in SCARC (cf. Section 2.3.2), it does not make sense to solve the local problems exactly and thus to generate a more favourable ratio between local and global work: Due to the elliptic nature of many problems we are concerned with, the additional local work often does not lead to an improvement of the convergence behaviour of the global scheme.

Alternative techniques that have been discussed during the initial design phase of SCARC [130] are the so-called substructuring techniques of BPS and BPX type [35–37] that can be summarised as non-overlapping domain decomposition with global coarse grid information. Since these technique have, at least in theory, a very favourable ratio of local to global work, revisiting them in future work might become necessary.

'Optimal' convergence rates are not the ultimate goal in practice. All schemes should scale well in the weak and strong sense, and converge independently of the number of subdomains and reasonably independent of the local mesh width, but a method that has slightly worse convergence rates (that are independent of the number of subdomains) might very well be much faster in practice.

Future Work (III): Assembly

For the nonlinear flow solver and for other (nonlinear) problems where the (repeated) assembly of the system matrices and right hand sides requires a substantial fraction of the total time to solution, a feasible idea is to execute not only (parts of) the linear solver on the device, but also (parts of) the system assembly. This task requires co-processors that natively support double precision, such as the latest generation of graphics processors. A lot of this work is trivially parallel on the level of elements, but nonetheless, this task is challenging to implement efficiently. Filipovič et al. [73] and in particular Cecka et al. [47] have started to work on this problem, as well as Komatitsch et al. [136]. These research efforts can be used as a starting point to integrate efficient assembly routines on GPUs into FEAST in future work.

6.6.5. Critical Discussion in Terms of Future Scalability

Detailed measurements reveal that the local speedup, i. e., the speedup measured inside the accelerable parts of the solution process, are very high for sufficiently large local problem sizes. However, even a speedup by a factor of 25 in 90 % of the code only translates to a total speedup by a factor of eight. We have developed an analytical performance model that relates the accelerable fraction of the solver and the measured local acceleration, to predict the total speedup. Our measured values are in close range to the predicted ones. Figure 6.8 on page 234 illustrates this relation. An important consequence is that as soon as an imbalance between accelerable and fixed components of a given code exists, it has a substantial impact. Even if the current limitation might not be too large (only 10 % of the code are 'fixed'), it means that future increases of the local speedup will at some point yield only diminishing returns. As GPUs are increasing performance faster than CPUs and CPUs in turn faster than the interconnect, this is a practically relevant issue. It is thus tantamount to increase the accelerable fraction of the code with techniques such as the ones outlined in Section 6.6.4 as much as possible. Nonetheless, in the long run a well-balanced interplay of the available hardware resources and software components is of utmost importance in order to maintain future scalability and speedup with newer hardware generations.

Heterogenisation and Hybridisation of Resources

Hardware resources within a compute cluster are becoming increasingly heterogeneous. A GPU-enhanced cluster is a current example, and can often include imbalances between the number of CPU cores and the number of GPUs per node. The ultimate goal is thus to balance resource usage as good as possible. Some tasks can only be performed efficiently on CPUs due to lack of fine-grained parallelism or highly irregular memory access and communication patterns, some tasks are better suited for accelerators. The CPU is always required, as accelerators are not expected to include fast interconnect hardware (e. g., Infiniband) in the near future. There may be different types of accelerators in each node. Furthermore, CPUs are becoming heterogeneous themselves (non-uniform memory access, NUMA), and cores within the same chip may not be able to access data that is located close to other cores as efficiently as their own data. This is a similar situation as we face now, as one GPU cannot efficiently communicate with another one even in the same cluster node, the data has to travel twice through the PCIe bottleneck, and the CPU and main memory are involved in these transfers. In summary, we are facing highly asymmetric performance, memory and communication characteristics in a heterogeneous environment.

The current approach in FEAST is to assign one MPI process per CPU core (potentially leaving other CPU cores in the same node idle due to memory bus contention), that eventually orchestrates only the computations on an accelerator and performs MPI communication. We have demonstrated that this approach introduces an unfair disadvantage on imbalanced heterogeneous nodes, at least when only parts of the code are accelerated. One idea to alleviate this problem is to introduce a hybrid parallelisation, i. e., to assign one MPI process per CPU or even per cluster node. Within each node, parallelisation among the different resources can then be realised via manual threading (pThreads on Linux systems, etc.). This approach has the potential to deliver more flexibility within the node, but the performance benefits of such a hybrid three-level parallelism (see below) are unclear at this point.

Furthermore, load balancing has to adaptively consider the heterogeneities, because (at least in production environments) the resources a large job is scheduled to are not necessarily known a priori.

Vision

Current and future computer systems employ parallelism on different scales, fine-grained parallelism within each processor (up to 30,000 concurrently executing threads in todays high-end graphics processors), medium-grained parallelism within each cluster node (cores in a CPU, CPU sockets in the node) and coarse-grained parallelism between the nodes in the cluster. The ultimate goal of any software package that executes on such a cluster is an *even scalability* on all levels:

- strong scaling inside fine-grained parallel devices that increase the degree of parallelism from one hardware generation to the next
- strong and weak scaling between resources with asymmetric improvements of performance characteristics
- load balancing between the heterogeneous resources
- 'classical' weak scaling when all resources in a cluster are used
- 'classical' strong scaling when more and more resources in a cluster are used

These aspects embrace, in the author's opinion, the most important challenges in the near and medium future, and a lot of research efforts are required to reach this goal.

A

Appendix

1.1. List of Symbols

$a \dots z$	(continuous) scalar functions
$\mathrm{a}_h \dots \mathrm{z}_h$	(discrete) scalar functions
$\boldsymbol{a} \dots \boldsymbol{z}$	(continuous) vectors
$\mathbf{A} \dots \mathbf{Z}$	(discrete) matrices
$\mathbf{a} \dots \mathbf{z}$	(discrete) vectors
$\alpha \dots \omega$	scalars
$\mathbf{A}^\mathsf{T}$	matrix transpose
$\mathbf{A}^{-1}$	matrix inverse
$\hat{\mathbf{x}}$	exact solution
$a_{i,j}$	matrix element
$\mathbf{a}_i$	vector element
$\mathbf{x}^{(k)}$	vector in the k-th iteration
$\mathbf{a} \cdot \mathbf{b} := (\mathbf{a}, \mathbf{b}) := \mathbf{a}^\mathsf{T}\mathbf{b}$	inner (dot) product
$\partial_i u := \frac{\partial u}{\partial x_i}$	partial derivative with respect to the i-th variable
$\mathrm{grad}\, x := \nabla x := (\partial_1 x, \dots, \partial_d x)^\mathsf{T}$	gradient of a scalar function
$\mathrm{div}\, \boldsymbol{x} := \nabla \cdot \boldsymbol{x} := \partial_1 x_1 + \ldots + \partial_d x_d$	divergence of a vector function
$\Delta u := \nabla \cdot (\nabla u) = \partial_1^2 u + \cdots + \partial_d^2 u$	Laplacian of a scalar function
$\partial_r u := \frac{\partial u}{\partial \boldsymbol{r}} := \boldsymbol{r} \cdot \nabla u$	directional derivative
$\mathbb{R}$	set of real numbers
$\lVert \cdot \rVert$	norm
$\mathcal{O}(\cdot)$	Landau symbol, asymptotic bound
Ω	domain, bounded open set
Ω_i	subdomain
Γ	domain boundary
$\boldsymbol{n}$	outer normal vector of Γ
$\mathbf{x}^{[i]}$	restriction to subdomain
$\mathbf{x}^{(L)}$	vector on refinement level L

Bibliography

[1] Mike Altieri. *Robuste und effiziente Mehrgitter-Verfahren auf verallgemeinerten Tensorprodukt-Gittern*. Diploma thesis, Universität Dortmund, Fachbereich Mathematik, May 2001.

[2] Mike Altieri, Christian Becker, and Stefan Turek. On the realistic performance of linear algebra components in iterative solvers. In Hans-Joachim Bungartz, Franz Durst, and Christoph Zenger, editors, *High Performance Scientific and Engineering Computing*, volume 8 of *Lecture Notes in Computational Science and Engineering*, pages 3–12. Springer, January 1999.

[3] Gene M. Amdahl. Validity of the single processor approach to achieving large scale computing capabilities. In *Proceedings of the 1967 Spring Joint Computer Conference (AFIPS '67)*, pages 483–485, April 1967. DOI: 10.1145/1465482.1465560.

[4] Joshua A. Anderson, Chris D. Lorenz, and Alex Travesset. General purpose molecular dynamics simulations fully implemented on graphics processing units. *Journal of Computational Physics*, 227(10):5342–5359, May 2008. DOI: 10.1016/j.jcp.2008.01.047.

[5] Jeff Andrews and Nick Baker. Xbox 360 system architecture. *IEEE Micro*, 26(2):25–37, March 2006. DOI: 10.1109/MM.2006.45.

[6] Thomas Apel, Tobias Knopp, and Gert Lube. Stabilized finite element methods with anisotropic mesh refinement for the Oseen problem. *Applied Numerical Mathematics*, 58 (12):1830–1843, November 2007. DOI: 10.1016/j.apnum.2007.11.016.

[7] Mario Arioli, Daniel Loghin, and Andrew J. Warthen. Stopping criteria for iterations in finite element methods. *Numerische Mathematik*, 99(3):381–410, January 2005. DOI: 10.1007/s00211-004-0568-z.

[8] Krste Asanovic, Ras Bodik, Bryan Christopher Catanzaro, Joseph James Gebis, Parry Husbands, Kurt Keutzer, David A. Patterson, William Lester Plishker, John Shalf, Samuel Webb Williams, and Katherine A. Yelick. The landscape of parallel computing research: A view from Berkeley. Technical Report UCB/EECS-2006-183, EECS Department, University of California, Berkeley, December 2006.

[9] Owe Axelsson. On iterative solvers in structural mechanics; separate displacement orderings and mixed variable methods. *Mathematics and Computers in Simulations*, 50(1–4): 11–30, November 1999. DOI: 10.1016/S0378-4754(99)00058-0.

[10] Owe Axelsson and Vincent A. Barker. *Finite Element Solution of Boundary Value Problems*, volume 35 of *Classics in Applied Mathematics*. SIAM, July 2001.

[11] David H. Bailey, Yoza Hida, Karthik Jeyabalan, Xiaoye S. Li, and Brandon J. Thompson. High-precision software directory. `http://crd.lbl.gov/~dhbailey/mpdist/`, August 2008.

[12] Ali Bakhoda, George L. Yuan, Wilson W. L. Fung, Henry Wong, and Tor M. Aamondt. Analyzing CUDA workloads using a detailed GPU simulator. In *2009 IEEE International Symposium on Performance Analysis of Systems and Software (ISPASS)*, pages 163–174, April 2009. DOI: 10.1109/ISPASS.2009.4919648.

[13] Richard Barrett, Michael Berry, Tony F. Chan, James W. Demmel, June Donato, Jack J. Dongarra, Victor Eijkhout, Roldan Pozo, Charles Romine, and Henk A. van der Vorst. *Templates for the Solution of Linear Systems: Building Blocks for Iterative Methods*. SIAM, 2nd edition, November 1994.

[14] Muthu Manikandan Baskaran and Rajesh Bordawekar. Optimizing sparse matrix-vector multiplication on GPUs. Technical Report RC24704, IBM Research, December 2008.

[15] Christian Becker. *Strategien und Methoden zur Ausnutzung der High-Performance-Computing-Ressourcen moderner Rechnerarchitekturen für Finite Element Simulationen und ihre Realisierung in FEAST (Finite Element Analysis & Solution Tools)*. PhD thesis, Universität Dortmund, May 2007. `http://www.logos-verlag.de/cgi-bin/buch?isbn=1637`.

[16] Christian Becker, Susanne Kilian, and Stefan Turek. Consequences of modern hardware design for numerical simulations and their realization in FEAST. In Patrick Amestoy, Philippe Berger, Michel Daydé, Iain Duff, Valérie Frayssé, Luc Giraud, and Daniel Ruiz, editors, *Euro-Par'99 Parallel Processing*, volume 1685 of *Lecture Notes in Computer Science*, pages 643–650. Springer, January 1999. DOI: 10.1007/3-540-48311-X_90.

[17] Christian Becker, Susanne Kilian, and Stefan Turek. Some concepts of the software package FEAST. In José M. L. M. Palma, Jack J. Dongarra, and Vicente Hernándes, editors, *Vector and Parallel Processing—VECPAR'98*, volume 1573 of *Lecture Notes in Computer Science*, pages 271–284. Springer, June 1999. DOI: 10.1007/10703040_22.

[18] Christian Becker, Sven H.M. Buijssen, Susanne Kilian, and Stefan Turek. High performance FEM simulation via FEAST and application to parallel CFD via FeatFlow. In Horst Rollnik and Dietrich Wolf, editors, *NIC Symposium 2001*, volume 9 of *NIC Series*, pages 493–502, May 2002.

[19] Christian Becker, Sven H.M. Buijssen, Hilmar Wobker, and Stefan Turek. FEAST: Development of HPC technologies for FEM applications. In Gernot Münster, Dietrich Wolf, and Manfred Kremer, editors, *NIC Symposium 2006*, volume 32 of *NIC Series*, pages 299–306, February 2006.

[20] Christian Becker, Sven H.M. Buijssen, and Stefan Turek. FEAST: Development of HPC technologies for FEM applications. In Wolfgang Nagel, Dietmar Kröner, and Michael Resch, editors, *High Performance Computing in Science and Engineering '07*, volume 2007 of *Transactions of the High Performance Computing Center, Stuttgart (HLRS)*, pages 503–516, December 2007. DOI: 10.1007/978-3-540-74739-0_34.

[21] Nathan Bell and Michael Garland. Efficient sparse matrix-vector multiplication using CUDA. Technical Report NVR-2008-4, NVIDIA, December 2008.

[22] Nathan Bell and Michael Garland. Implementing sparse matrix-vector multiplication on throughput-oriented processors. In *SC '09: Proceedings of the 2009 ACM/IEEE conference on Supercomputing*, November 2009. DOI: 10.1145/1654059.1654078. Article No. 18.

[23] Benjamin Bergen, Frank Hülsemann, and Ulrich Rüde. Is 1.7×10^{10} unknowns the largest finite element system that can be solved today? In *SC '05: Proceedings of the 2005 ACM/IEEE conference on Supercomputing*, November 2005. DOI: 10.1109/SC.2005.38. Article No. 5.

[24] Benjamin Bergen, Gerhard Wellein, Frank Hülsemann, and Ulrich Rüde. Hierachical hybrid grids: Achieving TFLOPS performance on large scale finite element simulations. *International Journal of Parallel, Emergent and Distributed Systems*, 22(4):311–329, January 2007. DOI: 10.1080/17445760701442218.

[25] Markus Billeter, Ola Olsson, and Ulf Assarsson. Efficient stream compaction on wide SIMD many-core architectures. In *High Performance Graphics 2009*, August 2009. DOI: 10.1145/1572769.1572795.

[26] Guy E. Blelloch. Prefix sums and their applications. Technical Report CMU-CS-90-190, School of Computer Science, Carnegie Mellon University, November 1990.

[27] Heribert Blum, Joachim Harig, Steffen Müller, and Stefan Turek. FEAT2D: Finite element analysis tools. user manual. release 1.3. Technical report, Universität Heidelberg, 1992.

[28] David Blythe. The Direct3D 10 system. *ACM Transactions on Graphics*, 25(3):724–734, July 2006. DOI: 10.1145/1141911.1141947.

[29] David Blythe. Rise of the graphics processor. *Proceedings of the IEEE*, 96(5):761–778, May 2008. DOI: 10.1109/JPROC.2008.917718.

[30] Jeff Bolz, Ian Farmer, Eitan Grinspun, and Peter Schröder. Sparse matrix solvers on the GPU: Conjugate gradients and multigrid. *ACM Transactions on Graphics*, 22(3):917–924, July 2003. DOI: 10.1145/882262.882364.

[31] Kiran Bondalapati and Viktor K. Prasanna. Reconfigurable computing systems. *Proceedings of the IEEE*, 90(7):1201–1217, July 2002. DOI: 10.1109/JPROC.2002.801446.

[32] Hilary J. Bowdler, Roger S. Martin, G. Peters, and James H. Wilkinson. Solution of real and complex systems of linear equations. *Numerische Mathematik*, 8(3):217–234, May 1966. DOI: 10.1007/BF02162559.

[33] Chas. Boyd. Data-parallel computing. *ACM Queue*, 6(2):30–39, March/April 2008. DOI: 10.1145/1365490.1365499.

[34] Dietrich Braess. *Finite Elements – Theory, Fast Solvers and Applications in Solid Mechanics*. Cambridge University Press, 2nd edition, April 2001. DOI: 10.2277/0521011957.

[35] James H. Bramble, Joseph E. Pasciak, and Alfred H. Schatz. The construction of preconditioners for elliptic problems by substructuring. I. *Mathematics of Computation*, 47(175): 103–134, July 1986. DOI: 10.2307/2008084.

[36] James H. Bramble, Joseph E. Pasciak, and Alfred H. Schatz. The construction of preconditioners for elliptic problems by substructuring. II. *Mathematics of Computation*, 49(179): 1–16, July 1987. DOI: 10.2307/2008246.

[37] James H. Bramble, Joseph E. Pasciak, and Jinchao Xu. Parallel multilevel preconditioners. *Mathematics of Computation*, 55(191):1–22, July 1990. DOI: 10.2307/2008789.

[38] Tobias Brandvik and Graham Pullan. Acceleration of a two-dimensional Euler flow solver using commodity graphics hardware. *Proceedings of the Institution of Mechanical Engineers, Part C: Journal of Mechanical Engineering Science*, 221(12):1745–1748, December 2007. DOI: 10.1243/09544062JMES813FT.

[39] Susanne C. Brenner. Convergence of the multigrid V-cycle algorithm for second-order boundary value problems without full elliptic regularity. *Mathematics of Computation*, 71 (238):507–525, April 2002. DOI: 10.1090/S0025-5718-01-01361-8.

[40] Alexander N. Brooks and Thomas J. R. Hughes. Streamline upwind/Petrov-Galerkin formulations for convection dominated flows with particular emphasis on the incompressible Navier-Stokes equations. *Computer Methods in Applied Mechanics and Engineering*, 32 (1–3):199–259, September 1982. DOI: 10.1016/0045-7825(82)90071-8.

[41] Luc Buatois, Guillaume Caumon, and Bruno Lévy. Concurrent number cruncher: An efficient sparse linear solver on the GPU. In *High Performance Computing and Communications*, volume 4782 of *Lecture Notes In Computer Science*, pages 358–371. Springer, September 2007. DOI: 10.1007/978-3-540-75444-2_37.

[42] Luc Buatois, Guillaume Caumon, and Bruno Lévy. Concurrent number cruncher – A GPU implementation of a general sparse linear solver. *International Journal of Parallel, Emergent and Distributed Systems*, 24(9):205–223, June 2009. DOI: 10.1080/17445760802337010.

[43] Ian Buck, Tim Foley, Daniel R. Horn, Jeremy Sugerman, Kayvon Fatahalian, Mike Houston, and Pat Hanrahan. Brook for GPUs: Stream computing on graphics hardware. *ACM Transactions on Graphics*, 23(3):777–786, August 2004. DOI: 10.1145/1015706.1015800.

[44] Sven H.M. Buijssen. *Efficient Multilevel Solvers and High Performance Computing Techniques for the Finite Element Simulation of the Transient, Incompressible Navier–Stokes Equations*. PhD thesis, TU Dortmund, Fakultät für Mathematik, 2010. in preparation.

[45] Sven H.M. Buijssen, Hilmar Wobker, Dominik Göddeke, and Stefan Turek. FEASTSolid and FEASTFlow: FEM applications exploiting FEAST's HPC technologies. In Wolfgang Nagel, Dietmar Kröner, and Michael Resch, editors, *High Performance Computing in Science and Engineering '08*, volume 2008 of *Transactions of the High Performance Computing Center, Stuttgart (HLRS)*, pages 425–440. Springer, December 2008. DOI: 10.1007/978-3-540-88303-6_30.

[46] Bill L. Buzbee, Gene H. Golub, and Clair W. Nielson. On direct methods for solving Poisson's equations. *SIAM Journal on Numerical Analysis*, 7(4):627–656, December 1970. DOI: 10.1137/0707049.

[47] Cris Cecka, Adrian J. Lew, and Eric Darve. Assembly of finite element methods on graphics processors. *submitted to International Journal Numerical Methods in Engineering, draft made available by the authors*, 2010.

[48] Shuai Che, Michael Boyer, Jiayuan Meng, David Tarjan, Jeremy W. Sheaffer, and Kevin Skadron. A performance study of general-purpose applications on graphics processors using CUDA. *Journal of Parallel and Distributed Computing*, 68(10):1370–1380, June 2008. DOI: 10.1016/j.jpdc.2008.05.014.

[49] Jonathan M. Cohen and M. Jeroen Molemaker. A fast double precision CFD code using CUDA. In *ParCFD'2009: 21st International Conference on Parallel Computational Fluid Dynamics*, May 2009.

[50] Phillip Colella, Thom H. Dunning Jr., William D. Gropp, and David E. Keyes. A science–based case for large–scale simulation. Technical report, Office of Science, US Department of Energy, July 2003. `http://www.pnl.gov/scales`.

[51] Katherine Compton and Scott Hauck. Reconfigurable computing: A survey of systems and software. *ACM Computing Surveys*, 34(2):171–210, June 2002. DOI: 10.1145/508352.508353.

[52] Andrew Corrigan, Fernando Camelli, Rainald Löhner, and John Wallin. Running unstructured grid based CFD solvers on modern graphics hardware. In *19th AIAA Computational Fluid Dynamics Conference*, June 2009. AIAA 2009-4001.

[53] Guillaume Da Graça and David Defour. Implementation of float-float operators on graphics hardware. In *RNC7: 7th Conference on Real Numbers and Computers*, pages 23–32, July 2006.

[54] Kevin Dale, Jeremy W. Sheaffer, Vinu Vijay Kumar, David P. Luebke, Greg Humphreys, and Kevin Skadron. Applications of small-scale reconfigurability to graphics processors. In Koen Bertels, João M. P. Cardoso, and Stamatis Vassiliadis, editors, *Reconfigurable Computing: Architectures and Applications (Selected revised papers from ARC 2006)*, volume 3985 of *Lecture Notes in Computer Science*, pages 99–108. Springer, August 2006. DOI: 10.1007/11802839_14.

[55] William J. Dally, Pat Hanrahan, Mattan Erez, Timothy J. Knight, François Labonté, Jung-Ho Ahn, Nuwan Jayasena, Ujval J. Kapasi, Abhishek Das, Jayanth Gummaraju, and Ian Buck. Merrimac: Supercomputing with streams. In *SC '03: Proceedings of the 2003 ACM/IEEE conference on Supercomputing*, page 35, November 2003. DOI: 10.1109/SC.2003.10043.

[56] Marc Daumas, Guillaume Da Graça, and David Defour. Caractéristiques arithmétiques des processeurs graphiques. In *SympA: Symposium en Architecture de Machines*, pages 86–95, October 2006.

[57] Timothy A. Davis. A column pre-ordering strategy for the unsymmetric-pattern multifrontal method. *ACM Transactions on Mathematical Software*, 30(2):165–195, June 2004. DOI: 10.1145/992200.992205.

[58] Theodorus J. Dekker. A floating-point technique for extending the available precision. *Numerische Mathematik*, 18(3):224–242, June 1971. DOI: 10.1007/BF01397083.

[59] James W. Demmel, Stanley C. Eisenstat, John R. Gilbert, Xiaoye S. Li, and Joseph W. H. Liu. A supernodal approach to sparse partial pivoting. *SIAM Journal on Matrix Analysis and Applications*, 20(3):720–755, July 1999. DOI: 10.1137/S0895479895291765.

[60] James W. Demmel, Yozo Hida, William M. Kahan, Xiaoye S. Li, Sonil Mukherjee, and E. Jason Riedy. Error bounds from extra-precise iterative refinement. *ACM Transactions on Mathematical Software*, 32(2):325–351, June 2006. DOI: 10.1145/1141885.1141894.

[61] Jean Donea and Antonio Huerta. *Finite Element Methods for Flow Problems*. John Wiley & Sons, April 2003.

[62] Yuri Dotsenko, Naga K. Govindaraju, Peter-Pike Sloan, Charles Boyd, and John Manferdelli. Fast scan algorithms on graphics processors. In *ICS'08: Proceedings of the 22nd annual International Conference on Supercomputing*, pages 205–213, June 2008. DOI: 10.1145/1375527.1375559.

[63] Craig C. Douglas, Jonathan Hu, Markus Kowarschik, Ulrich Rüde, and Christian Weiß. Cache optimization for structured and unstructured grid multigrid. *Electronic Transactions on Numerical Analysis*, 10:21–40, February 2000.

[64] Ömer Eğecioğlu, Cetin K. Koc, and Alan J. Laub. A recursive doubling algorithm for solution of tridiagonal systems on hypercube multiprocessors. *Journal of Computational and Applied Mathematics*, 27(1+2):95–108, September 1989. DOI: 10.1016/0377-0427(89)90362-2.

[65] Erich Elsen, Patrick LeGresley, and Eric Darve. Large calculation of the flow over a hypersonic vehicle using a GPU. *Journal of Computational Physics*, 227(24):10148–10161, December 2008. DOI: 10.1016/j.jcp.2008.08.023.

[66] Mattan Erez, Jung Ho Ahn, Jayanth Gummaraju, Mendel Rosenblum, and William J. Dally. Executing irregular scientific applications on stream architectures. In *ICS'07: Proceedings of the 21st annual International Conference on Supercomputing*, pages 93–104, June 2007. DOI: 10.1145/1274971.1274987.

[67] Zhe Fan, Feng Qiu, Arie Kaufman, and Suzanne Yoakum-Stover. GPU cluster for high performance computing. In *SC '04: Proceedings of the 2004 ACM/IEEE conference on Supercomputing*, page 47, November 2004. DOI: 10.1109/SC.2004.26.

[68] Kayvon Fatahalian and Mike Houston. A closer look at GPUs. *Communications of the ACM*, 51(10):50–57, October 2008. DOI: 10.1145/1400181.1400197.

[69] Kayvon Fatahalian, Timothy J. Knight, Mike Houston, Mattan Erez, Daniel R. Horn, Larkhoon Leem, Ji Young Park, Manman Ren, Alex Aiken, William J. Dally, and Pat Hanrahan. Sequoia: Programming the memory hierarchy. In *SC '06: Proceedings of the 2006 ACM/IEEE conference on Supercomputing*, November 2006. DOI: 10.1145/1188455.1188543. Article No. 83.

[70] Massimiliano Fatica. Accelerating Linpack with CUDA on heterogenous clusters. In David Kaeli and Miriam Leeser, editors, *GPGPU-2: Proceedings of 2nd Workshop on General Purpose Processing on Graphics Processing Units*, number 383 in ACM International Conference Proceeding Series, pages 46–51, March 2009. DOI: 10.1145/1513895.1513901.

[71] Zhuo Feng and Peng Li. Multigrid on GPU: Tackling power grid analysis on parallel SIMT platforms. In *ICCAD 2008: IEEE/ACM International Conference on Computer-Aided Design*, pages 647–654, November 2008. DOI: 10.1109/ICCAD.2008.4681645.

[72] Randima Fernando, editor. *GPU Gems*. Addison-Wesley, April 2004.

[73] Jiří Filipovič, Igor Peterlík, and Jan Fousek. GPU acceleration of equations assembly in finite elements method – preliminary results. In *2009 Symposium on Application Accelerators in High Performance Computing (SAAHPC'09)*, July 2009.

[74] Michael J. Flynn. Some computer organizations and their effectiveness. *IEEE Transactions on Computers*, C-21(9):948–960, September 1972. DOI: 10.1109/TC.1972.5009071.

[75] Matteo Frigo and Volker Strumpen. The cache complexity of multithreaded cache oblivious algorithms. In *SPAA 2006: Eighteenth Annual ACM Symposium on Parallelism in Algorithms and Architectures*, pages 271–280, August 2006. DOI: 10.1145/1148109.1148157.

[76] Nico Galoppo, Naga K. Govindaraju, Michael Henson, and Dinesh Manocha. LU-GPU: Efficient algorithms for solving dense linear systems on graphics hardware. In *SC '05: Proceedings of the 2005 ACM/IEEE conference on Supercomputing*, November 2005. DOI: 10.1109/SC.2005.42.

[77] Michael Garland, Scott Le Grand, John Nickolls, Joshua A. Anderson, Jim Hardwick, Scott Morton, Everett H. Phillips, Yao Zhang, and Vasily Volkov. Parallel computing experiences with CUDA. *IEEE Micro*, 28(4):13–27, July 2008. DOI: 10.1109/MM.2008.57.

[78] Keith O. Geddes and Wei Wei Zheng. Exploiting fast hardware floating point in high precision computation. In *ISSAC '03: Proceedings of the 2003 International Symposium on Symbolic and Algebraic Computation*, pages 111–118, August 2003. DOI: 10.1145/860854.860886.

[79] Dominik Göddeke and Robert Strzodka. Performance and accuracy of hardware-oriented native, emulated- and mixed-precision solvers in FEM simulations (part 2: Double precision GPUs). Technical report, Fakultät für Mathematik, Technische Universität Dortmund, August 2008. Ergebnisberichte des Instituts für Angewandte Mathematik, Nummer 370.

[80] Dominik Göddeke and Robert Strzodka. Mixed precision GPU-multigrid solvers with strong smoothers. In Jakub Kurzak, David A. Bader, and Jack J. Dongarra, editors, *Scientific Computing with Multicore and Accelerators*, chapter 7. CRC Press, December 2010.

[81] Dominik Göddeke and Robert Strzodka. Cyclic reduction tridiagonal solvers on GPUs applied to mixed precision multigrid. *IEEE Transactions on Parallel and Distributed Systems*, 22(1):22–32, January 2011. DOI: 10.1109/TPDS.2010.61.

[82] Dominik Göddeke, Robert Strzodka, and Stefan Turek. Accelerating double precision FEM simulations with GPUs. In Frank Hülsemann, Matthias Kowarschik, and Ulrich Rüde, editors, *18th Symposium Simulationstechnique (ASIM'05)*, Frontiers in Simulation, pages 139–144, September 2005.

[83] Dominik Göddeke, Christian Becker, and Stefan Turek. Integrating GPUs as fast coprocessors into the parallel FE package FEAST. In Matthias Becker and Helena Szczerbicka, editors, *19th Symposium Simulationstechnique (ASIM'06)*, Frontiers in Simulation, pages 277–282, September 2006.

[84] Dominik Göddeke, Robert Strzodka, Jamaludin Mohd-Yusof, Patrick S. McCormick, Sven H.M. Buijssen, Matthias Grajewski, and Stefan Turek. Exploring weak scalability for FEM calculations on a GPU-enhanced cluster. *Parallel Computing*, 33(10–11):685–699, September 2007. DOI: 10.1016/j.parco.2007.09.002.

[85] Dominik Göddeke, Robert Strzodka, and Stefan Turek. Performance and accuracy of hardware-oriented native-, emulated- and mixed-precision solvers in FEM simulations. *International Journal of Parallel, Emergent and Distributed Systems*, 22(4):221–256, January 2007. DOI: 10.1080/17445760601122076.

[86] Dominik Göddeke, Hilmar Wobker, Robert Strzodka, Jamaludin Mohd-Yusof, Patrick S. McCormick, and Stefan Turek. Co-processor acceleration of an unmodified parallel structural mechanics code with FEAST-GPU, November 2007. Supercomputing 2007 Posters.

[87] Dominik Göddeke, Robert Strzodka, Jamaludin Mohd-Yusof, Patrick S. McCormick, Hilmar Wobker, Christian Becker, and Stefan Turek. Using GPUs to improve multigrid solver performance on a cluster. *International Journal of Computational Science and Engineering*, 4(1):36–55, November 2008. DOI: 10.1504/IJCSE.2008.021111.

[88] Dominik Göddeke, Sven H.M. Buijssen, Hilmar Wobker, and Stefan Turek. GPU acceleration of an unmodified parallel finite element Navier-Stokes solver. In Waleed W. Smari and John P. McIntire, editors, *High Performance Computing & Simulation 2009*, pages 12–21, June 2009. DOI: 10.1109/HPCSIM.2009.5191718.

[89] Dominik Göddeke, Sven H.M. Buijssen, Hilmar Wobker, and Stefan Turek. GPU cluster computing for finite element applications, March 2009. SIAM Conference on Computational Science and Engineering, Miami, Florida.

[90] Dominik Göddeke, Hilmar Wobker, Robert Strzodka, Jamaludin Mohd-Yusof, Patrick S. McCormick, and Stefan Turek. Co-processor acceleration of an unmodified parallel solid mechanics code with FEASTGPU. *International Journal of Computational Science and Engineering*, 4(4):254–269, October 2009. DOI: 10.1504/IJCSE.2009.029162.

[91] David Goldberg. What every computer scientist should know about floating-point arithmetic. *ACM Computing Surveys*, 23(1):5–48, March 1991. DOI: 10.1145/103162.103163.

[92] Nolan Goodnight, Cliff Woolley, Gregory Lewin, David P. Luebke, and Greg Humphreys. A multigrid solver for boundary value problems using programmable graphics hardware. In Michael Doggett, Wolfgang Heidrich, William R. Mark, and Andreas Schilling, editors, *Graphics Hardware 2003*, pages 102–111, July 2003.

[93] Georgios Goumas, Kornilios Kourtis, Nikos Anastopoulos, Vasileios Karakasis, and Nectarios Koziris. Understanding the performance of sparse matrix-vector-multiplication. In *16th Euromicro Conference on Parallel, Distributed and Network-Based Processing (PDP 2008)*, pages 283–292, February 2008. DOI: 10.1109/PDP.2008.41.

[94] Naga K. Govindaraju, Jim Gray, Ritesh Kumar, and Dinesh Manocha. GPUTeraSort: High performance graphics co-processor sorting for large database management. In *SIGMOD '06: Proceedings of the 2006 ACM SIGMOD international conference on Management of Data*, pages 325–336, June 2006. DOI: 10.1145/1142473.1142511.

[95] Naga K. Govindaraju, Scott Larsen, Jim Gray, and Dinesh Manocha. A memory model for scientific algorithms on graphics processors. In *SC '06: Proceedings of the 2006 ACM/IEEE conference on Supercomputing*, November 2006. DOI: 10.1145/1188455.1188549. Article No. 89.

[96] Tobias Gradl and Ulrich Rüde. High performance multigrid in current large scale parallel computers. In Wolfgang E. Nagel, Rolf Hoffmann, and Andreas Koch, editors, *9th Workshop on Parallel Systems and Algorithms (PASA)*, volume 124 of *GI Edition: Lecture Notes in Informatics*, pages 37–45, February 2008.

[97] Susan L. Graham, Marc Snir, and Cynthia A. Patterson (eds.). *Getting up to Speed - The Future of Supercomputing*. The National Academies Press, Washington, D.C., November 2004.

[98] Matthias Grajewski. *A new fast and accurate grid deformation method for r–adaptivity in the context of high performance computing*. PhD thesis, TU Dortmund, Fakultät für Mathematik, March 2008. `http://www.logos-verlag.de/cgi-bin/buch?isbn=1903`.

[99] Ananth Grama, Anshul Gupta, George Karypis, and Vipin Kumar. *Introduction to Parallel Computing: Design and Analysis of Algorithms*. Addison-Wesley, 2nd edition, February 2003.

[100] Alexander Greß, Michael Guthe, and Reinhard Klein. GPU-based collision detection for deformable parameterized surfaces. *Computer Graphics Forum*, 25(3):497–506, September 2006. DOI: 10.1111/j.1467-8659.2006.00969.x.

[101] Marcus J. Grote and Thomas Huckle. Parallel preconditioning with sparse approximate inverses. *SIAM Journal on Scientific Computing*, 18(3):838–853, May 1997. DOI: 10.1137/S1064827594276552.

[102] Wolfgang Hackbusch. *Multi-grid methods and applications*. Springer, October 1985.

[103] Wolfgang Hackbusch. *Iterative Lösung großer schwachbesetzer Gleichungssysteme*. Teubner, 2nd edition, January 1993.

[104] Wolfgang Hackbusch. *Iterative Solution of Large Sparse Systems*. Springer, November 1994.

[105] Eric Haines. An introductory tour of interactive rendering. *IEEE Computer Graphics and Applications*, 26(1):76–87, January/February 2006. DOI: 10.1109/MCG.2006.9.

[106] Tom R. Halfhill. Parallel processing with CUDA. *Microprocessor Report*, 01/28/08-01, January 2008.

[107] Mark J. Harris. *Real-Time Cloud Simulation and Rendering*. PhD thesis, University of North Carolina at Chapel Hill, September 2003.

[108] Mark J. Harris. Fast fluid dynamics simulation on the GPU. In Randima Fernando, editor, *GPU Gems*, pages 637–665. Addison-Wesley, March 2004.

[109] Mark J. Harris. Mapping computational concepts to GPUs. In Matt Pharr, editor, *GPU Gems 2*, chapter 31, pages 493–508. Addison-Wesley, March 2005.

[110] Mark J. Harris and Ian Buck. GPU flow control idioms. In Matt Pharr, editor, *GPU Gems 2*, chapter 34, pages 547–555. Addison-Wesley, March 2005.

[111] Mark J. Harris, Greg Coombe, Thorsten Scheuermann, and Anselmo Lastra. Physically-based visual simulation on graphics hardware. In Thomas Ertl, Wolfgang Heidrich, and Michael Doggett, editors, *Graphics Hardware 2002*, pages 109–118, September 2002.

[112] Mark J. Harris, William V. Baxter III, Thorsten Scheuermann, and Anselmo Lastra. Simulation of cloud dynamics on graphics hardware. In Michael Doggett, Wolfgang Heidrich, William R. Mark, and Andreas Schilling, editors, *Graphics Hardware 2003*, pages 92–101, July 2003.

[113] Mark J. Harris, Shubhabrata Sengupta, and John D. Owens. Parallel prefix sum (scan) with CUDA. In Hubert Nguyen, editor, *GPU Gems 3*, chapter 39, pages 851–876. Addison-Wesley, August 2007.

[114] Yoza Hida, Xiaoye S. Li, and David H. Bailey. Algorithms for quad-double precision floating point arithmetic. In Neil Burgess and Luigi Ciminiera, editors, *ARITH'01: Proceedings of the 15th IEEE Symposium on Computer Arithmetic*, pages 155–162, June 2001. DOI: 10.1109/ARITH.2001.930115.

[115] Nicolas J. Higham. *Accuracy and Stability of Numerical Algorithms*. SIAM, 2nd edition, August 2002.

[116] Karl E. Hillesland and Anselmo Lastra. GPU floating-point paranoia. In *2004 ACM Workshop on General Purpose Computing on Graphics Processors*, pages C–8, August 2004.

[117] Roger W. Hockney. A fast direct solution of Poisson's equation using Fourier analysis. *Journal of the ACM*, 12(1):95–113, January 1965. DOI: 10.1145/321250.321259.

[118] Roger W. Hockney and Chris R. Jesshope. *Parallel Computers*. Adam Hilger, November 1981.

[119] Daniel R. Horn. Stream reduction operations for GPGPU applications. In Matt Pharr, editor, *GPU Gems 2*, chapter 36, pages 573–589. Addison-Wesley, March 2005.

[120] Wen-mei Hwu, Shane Ryoo, Sain-Zee Ueng, John H. Kelm, Isaac Gelado, Sam S. Stone, Robert E. Kidd, Sara S. Baghsorkhi, Aqeel A. Mahesri, Stephanie C. Tsao, Nacho Navarro, Steve S. Lumetta, Matthew I. Frank, and Sanjay J. Patel. Implicitly parallel programming models for thousand-core microprocessors. In *DAC'07: Proceedings of the 44th annual conference on Design Automation*, pages 754–759, June 2007. DOI: 10.1145/1278480.1278669.

[121] Wen-mei Hwu, Kurt Keutzer, and Timothy G. Mattson. The concurrency challenge. *IEEE Design & Test of Computers*, 25(4):312–320, May 2008. DOI: 10.1109/MDT.2008.110.

[122] IEEE Computer Society. ANSI/IEEE Std 754-1985 for binary floating point arithmetic, March 1985.

[123] William M. Kahan. Further remarks on reducing truncation errors. *Communications of the ACM*, 8(1):40, January 1965. DOI: 10.1145/363707.363723.

[124] Michael Kass, Aaron E. Lefohn, and John D. Owens. Interactive depth of field using simulated diffusion. Technical Report 06-01, Pixar Animation Studios, January 2006.

[125] Michael Kazhdan and Hugues Hoppe. Streaming multigrid for gradient-domain operations on large images. *ACM Transactions on Graphics*, 27(3):1–10, August 2008. DOI: 10.1145/1360612.1360620.

[126] David E. Keyes. Terascale implicit methods for partial differential equations. In Xiaobing Feng and Tim P. Schulze, editors, *Recent Advances in Numerical Methods for Partial Differential Equations and Applications*, volume 306 of *Contemporary Mathematics*, pages 29–84. American Mathematical Society, January 2002.

[127] Brucek Khailany, William J. Dally, Scott Rixner, Ujval J. Kapasi, John D. Owens, and Brian Towles. Exploring the VLSI scalability of stream processors. In *HPCA '03: Proceedings of the Ninth International Symposium on High Performance Computer Architecture*, pages 153–164, February 2003. DOI: 10.1109/HPCA.2003.1183534.

[128] Khronos OpenCL Working Group. The OpenCL Specification, version 1.0. `http://www.khronos.org/opencl`, December 2008.

[129] Emmett Kilgariff and Randima Fernando. The GeForce 6 series GPU architecture. In Matt Pharr, editor, *GPU Gems 2*, chapter 30, pages 471–491. Addison-Wesley, March 2005.

[130] Susanne Kilian. *ScaRC: Ein verallgemeinertes Gebietszerlegungs-/Mehrgitterkonzept auf Parallelrechnern*. PhD thesis, Universität Dortmund, Fachbereich Mathematik, January 2001. http://www.logos-verlag.de/cgi-bin/buch?isbn=0092.

[131] Susanne Kilian and Stefan Turek. An example for parallel ScaRC and its application to the incompressible Navier-Stokes equations. In Hans Georg Bock, Guido Kanschat, Rolf Rannacher, Franco Brezzi, Roland Glowinski, Yuri A. Kuznetsov, and Jacques Periaux, editors, *ENUMATH 97: Proceedings of the 2nd European Conference on Numerical Mathematics and Advanced Applications*, pages 389–396, September 1997.

[132] Volodymyr V. Kindratenko, Jeremy J. Enos, Guochun Shi, Michael T. Showerman, Galen W. Arnold, John E. Stone, James C. Phillips, and Wen-mei Hwu. GPU clusters for high-performance computing. In *Proceedings of the Workshop on Parallel Programming on Accelerator Clusters (PPAC'09)*, pages 1–8, August 2009.

[133] Donald E. Knuth. *The Art of Computer Programming Volume 2: Seminumerical Algorithms*. Addison-Wesley, 3rd edition, October 1997.

[134] Peter Kogge, Keren Bergman, Shekhar Borkar, Dan Campbell, William Carlson, William Dally, Monty Denneau, Paul Franzon, William Harrod, Kerry Hill, Jon Hiller, Sherman Karp, Stephen Keckler, Dean Klein, Robert Lucas, Mark Richards, Al Scarpelli, Steven Scott, Allan Snavely, Thomas Sterling, R. Stanley Williams, and Katherine Yelick. Exascale computing study: Technology challenges in achieving exascale systems. Technical report, DARPA IPTO, September 2008.

[135] Dimitri Komatitsch, David Michéa, and Gordon Erlebacher. Porting a high-order finite-element earthquake modeling application to NVIDIA graphics cards using CUDA. *Journal of Parallel and Distributed Computing*, 69(5):451–460, May 2009. DOI: 10.1016/j.jpdc.2009.01.006.

[136] Dimitri Komatitsch, Gordon Erlebacher, Dominik Göddeke, and David Michéa. High-order finite-element seismic wave propagation modeling with MPI on a large GPU cluster. *Journal of Computational Physics*, 229:7692–7714, October 2010. DOI: 10.1016/j.jcp.2010.06.024.

[137] Michael Köster, Dominik Göddeke, Hilmar Wobker, and Stefan Turek. How to gain speedups of 1000 on single processors with fast FEM solvers — Benchmarking numerical and computational efficiency. Technical report, Fakultät für Mathematik, TU Dortmund, October 2008. Ergebnisberichte des Instituts für Angewandte Mathematik, Nummer 382.

[138] Markus Kowarschik and Christian Weiß. An overview of cache optimization techniques and cache-aware numerical algorithms. In Ulrich Meyer, Peter Sanders, and Jop Sibeyn, editors, *Algorithms for Memory Hierarchies*, volume 2625 of *Lecture Notes In Computer Science*, pages 213–232. Springer, July 2003. DOI: 10.1007/3-540-36574-5_10.

[139] Jens Krüger. *A GPU Framework for Interactive Simulation and Rendering of Fluid Effects*. PhD thesis, Technische Universität München, December 2006.

[140] Jens Krüger and Rüdiger Westermann. Linear algebra operators for GPU implementation of numerical algorithms. *ACM Transactions on Graphics*, 22(3):908–916, July 2003. DOI: 10.1145/882262.882363.

[141] Jens Krüger and Rüdiger Westermann. A GPU framework for solving systems of linear equations. In Matt Pharr, editor, *GPU Gems 2*, chapter 44, pages 703–718. Addison-Wesley, March 2005.

[142] Julie Langou, Julien Langou, Piotr Luszczek, Jakub Kurzak, Alfredo Buttari, and Jack J. Dongarra. Exploiting the performance of 32 bit floating point arithmetic in obtaining 64 bit accuracy (revisiting iterative refinement for linear systems). In *SC '06: Proceedings of the 2006 ACM/IEEE conference on Supercomputing*, November 2006. DOI: 10.1145/1188455.1188573. Article No. 113.

[143] Aaron E. Lefohn. *Glift: Generic Data Structures for Graphics Hardware*. PhD thesis, Computer Science, University of California, Davis, September 2006.

[144] Aaron E. Lefohn, Joe M. Kniss, and John D. Owens. Implementing efficient parallel data structures on GPUs. In Matt Pharr, editor, *GPU Gems 2*, chapter 33, pages 521–545. Addison-Wesley, March 2005.

[145] Aaron E. Lefohn, Joe M. Kniss, Robert Strzodka, Shubhabrata Sengupta, and John D. Owens. Glift: Generic, efficient, random-access GPU data structures. *ACM Transactions on Graphics*, 25(1):60–99, January 2006. DOI: 10.1145/1122501.1122505.

[146] Jed Lengyel, Mark Reichert, Bruce R. Donald, and Donald P. Greenberg. Real-time robot motion planning using rasterizing computer graphics hardware. *Computer Graphics (Proceedings of SIGGRAPH 90)*, 24(4):327–335, August 1990. DOI: 10.1145/97880.97915.

[147] Xiaoye S. Li, James W. Demmel, David H. Bailey, Greg Henry, Yozo Hida, Jimmy Iskandar, William M. Kahan, Suh Y. Kang, Anil Kapur, Michael C. Martin, Brandon J. Thompson, Teresa Tung, and Daniel J. Yoo. Design, implementation and testing of extended and mixed precision BLAS. *ACM Transactions on Mathematical Software*, 28(2):152–205, June 2002. DOI: 10.1145/567806.567808.

[148] Erik Lindholm, Mark J. Kilgard, and Henry Moreton. A user-programmable vertex engine. In *Proceedings of SIGGRAPH 2001*, pages 149–158, August 2001. DOI: 10.1145/383259.383274.

[149] Erik Lindholm, John Nickolls, Stuart Oberman, and John Montrym. NVIDIA Tesla: A unified graphics and computing architecture. *IEEE Micro*, 28(2):39–55, March/April 2008. DOI: 10.1109/MM.2008.31.

[150] Eugene Loh and G. William Walster. Rump's example revisited. *Reliable Computing*, 8(3): 245–248, June 2002. DOI: 10.1023/A:1015569431383.

[151] Rainald Löhner. *Applied Computational Fluid Dynamics Techniques: An Introduction Based on Finite Elements*. John Wiley & Sons, 2nd edition, March 2008.

[152] Philipp Lucas, Nicolas Fritz, and Reinhard Wilhelm. The CGiS compiler – A tool demonstration. In *Proceedings of the 15th International Conference on Compiler Construction*, volume 3923 of *Lecture Notes in Computer Science*, pages 105–108. Springer, March 2006. DOI: 10.1007/11688839_10.

[153] William R. Mark. Future graphics architectures. *ACM Queue*, 6(2):54–64, March/April 2008. DOI: 10.1145/1365490.1365501.

[154] William R. Mark, R. Steven Glanville, Kurt Akeley, and Mark J. Kilgard. Cg: A system for programming graphics hardware in a C-like language. *ACM Transactions on Graphics*, 22 (3):896–907, July 2003. DOI: 10.1145/882262.882362.

[155] Roger S. Martin, G. Peters, and James H. Wilkinson. Iterative refinement of the solution of a positive definite system of equations. *Numerische Mathematik*, 8(3):203–216, May 1966. DOI: 10.1007/BF02162558.

[156] Michael McCool, Stefanus Du Toit, Tiberiu S. Popa, Bryan Chan, and Kevin Moule. Shader algebra. *ACM Transactions on Graphics*, 23(3):787–795, August 2004. DOI: 10.1145/1015706.1015801.

[157] Michael D. McCool, Zheng Qin, and Tiberiu S. Popa. Shader metaprogramming. In Thomas Ertl, Wolfgang Heidrich, and Michael Doggett, editors, *Graphics Hardware 2002*, pages 57–68, September 2002.

[158] Patrick S. McCormick, Jeff Inman, James P. Ahrens, Chuck Hansen, and Greg Roth. Scout: A hardware-accelerated system for quantitatively driven visualization and analysis. In *IEEE Visualization 2004*, pages 171–178, October 2004. DOI: 10.1109/VIS.2004.95.

[159] Patrick S. McCormick, Jeff Inman, James P. Ahrens, Jamaludin Mohd-Yusof, Greg Roth, and Sharen Cummins. Scout: A data-parallel programming language for graphics processors. *Parallel Computing*, 33(10–11):648–662, November 2007. DOI: 10.1016/j.parco.2007.09.001.

[160] Paulius Micikevicius. 3D finite-difference computation on GPUs using CUDA. In *GPGPU-2: Proceedings of the 2nd Workshop on General Purpose Processing on Graphics Processing Units*, pages 79–84, March 2009. DOI: 10.1145/1513895.1513905.

[161] M. Jeroen Molemaker, Jonathan M. Cohen, Sanjit Patel, and Jonyong Noh. Low viscosity flow simulations for animations. In Markus Gross and Doug James, editors, *Eurographics / ACM SIGGRAPH Symposium on Computer Animation*, July 2008.

[162] Cleve B. Moler. Iterative refinement in floating point. *Journal of the ACM*, 14(2):316–321, April 1967. DOI: 10.1145/321386.321394.

[163] Ole Møller. Quasi double-precision in floating point addition. *BIT Numerical Mathematics*, 5(1):37–50, March 1965. DOI: 10.1007/BF01975722.

[164] John Montrym and Henry Moreton. The GeForce 6800. *IEEE Micro*, 25(2):41–51, March/April 2005. DOI: 10.1109/MM.2005.37.

[165] Christoph Müller, Steffen Frey, Magnus Strengert, Carsten Dachsbacher, and Thomas Ertl. A compute unified system architecture for graphics clusters incorporating data locality. *IEEE Transactions on Visualization and Computer Graphics*, 15(4):605–617, July/August 2009. DOI: 10.1109/TVCG.2008.188.

[166] Malcolm F. Murphy, Gene H. Golub, and Andrew J. Wathen. A note on preconditioning for indefinite linear systems. *SIAM Journal on Scientific Computing*, 21(6):1969–1972, May 2000. DOI: 10.1137/S1064827599355153.

[167] Theodore H. Myer and Ivan E. Sutherland. On the design of display processors. *Communications of the ACM*, 11(6):410–414, June 1968. DOI: 10.1145/363347.363368.

[168] Hubert Nguyen, editor. *GPU Gems 3*. Addison-Wesley, August 2007.

[169] John Nickolls, Ian Buck, Michael Garland, and Kevin Skadron. Scalable parallel programming with CUDA. *ACM Queue*, 6(2):40–53, March/April 2008. DOI: 10.1145/1365490.1365500.

[170] NVIDIA Corporation. NVIDIA CUDA best practice guide version 2.3. `http://www.nvidia.com/cuda`, July 2009.

[171] NVIDIA Corporation. NVIDIA CUDA programming guide version 2.3. `http://www.nvidia.com/cuda`, July 2009.

[172] NVIDIA Corporation. Whitepaper: NVIDIA's next generation CUDA compute architecture: Fermi. `http://www.nvidia.com/object/fermi_architecture.html`, September 2009.

[173] Evgueni E. Ovtchinnikov and Leonidas S. Xanthis. Iterative subspace correction methods for thin elastic structures and Korn's type inequality in subspaces. *Proceedings of the Royal Society*, 454(1976):2023–2039, August 1998. DOI: 10.1098/rspa.1998.0247.

[174] John D. Owens, David P. Luebke, Naga K. Govindaraju, Mark J. Harris, Jens Krüger, Aaron E. Lefohn, and Timothy J. Purcell. A survey of general-purpose computation on graphics hardware. In *Eurographics 2005, State of the Art Reports*, pages 21–51, September 2005.

[175] John D. Owens, David P. Luebke, Naga K. Govindaraju, Mark J. Harris, Jens Krüger, Aaron E. Lefohn, and Timothy J. Purcell. A survey of general-purpose computation on graphics hardware. *Computer Graphics Forum*, 26(1):80–113, March 2007. DOI: 10.1111/j.1467-8659.2007.01012.x.

[176] John D. Owens, Mike Houston, David P. Luebke, Simon Green, John E. Stone, and James C. Phillips. GPU computing. *Proceedings of the IEEE*, 96(5):879–899, May 2008. DOI: 10.1109/JPROC.2008.917757.

[177] Alexandros Papakonstantinou, Karthik Gururaj, John A. Stratton, Deming Chen, Jason Cong, and Wen-mei Hwu. FCUDA: Enabling efficient compilation of CUDA kernels on FPGAs. In *7th IEEE Symposium on Application Specific Processors*, pages 35–42, July 2009.

[178] Bryson R. Payne and Markus A. Hitz. Implementation of residue number systems on GPUs. In *ACM SIGGRAPH 2006 Research posters*, page 57, August 2006. DOI: 10.1145/1179622.1179687.

[179] Donald W. Peaceman and Henry H. Rachford Jr. The numerical solution of parabolic and elliptic differential equations. *Journal of the Society for Industrial and Applied Mathematics*, 3(1):28–41, March 1955. DOI: 10.1137/0103003.

[180] Mark Peercy, Mark Segal, and Derek Gerstmann. A performance-oriented data parallel virtual machine for GPUs. In *ACM SIGGRAPH 2006 Conference Abstracts and Applications*, page 184, August 2006. DOI: 10.1145/1179849.1180079.

[181] Dac C. Pham, Shigehiro Asano, Mark Bolliger, Michael N. Day, H. Peter Hofstee, Charles R. Johns, James A. Kahle, Atsushi Kameyama, John Keaty, Yoshio Masubuchi, Mack Riley, David Shippy, Daniel L. Stasiak, Masakazu Suzuoki, M. Wang, James

Warnock, Steve Weitzel, Dieter Wendel, Takeshi Yamazaki, and Kazuaki Yazawa. The design and implementation of a first-generation CELL processor. In *Solid-State Circuits Conference, ISSCC 2005, Digest of Technical Papers*, pages 184–592 Vol. 1, February 2005. DOI: 10.1109/ISSCC.2005.1493930.

[182] Everett H. Phillips, Yao Zhang, Roger L. Davis, and John D. Owens. Rapid aerodynamic performance prediction on a cluster of graphics processing units. In *Proceedings of the 47th AIAA Aerospace Sciences Meeting*, January 2009. AIAA 2009-565.

[183] James C. Phillips, John E. Stone, and Klaus Schulten. Adapting a message-driven parallel application to GPU-accelerated clusters. In *SC '08: Proceedings of the 2008 ACM/IEEE conference on Supercomputing*, November 2008. DOI: 10.1145/1413370.1413379. Article No. 8.

[184] Douglas M. Priest. Algorithms for arbitrary precision floating point arithmetic. In *10th IEEE Symposium on Computer Arithmetic*, pages 132–143, June 1991. DOI: 10.1109/ARITH.1991.145549.

[185] Stefanie Reese, Martin Küssner, and Batmanathan Dayanand Reddy. A new stabilization technique for finite elements in finite elasticity. *International Journal for Numerical Methods in Engineering*, 44(11):1617–1652, March 1999. DOI: 10.1002/(SICI)1097-0207(19990420)44:11<1617::AID-NME557>3.0.CO;2-X.

[186] Randy J. Rost. *OpenGL Shading Language*. Addison-Wesley, 2nd edition, February 2006.

[187] Siegfried M. Rump. Algorithms for verified inclusions: Theory and practice. In Ramon E. Moore, editor, *Reliability in Computing: The Role of Interval Methods in Scientific Computing*, volume 19 of *Perspectives in Computing*, pages 109–126. Academic Press, 1988.

[188] Martin Rumpf and Robert Strzodka. Nonlinear diffusion in graphics hardware. In *Proceedings of EG/IEEE TCVG Symposium on Visualization (VisSym '01)*, pages 75–84, May 2001.

[189] Shane Ryoo, Christopher I. Rodrigues, Sam S. Stone, Sara S. Baghsorkhi, Sain-Zee Ueng, John A. Stratton, and Wen-mei Hwu. Program optimization space pruning for a multithreaded GPU. In *IEEE/ACM international symposium on Code generation and optimization (CGO'08)*, pages 195–204, April 2008. DOI: 10.1145/1356058.1356084.

[190] Shane Ryoo, Christopher I. Rodrigues, Sam S. Stone, John A. Stratton, Sain-Zee Ueng, Sara S. Baghsorkhi, and Wen-mei Hwu. Program optimization carving for GPU computing. *Journal of Parallel and Distributed Computing*, 68(10):1389–1401, October 2008. DOI: 10.1016/j.jpdc.2008.05.011.

[191] Yousef Saad. *Iterative Methods for Sparse Linear Systems*. SIAM, 2nd edition, January 2000.

[192] Nadathur Satish, Mark J. Harris, and Michael Garland. Designing efficient sorting algorithms for manycore GPUs. In *Proceedings of the 23rd IEEE International Parallel and Distributed Processing Symposium*, pages 1–10, May 2009. DOI: 10.1109/IPDPS.2009.5161005.

[193] Bianca Schroeder, Eduardo Pinheiro, and Wolf-Dietrich Weber. DRAM errors in the wild: A large-scale field study. In *SIGMETRICS/Performance '09: Proceedings of the 2009 Joint International Conference on Measurement & Modeling of Computer Systems*, pages 193–204, June 2009. DOI: 10.1145/1555349.1555372.

[194] Larry Seiler, Doug Carmean, Eric Sprangle, Tom Forsyth, Michael Abrash, Pradeep Dubey, Stephen Junkins, Adam Lake, Jeremy Sugerman, Robert Cavin, Roger Espasa, Ed Grochowski, Toni Juan, and Pat Hanrahan. Larrabee: A many-core x86 architecture for visual computing. *ACM Transactions on Graphics*, 27(3):1–15, August 2008. DOI: 10.1145/1360612.1360617.

[195] Sriram Sellappa and Siddhartha Chatterjee. Cache–efficient multigrid algorithms. *International Journal of High Performance Computing Applications*, 18(1):115–133, February 2004. DOI: 10.1177/1094342004041295.

[196] Shubhabrata Sengupta, Aaron E. Lefohn, and John D. Owens. A work-efficient step-efficient prefix sum algorithm. In *Proceedings of the Workshop on Edge Computing Using New Commodity Architectures*, pages D–26–27, May 2006.

[197] Shubhabrata Sengupta, Mark J. Harris, Yao Zhang, and John D. Owens. Scan primitives for GPU computing. In Timo Aila and Mark Segal, editors, *Graphics Hardware 2007*, pages 97–106, August 2007.

[198] Jeremy W. Sheaffer, David P. Luebke, and Kevin Skadron. The visual vulnerability spectrum: Characterizing architectural vulnerability for graphics hardware. In Mark Olano and Philipp Slusallek, editors, *Graphics Hardware 2006*, pages 9–16, September 2006. DOI: 10.1145/1283900.1283902.

[199] Jeremy W. Sheaffer, David P. Luebke, and Kevin Skadron. A hardware redundancy and recovery mechanism for reliable scientific computation on graphics processors. In Timo Aila and Mark Segal, editors, *Graphics Hardware 2007*, pages 55–64, August 2007.

[200] Jonathan R. Shewchuk. Adaptive precision floating-point arithmetic and fast robust geometric predicates. *Discrete & Computational Geometry*, 18(3):305–363, October 1997. DOI: 10.1007/PL00009321.

[201] Barry F. Smith, Petter E. Bjørstad, and William D. Gropp. *Domain Decomposition: Parallel Multilevel Methods for Elliptic Partial Differential Equations*. Cambridge University Press, June 1996.

[202] Jos Stam. Stable fluids. In *Proceedings of SIGGRAPH 1999*, pages 121–128, August 1999. DOI: 10.1145/311535.311548.

[203] Gilbert W. Stewart. *Introduction to Matrix Computations*. Academic Press, May 1973.

[204] Harold S. Stone. An efficient parallel algorithm for the solution of a tridiagonal linear system of equations. *Journal of the ACM*, 20(1):27–38, January 1973. DOI: 10.1145/321738.321741.

[205] John E. Stone, James C. Phillips, Peter L. Freddolino, David J. Hardy, Leonardo G. Trabuco, and Klaus Schulten. Accelerating molecular modeling applications with graphics processors. *Journal of Computational Chemistry*, 28(16):1618–2640, September 2007. DOI: 10.1002/jcc.20829.

[206] John A. Stratton, Sam S. Stone, and Wen-mei Hwu. MCUDA: An efficient implementation of CUDA kernels for multi-core CPUs. In José Nelson Amaral, editor, *Languages and Compilers for Parallel Computing*, volume 5335 of *Lecture Notes in Computer Science*, pages 16–30. Springer, November 2008. DOI: 10.1007/978-3-540-89740-8_2.

[207] Magnus Strengert, Christoph Müller, Carsten Dachsbacher, and Thomas Ertl. CUDASA: Compute unified device and systems architecture. In Jean Favre, Kwan-Liu Ma, and Daniel Weiskopf, editors, *Eurographics Symposium on Parallel Graphics and Visualisation*, pages 49–56, April 2008.

[208] Robert Strzodka. Virtual 16 bit precise operations on RGBA8 textures. In Günther Greiner, Heinrich Niemann, Thomas Ertl, Bernd Girod, and Hans-Peter Seidel, editors, *Proceedings of Vision, Modeling, and Visualization (VMV'02)*, pages 171–178, November 2002.

[209] Robert Strzodka. *Hardware Efficient PDE Solvers in Quantized Image Processing*. PhD thesis, University of Duisburg-Essen, December 2004.

[210] Robert Strzodka and Dominik Göddeke. Mixed precision methods for convergent iterative schemes. In *Proceedings of the Workshop on Edge Computing Using New Commodity Architectures*, pages D–59–60, May 2006.

[211] Robert Strzodka and Dominik Göddeke. Pipelined mixed precision algorithms on FPGAs for fast and accurate PDE solvers from low precision components. In *Proceedings of the 14th Annual IEEE Symposium on Field-Programmable Custom Computing Machines (FCCM'06)*, pages 259–270, April 2006. DOI: 10.1109/FCCM.2006.57.

[212] Robert Strzodka, Marc Droske, and Martin Rumpf. Fast image registration in DX9 graphics hardware. *Journal of Medical Informatics and Technologies*, 6:43–49, November 2003.

[213] Robert Strzodka, Marc Droske, and Martin Rumpf. Image registration by a regularized gradient flow – A streaming implementation in DX9 graphics hardware. *Computing*, 73(4): 373–389, November 2004. DOI: 10.1007/s00607-004-0087-x.

[214] Jeff A. Stuart and John D. Owens. Message passing on data-parallel architectures. In *Proceedings of the 23rd IEEE International Parallel and Distributed Processing Symposium*, pages 1–12, May 2009. DOI: 10.1109/IPDPS.2009.5161065.

[215] Jeremy Sugerman, Kayvon Fatahalian, Solomon Boulos, Kurt Akeley, and Pat Hanrahan. GRAMPS: A programming model for graphics pipelines. *ACM Transactions on Graphics*, 28(1):4:1–4:11, January 2009. DOI: 10.1145/1477926.1477930.

[216] Junqing Sun, Gregory D. Peterson, and Olaf O. Storaasli. High-performance mixed-precision linear solver for FPGAs. *IEEE Transactions on Computers*, 57(12):1614–1623, December 2008. DOI: 10.1109/TC.2008.89.

[217] Sascha Mirko Sykorra. *Lösung partieller Differentialgleichungen mit gemischter Genauigkeit auf field programmable gate arrays (FPGAs)*. Diploma thesis, Universität Dortmund, Fachbereich Informatik, July 2007.

[218] David Tarditi, Sidd Puri, and Jose Oglesby. Accelerator: Using data parallelism to program GPUs for general-purpose uses. *ACM SIGARCH Computer Architecture News*, 34(5):325–335, December 2006. DOI: 10.1145/1168919.1168898.

[219] Andrew Thall. Extended-precision floating-point numbers for GPU computation. In *ACM SIGGRAPH 2006 Research posters*, page 52, August 2006. DOI: 10.1145/1179622.1179682.

[220] Julien C. Thibault and Inanc Senocak. CUDA implementation of a Navier-Stokes solver on multi-GPU desktop platforms for incompressible flows. In *Proceedings of the 47th AIAA Aerospace Sciences Meeting*, January 2009. AIAA 2009-0758.

[221] Llewellyn Hilleth Thomas. Elliptic problems in linear difference equations over a network, 1949. Watson Scientific Computing Laboratory Report, Columbia University, New York.

[222] Jonas Tölke and Manfred Krafczyk. TeraFLOP computing on a desktop PC with GPUs for 3D CFD. *International Journal of Computational Fluid Dynamics*, 22(7):443–456, August 2008. DOI: 10.1080/10618560802238275.

[223] Andrea Toselli and Olof B. Widlund. *Domain Decomposition Methods – Algorithms and Theory*, volume 34 of *Springer Series in Computational Mathematics*. Springer, November 2005.

[224] Stefan Turek. On ordering strategies in a multigrid algorithm. In *Proc. 8th GAMM–Seminar*, volume 41 of *Notes on Numerical Fluid Mechanics*. Vieweg, 1992.

[225] Stefan Turek. *Efficient Solvers for Incompressible Flow Problems: An Algorithmic and Computational Approach*. Springer, June 1999.

[226] Stefan Turek and Michael Schäfer. Benchmark computations of laminar flow around a cylinder. In Ernst H. Hirschel, editor, *Flow Simulation with High-Performance Computers II*, volume 52 of *Notes on Numerical Fluid Mechanics*, pages 547–566. Vieweg, 1996.

[227] Stefan Turek, Mike Altieri, Christian Becker, Susanne Kilian, Hubertus Oswald, and John Wallis. Some basic concepts of FEAST. In *Proceedings 14th GAMM Seminar 'Concepts of Numerical Software'*, Notes on Numerical Fluid Mechanics, May 1998.

[228] Stefan Turek, Christian Becker, and Alexander Runge. Performance rating via the FEAST indices. *Computing*, 63(3):283–297, November 1999. DOI: 10.1007/s006070050035.

[229] Stefan Turek, Alexander Runge, and Christian Becker. The FEAST indices – Realistic evaluation of modern software components and processor technologies. *Computers and Mathematics with Applications*, 41(10):1431–1464, May 2001. DOI: 10.1016/S0898-1221(01)00108-0.

[230] Stefan Turek, Christian Becker, and Susanne Kilian. Hardware–oriented numerics and concepts for PDE software. *Future Generation Computer Systems*, 22(1-2):217–238, February 2004. DOI: 10.1016/j.future.2003.09.007.

[231] Stefan Turek, Dominik Göddeke, Christian Becker, Sven H.M. Buijssen, and Hilmar Wobker. UCHPC – Unconventional high-performance computing for finite element simulations. In *International Supercomputing Conference (ISC'08)*, June 2008.

[232] Stefan Turek, Abderrahim Ouazzi, and Jaroslav Hron. On pressure separation algorithms (PSepA) for improving the accuracy of incompressible flow simulations. *International Journal for Numerical Methods in Fluids*, 59(4):387–403, April 2008. DOI: 10.1002/fld.1820.

[233] Stefan Turek, Dominik Göddeke, Christian Becker, Sven H.M. Buijssen, and Hilmar Wobker. FEAST – Realisation of hardware-oriented numerics for HPC simulations with finite elements. *Concurrency and Computation: Practice and Experience*, 22(6):2247–2265, November 2010. DOI: 10.1002/cpe.1584.

[234] Stefan Turek, Dominik Göddeke, Sven H.M. Buijssen, and Hilmar Wobker. Hardware-oriented multigrid finite element solvers on GPU-accelerated clusters. In Jakub Kurzak, David A. Bader, and Jack J. Dongarra, editors, *Scientific Computing with Multicore and Accelerators*, chapter 6. CRC Press, December 2010.

[235] Kathryn Turner and Homer F. Walker. Efficient high accuracy solutions with GMRES(m). *SIAM Journal on Scientific and Statistical Computing*, 13(3):815–825, May 1992. DOI: 10.1137/0913048.

[236] Danny van Dyk, Markus Geveler, Sven Mallach, Dirk Ribbrock, Dominik Göddeke, and Carsten Gutwenger. HONEI: A collection of libraries for numerical computations targeting multiple processor architectures. *Computer Physics Communications*, 180(12):2534–2543, December 2009. DOI: 10.1016/j.cpc.2009.04.018.

[237] Vasily Volkov and James W. Demmel. Benchmarking GPUs to tune dense linear algebra. In *SC '08: Proceedings of the 2008 ACM/IEEE conference on Supercomputing*, November 2008. DOI: 10.1145/1413370.1413402. Article No. 31.

[238] Pieter Wesseling. *An Introduction to Multigrid Methods*. John Wiley & Sons, 1992.

[239] James H. Wilkinson. *Rounding errors in algebraic processes*. Prentice-Hall, January 1963.

[240] Samuel Williams, John Shalf, Leonid Oliker, Shoaib Kamil, Parry Husbands, and Katherine Yelick. The potential of the Cell processor for scientific computing. In *Computing Frontiers 2006*, pages 9–20, May 2006. DOI: 10.1145/1128022.1128027.

[241] Samuel Williams, Leonid Oliker, Richard Vuduc, John Shalf, Katherine Yelick, and James W. Demmel. Optimization of sparse matrix-vector multiplication on emerging multicore platforms. In *SC '07: Proceedings of the 2007 ACM/IEEE conference on Supercomputing*, November 2007. DOI: 10.1145/1362622.1362674. Article No. 38.

[242] Hilmar Wobker. *Efficient Multilevel Solvers and High Performance Computing Techniques for the Finite Element Simulation of Large-Scale Elasticity Problems*. PhD thesis, TU Dortmund, Fakultät für Mathematik, March 2010. `http://hdl.handle.net/2003/26998`.

[243] Jack M. Wolfe. Reducing truncation errors by programming. *Communications of the ACM*, 7(6):355–356, June 1964. DOI: 10.1145/512274.512287.

[244] Cliff Woolley. GPU program optimization. In Matt Pharr, editor, *GPU Gems 2*, chapter 35, pages 557–571. Addison-Wesley, March 2005.

[245] Junchao Xu. Iterative methods by space decomposition and subspace correction. *SIAM Review*, 34(4):581–613, December 1992. DOI: 10.1137/1034116.

[246] David Yeh, Li-Shiuan Peh, Shekhar Borkar, John Darringer, Anant Agarwal, and Wen-mei Hwu. Thousand-core chips. *IEEE Design & Test of Computers*, 25(3):272–278, May 2008. DOI: 10.1109/MDT.2008.85.

[247] Yao Zhang, Jonathan Cohen, and John D. Owens. Fast tridiagonal solvers on the GPU. In *Proceedings of the 15th ACM SIGPLAN Symposium on Principles and Practice of Parallel Programming (PPoPP 2010)*, pages 127–136, January 2010. DOI: 10.1145/1693453.1693472.

[248] Gerhard Zielke and Volker Drygalla. Genaue Lösung linearer Gleichungssysteme. *GAMM-Mitteilungen*, 26(1–2):7–107, December 2003.